WAVES IN DUSTY, SOLAR, AND SPACE PLASMAS

Related Titles from AIP Conference Proceedings

WAVES IN DUSTY, SOLAR, AND SPACE PLASMAS

Leuven, Belgium   22–26 May 2000

EDITORS

F. Verheest
Universiteit Gent, Belgium

M. Goossens
Katholieke Universiteit Leuven, Belgium

M. A. Hellberg
University of Natal, Durban, South Africa

R. Bharuthram
University of Durban-Westville, Durban, South Africa

AMERICAN INSTITUTE OF PHYSICS

Melville, New York, 2000
AIP CONFERENCE PROCEEDINGS ■ VOLUME 537

Editors:

F. Verheest
Sterrenkundig Observatorium
Universiteit Gent
Krijgslaan 281
B-9000 Gent, Belgium
E-mail: Frank.Verheest@rug.ac.be

M. Goossens
Centrum voor Plasma Astrofysica
Katholieke Universiteit Leuven
Celestijnenlaan 200B
B-3001 Leuven-Heverlee, Belgium
E-mail: Marcel.Goossens@wis.kuleuven.ac.be

M. A. Hellberg
School of Pure and Applied Physics
University of Natal
King George V Avenue
Durban 4041, South Africa
E-mail: hellberg@nu.ac.za

R. Bharuthram
M L Sultan Technikon
Durban 4000, South Africa
E-mail: RameshB@wpogate.mlsultan.ac.za

L.C. Catalog Card No. 00-107788
ISBN 1-56396-962-9
ISSN 0094-243X
Printed in the United States of America

CONTENTS

Preface

This volume contains the papers presented at the international *Workshop on Waves in Dusty, Solar and Space Plasmas* held in May, 2000, in the historic university city of Leuven, Belgium.

The Workshop had its roots in a collaborative research project involving the Workshop Conveners, as part of the Flemish-South African Bilateral Scientific and Technological Cooperation Agreement. This project was supported for three years by the Flemish Government (Department of Science and Technology) and the (South African) National Research Foundation, enabling regular visits between the two countries to take place. Both these bodies are thanked for their generous support over these years.

One of the specific requirements on the Flemish side was that we organize a workshop in Flanders, where the results of our scientific interaction would be presented. As true physicists we thought that it would be much more natural and stimulating to open up this workshop to the scientific community at large. As a result, the *Workshop on Waves in Dusty, Solar and Space Plasmas* was held as a truly open and international meeting. The response was gratifying, with more than 60 scientists participating in the meeting. There were more who submitted abstracts, but were unfortunately not able to attend because of financial constraints.

Papers were presented in all three areas of interest, and they covered a wide range. The studies of dusty plasmas were invariably related to applications in space and astrophysical plasmas. Papers included material on general aspects of collective processes, self-gravitational effects, and fugacity in dusty plasmas, Alfvén waves, linear and nonlinear studies, fluid and kinetic theory. In the solar physics area, the emphasis was on MHD waves and solar coronal heating, and was enhanced by some of the spectacular observations made possible by recent space missions.

In this volume the papers have been arranged according to the area of interest. Within each of the three sections the invited papers are included in the order of presentation, and they are then followed by the other contributions, in alphabetical order of first authors.

It is a great pleasure to thank those bodies without whose support the meeting could not have taken place: the Flemish Department of Science and Technology for the generous funding specifically for the Workshop as a whole; the National Research Foundation for support of the South African participants' travel; the Katholieke Universiteit Leuven, celebrating its 575th anniversary, for the stylish venue, conference bags and other support; and the City of Leuven, through its Mayor, for a guided tour through the old town and a civic reception in the magnificent medieval Town Hall.

The Conveners would like to single out two individuals for their efforts in making this a successful meeting. Gerald Jacobs, from the Sterrenkundig Observatorium, Universiteit Gent, provided vital support in the preparatory work on the scientific programme, while at Leuven, Leona Vandezande from the Department of Mathematics took excellent care of the social arrangements, ranging from housing to a welcome reception to a memorable farewell dinner.

The size of the meeting was such that the proceedings were enlivened by avid discussion of the papers, and participants therefore enjoyed benefits from the Workshop, which the reader of the papers in this volume will unfortunately not be able to share.

We trust that readers who were not able to attend the Workshop will nonetheless from this volume derive some of the intellectual stimulation that participants experienced, and glean some pleasure from reading papers which cover recent developments in a number of highly topical areas of the subject of waves in dusty, solar and space plasmas.

F Verheest

Sterrenkundig Observatorium, Universiteit Gent, Belgium

M Goossens

Centrum voor Plasma Astrofysica, Katholieke Universiteit Leuven, Belgium

M A Hellberg

School of Pure and Applied Physics, University of Natal, Durban, South Africa

R Bharuthram

Physics Department, University of Durban-Westville, Durban, South Africa
now at M L Sultan Technikon, Durban, South Africa

PART 1: DUSTY PLASMAS

Collective processes in complex plasmas

P. K. Shukla

Fakultät für Physik und Astronomie
Ruhr-Universität Bochum, D-44780 Bochum, Germany
and Department of Plasma Physics, Umeå University,
S-90187 Umeå, Sweden

Abstract. During the last decade, we have seen a very rapid growth of the dusty plasma physics with many discoveries. The latter include the dust acoustic wave, the wakefield and the ion focusing (which are responsible for the attraction of charged dust grains), the dusty plasma crystals as well as coherent nonlinear structures comprising the dust ion-acoustic shocks, the dust acoustic Mach cones, voids, and vortices. Since the dusty plasma physics involves the charging and the dynamics of extremely massive charged dust grains, it can be characterized as a complex plasma system whose characteristics are significantly different from the usual multi-ion plasmas. In this paper, we present the properties of a complex dusty plasma, as well as discuss two novel instabilities that are responsible for the excitation of low-frequency electrostatic modes in an unmagnetized dusty plasmas. The instabilities are associated with the combined effects of a dc electric field and the dust charge gradient on one hand, and the rotation of elongated dust grains on the other hand. The implications of the present investigation to laboratory and space plasmas are discussed.

INTRODUCTION

Two omnipresent ingredients of the Universe are plasmas and charged dust. The interplay between these two has opened up a new and fascinating research area, that of a dusty (or complex) plasma. A complex plasma is a normal electron-ion plasma with an additional charged component of small micron-sized particulates. This extra component, which increases the complexity of the system even further, is responsible for the name "complex plasma". Dusty or complex plasmas are ubiquitous in different parts of our solar system [1-2], namely, in planetary rings, in circumsolar dust rings, in the interplanetary medium, in cometary comae and tails, and in interstellar molecular clouds. The latter are seen in the Orion, Coalsack, Horsehead and Eagle nebulae. Dusty plasmas also occur in zodiacal light, in noctilucent clouds (in the form possibly of ice), in the arctic troposphere and polar mesosphere, in cloud-to-ground lightening in thunderstorms containing smoke-contaminated air over the United States, in rocket exhausts, in the flame of

CP537, *Waves in Dusty, Solar, and Space Plasmas,* edited by F. Verheest, et al.
© 2000 American Institute of Physics 1-56396-962-9/00/$17.00

humble candle, etc. Furthermore, meteoritic dust is thought to be present in the Earth's mesosphere at altitudes of $\sim 80 - 100$ km. It has been conjectured that in the cold summer mesopause, ice particles can form around meteoritic dust particles, with the icy dust particles possibly influencing the charge balance of the region. On the other hand, the presence of charged dust particles in the polar summer mesopause has been invoked to explain aspects of the very strong polar summer radar echoes referred to as PMSE (polar mesosphere summer echoes), which occur at altitudes of 80-93 km. Recently, the presence of charged dust in the mesosphere has been detected by direct rocket probe measurements, and both negatively and positively charged dust have been reported [3-4].

The physics of complex plasmas (dusty plasmas) has appeared as one of the most rapidly growing field of science, besides the field of Bose-Einstein condensate, as demonstrated by the number of published papers in scientific journals and conference proceedings. It has tremendous impact in astrophysics and low-temperature laboratory discharges including processing plasmas. In fact, the boost to dusty plasma physics came after the discovery of the dust acoustic wave [5,6], dust ion-acoustic waves [7] and the dusty plasma crystal [8,9]. We note that the theoretical idea of the dust acoustic wave (DAW) was put forward by Shukla [5] in the Capri Meeting on Dusty Plasmas in July 1989, while Ikezi [10] had theoretically predicted the Coulomb crystallization of charged dust grains in a strongly coupled dusty plasma system when the ratio between the Coulomb interaction and the dust thermal energies exceeds 170.

Dusty plasmas are fully or partially ionized low-temperature gases comprising electrons, ions, and micron-sized extremely massive charged dust grains. The latter, which are billion times heavier than the ions, acquire several thousands of the electron charge. The dust grain charging [2] occurs due to a variety of physical processes including the collection of the background plasma electrons and ions by dust grains, the photo electron emission, secondary emission and sputtering, etc. Dust grains can be charged both negatively and positively. The grains act like a source when they are charged positively due to the irradiation of the UV radiation. Both the positive and negative dust grains can coexist in laboratory and space plasmas. It appears that the dust grain charging is a new physical process in a dusty plasma, which marks a distinction between the latter and the usual multi component electron-ion plasma containing two ion species.

In this paper, we focus on two new instabilities that arise under the combined influence of the dc electric field and the dust charge gradient as well as when elongated dust grains have finite angular rotation. The manuscript is organized in the following fashion: In Section II, we briefly describe the general properties of dusty plasmas. In Section III, we present two novel mechanisms for the excitation of dust acoustic waves. Section IV contains a brief summary of our investigation.

PROPERTIES OF DUSTY PLASMAS

The constituents of dusty plasmas are electrons, ions, and extremely massive charged dust grains. There are three characteristic length scales for such a combined dust and plasma mixture. These are the dust grain radius R, the dusty plasma Debye radius λ_D, and an average intergrain distance d, roughly related to the dust number density n_d by $n_d d^3 \sim 1$. Here, the dusty plasma Debye radius [11] is given by $\lambda_D^{-2} = \lambda_{De}^{-2} + \lambda_{Di}^{-2}$, where $\lambda_{De}(\lambda_{Di})$ is the electron (ion) Debye radius. In dusty plasmas, we typically have $R << \lambda_D$. One can treat the dust from a particle dynamics point of view when $R << \lambda_D < d$, and in that case we have a plasma containing isolated screened dust grains, or dust-in-plasma. On the other hand, collective effects of charged dust grains become important when $R << d < \lambda_D$. Here, charged dust particulates, which are essential ingredients of the total plasma mixture, can be treated as massive point particles similar to multiply charged negative (or positive) ions in a multi-species plasma. The dusty plasma quasi-neutrality condition for negatively charged dust grains is $n_{i0} = n_{e0} + Z_{d0}n_{d0}$, where n_{j0} is the unperturbed number density of the particle species j (j equals e for the electrons, i for the ions, and d for the dust grains) and Z_{d0} is the number of charges residing on the dust grain surface. When most of the electrons from the ambient plasma are attached onto the dust grain surface, we may have $Z_{d0}n_{d0} >> n_{e0}$. Here, the dusty plasma may be regarded as a two component plasma composed of negatively charged dust grains and ions; the latter shield the dust grains. Such a situation is common in the Saturn rings as well as in low-temperature laboratory discharges. On the other hand, in thermal or UV irradiated dusty plasmas, the grains emit electrons and they are charged positively. The shielding of positive grains comes from the electrons, and at equilibrium we have $n_{e0} \approx Z_{d0}n_{d0}$, since the ion number density is completely depleted.

The dust grains in a dusty plasma could be either weakly or strongly correlated depending on the strength of the Coulomb coupling parameter $\Gamma = (Q^2/dT_d) \exp(-\kappa)$, where $Q = Z_d e$, e is the magnitude of the electron charge, and $\kappa = d/\lambda_D$. A dusty plasma can be considered as weakly coupled as long as $\Gamma \leq 1$. However, when $\Gamma >> 1$ and $\kappa >> 1$ charged dust microspheres strongly interact with each other, and we have the possibility of forming Coulomb lattices in a strongly coupled dusty plasma. According to Ikezi [10] the critical value of Γ for the Coulomb crystallization of charged dust grains is 170. Strongly coupled dusty plasmas are created in low-temperature dusty plasma discharges for studying the formation and dynamics of dusty crystals. They are also found in a highly evolved star, in a white dwarf, in planetary rings (narrow rings of Uranus, incomplete rings of Neptune, etc.), in the Jovian interior, in laser implosion experiments, as well as in colloidal systems.

There have been arguments that a dusty plasma is similar to a multi-ion plasma. However, this assertion has to be refuted because a dusty plasma is significantly different from a multi-ion plasma in that the presence of massive charged dust grains produce new collective phenomena on a completely new time and space scales. An example is the dust acoustic wave (DAW) in which the dust mass provides the inertia, while the restoring force comes from the pressure of the inertialess electrons and ions. In laboratory dusty plasma discharges, the DAW frequency is typically 10 Hz and the video images of the DAW wavefront is possible. Also the dust charge fluctuation dynamics [12,13] and dust-dust interactions [14,15] give rise to new effects. The dust mass and shape distributions [2] as well as the dust rotation [16] and the plasma boundary [17] introduce new effects in the dusty plasma as well. Furthermore, there is a dust lattice wave [18–20] whose counterpart exists only in solids. Finally, in a strongly coupled dusty plasma we have the possibility of new attractive forces (viz. the wakefield [21], the dipolar interaction [22] etc.) as well as the phase transition [23]. Thus, the knowledge of basic plasma physics, probe theory, statistical mechanics, as well as solid state physics and condensed matter physics is very essential for understanding collective processes in dusty plasmas.

NEW INSTABILITIES IN DUSTY PLASMAS

In the past, Rosenberg [24] discussed the possibility of dusty plasma wave excitation in the presence of equilibrium ion drifts in a collisionless dusty plasma which is uniform. Specifically, she has theoretically predicted the excitation of the dust acoustic and dust ion-acoustic waves due to the two-stream and kinetic instabilities. The kinetic results for the dust acoustic wave excitation are in excellent agreement with the experimental observations [25–28]. However, in a nonuniform dusty plasma sheath, there appears a dust charge gradient [29]. In the following, we show [13] that free energy stored in the latter can be coupled to the DAW. Consequently, the dusty plasma sheath becomes unstable [29]. In order to understand the physics of unstable dusty plasma sheath, we first consider its equilibrium properties which are governed by [13]

$$n_{d0} u_0 = \text{constant}, \tag{1}$$

$$u_0 u_0' + \frac{Q_0}{m_d} \phi_0' + g_x = 0, \tag{2}$$

$$u_0 Q_0' = I_{e0} + I_{i0} \equiv I_0, \tag{3}$$

and

$$e(n_{i0} - n_{e0}) + Q_0 n_{d0} = 0, \tag{4}$$

where u_0 is the component of the equilibrium dust fluid velocity along the x axis, $u_0' = \partial u_0 / \partial x$, $Q_0 (= V_0/R)$ is the unperturbed dust charge, V_0 is the unperturbed grain potential, $\phi_0' = \partial \phi_0 / \partial x \equiv -E_{0x}$ is the unperturbed sheath electric field, $m_d g_x$ is the x component of the gravity force, and m_d is the dust mass. The unperturbed OLM currents are

$$I_{e0} = -\pi R^2 e n_e(\phi_0)(8/\pi)^{1/2} v_{te} \exp(eQ_0/RT_e), \tag{5}$$

and

$$I_{i0} = \pi R^2 e n_i(\phi_0)(8/\pi)^{1/2} v_{ti}(1 - eQ_0/RT_i), \tag{6}$$

where $v_{tj} = (T_j/m_j)^{1/2}$ is the thermal velocity of the particle species j and m_j is the mass.

Equations (1) to (3) reveal that in the absence of the equilibrium dust fluid velocity, we have $Q_0 E_{0x} = m_d g_x$ and $I_{e0} + I_{i0} = 0$. The latter determines the equilibrium charge on the dust grain surface, while the former dictates that the balance between the sheath electric and gravity forces is responsible for the levitation of the dust grains. On the other hand, in the presence of a uniform dust flow, there appears a dust charge gradient $Q'_0 = I_0/u_0$, which can be expressed as [13]

$$Q'_0 = -\pi R^2 e(8/\pi u_0^2)^{1/2}\left[N_e v_{te}\exp(eQ_0/RT_e) - N_i v_{ti}(1 - eQ_0/RT_i)\right], \tag{7}$$

where $N_e = n_{e0}\exp(e\phi_0/T_e)$ and $N_i = n_{i0}\exp(-e\phi_0/T_i)$.

In order to study the instability of our equilibrium state, as described above, we let the number density $n_d = n_{d0} + n_1$, the dust fluid velocity $\mathbf{u}_d = u_0\hat{\mathbf{x}} + \mathbf{u}_1$, the potential $\phi = \phi_0(x) + \phi_1$, and the dust charge $q_d = Q_0(x) + Q_1$, where n_{d0} and u_0 are uniform, and n_1, $\mathbf{u}_1$, ϕ_1, and Q_1 are small perturbations in their equilibrium values. The relevant equations for perturbed quantities associated with the DAW in a nonuniform dusty plasma are [13] then

$$d_t n_1 + \nabla \cdot (n_{d0}\mathbf{u}_1) = 0, \tag{8}$$

$$d_t \mathbf{u}_1 + (Q_0/m_d)\nabla\phi_1 + \hat{\mathbf{x}}(\phi'_0/m_d)Q_1 = 0, \tag{9}$$

$$\nabla^2\phi_1 = k_D^2\phi_1 - 4\pi(Q_0 n_1 + n_{d0}Q_1), \tag{10}$$

and

$$(d_t + \nu_1)Q_1 + Q'_0 u_{1x} = -\nu_2 R\phi_1, \tag{11}$$

where $d_t = \partial_t + u_0\partial_x$, $k_D^2 = 4\pi e^2[(n_{e0}/T_e) + (n_{i0}/T_i)] \equiv \lambda_{De}^{-2} + \lambda_{Di}^{-2} = \lambda_D^{-2}$ is the square of the effective Debye wave number, $\lambda_{De} = (T_e/4\pi n_{e0}e^2)^{1/2} \equiv v_{te}/\omega_{pe}$ is the electron Debye radius, $\lambda_{Di} = (T_i n_{e0}/T_e n_{i0})^{1/2}\lambda_{De}$ is the ion Debye radius, and $\omega_{pe} = (4\pi n_{e0}e^2/m_e)^{1/2}$ is the electron plasma frequency. Furthermore, $\nu_1 = (R/\sqrt{2\pi})[(\omega_{pi}/\lambda_{Di}) + (\omega_{pe}/\lambda_{De})\exp(eV_0/T_e)]$ is the dust charge relaxation frequency arising from the dust grain surface potential changes, and $\nu_2 = (R/\sqrt{2\pi})[(\omega_{pi}/\lambda_{Di})(1 - eV_0/T_i) + (\omega_{pe}/\lambda_{De})\exp(eV_0/T_e)]$ is a frequency associated with changes in the OLM currents due to the presence of oscillating potential, and $\omega_{pi} = (4\pi n_{i0}e^2/m_i)^{1/2}$ is the ion plasma frequency. We notice that (9) and (11) contain the terms $(\phi'_0/m_d)Q_1$ and $Q'_0 u_{1x}$, which are associated with perturbed electrostatic forces involving the unperturbed sheath electric field and the convection of the equilibrium dust charge gradient, respectively. These two forces are responsible for the novel absolute instabilities, as discussed below.

The local dispersion relation can be obtained from (8) to (11) by supposing that the perturbed quantities are proportional to $\exp(iky - i\omega t)$, where k is the wave number and ω is the frequency. Accordingly, we have [13] for $k^2/k_D^2 << 1$

$$1 - \frac{\omega_D^2}{\omega^2} + \left(1 + \frac{\Omega_D^2}{\omega^2}\right)\frac{f\nu_2}{\nu_1 - i(\omega + \Omega)} = 0, \tag{13}$$

where $\omega_D = k\omega_{pd}/k_D$ is the dust acoustic frequency [6], $\omega_{pd} = (4\pi n_{d0}Q_0^2/m_d)^{1/2}$ is the dust plasma frequency, $\Omega_D^2 = Q_0 k_Q \phi_0'/m_d$, $k_Q = Q_0'/Q_0$, $f = 4\pi n_{d0}R/k_D^2$, and $\Omega = \Omega_D^2/\omega$. Equation (13) is a cubic polynomial in ω, which can be analyzed numerically. However, some useful analytical results, which exhibit novel instabilities, can be derived from (13) in several limiting cases.

We consider that the wave frequency is much smaller than the dust charge relaxation frequency ν_1, which is typically the case in low temperature laboratory dusty plasma discharges. Here, (13) takes the form [13]

$$(\omega^2 - \omega_D^2)(\nu_1 - i\Omega_D^2/\omega) + (\omega^2 + \Omega_D^2)f\nu_2 = 0. \tag{14}$$

Several useful results follow from (14). First, for $\Omega_D = 0$ (which ensures that there are no dc electric field and the dust charge gradient), we have from (14) the modified DAW frequency $\omega = \omega_D/(1 + f\nu_2/\nu_1) \equiv \Omega_0$. The term $f\nu_2/\nu_1$ is due to the dust charge fluctuation effect. Second, for $\omega \approx \omega_D + i\gamma$, where $\gamma < \omega_D, |\Omega_D|$, (14) gives the growth rate

$$\gamma = \frac{(\omega_D^2 + \Omega_D^2)\Omega_D^2 f\nu_2}{2(\nu_1^2\omega_D^2 + \Omega_D^4)}, \tag{15}$$

of a novel DAW instability for $\Omega_D^2 > 0$. The latter is fulfilled if $E_{0x}Q_0' < 0$. Clearly, the dc electric field and the dust charge gradient must oppose each other for the dusty plasma to become unstable. As an illustration, we mention that in laboratory experiments, we typically have $n_{e0} \sim n_{i0} \approx 10^8$ cm^{-3}, $T_e \sim 10T_i \approx 1$ eV, $R \sim 1 - 10\,\mu$m, $\lambda_D \sim 10^2 - 10^3\mu$m. Accordingly, for $f\nu_2/\nu_1 \sim 1$, $\nu_1 \sim 10^3$ s^{-1}, $\omega_D \sim 60$ s^{-1}, and $\Omega_D = 10\,$s^{-1} the growth time, deduced from (15), is a fraction of a second. This is consistent with the observation [29].

Next, we discuss the instability of the dust acoustic waves in a dusty plasma containing elongated and rotating dust grains. Here, the dipole moments of the dust grains are nonzero. The motion of the charged dust grain in the electromagnetic fields $[\mathbf{E} - \nabla\phi - c^{-1}\partial_t\mathbf{A}$ and $\mathbf{B} = \nabla \times \mathbf{A})$ is described by the Hamiltonian [16]

$$H = \frac{1}{2m_d}\left[\mathbf{P} - \frac{q}{c}\mathbf{A} + \frac{1}{c}(\mathbf{d} \times \mathbf{B})\right]^2 + \frac{P_\varphi^2}{2I} + q\phi - (\mathbf{d} \cdot \mathbf{E}), \tag{16}$$

where $\mathbf{P}$ is the generalized momentum, $P_\varphi = I\Omega$ is the angular momentum, I the moment of inertia of the elongated grain, Ω is the constant angular frequency of the dust grain rotation, $\mathbf{d}$ is the dipole moment, and q is the dust charge. The scalar and vector potentials of the electromagnetic fields are denoted by ϕ and $\mathbf{A}$, respectively. Mahmoodi et $al.$ [16] has derived the kinetic equation for the dust grain distribution in the presence of the Hamiltonian (16). Subsequently, the perturbed dust distribution function is obtained for the case where the dust grain size is much smaller than the wavelength of the perturbation, and that the wave phase velocity exceeds the thermal velocity of the dust particles. By choosing the unperturbed distribution function of the rotating dust grains of the form

$$f_{d0} = n_{d0}(2\pi m_d T_d)^{-3/2}(2\pi I T_d)^{-1/2} \exp\left[-\frac{p^2}{2m_d T_d} - \frac{(p_\varphi - p_{\varphi 0})^2}{2 I T_d}\right], \qquad (17)$$

where T_d is the dust temperature, $p_{\varphi 0} = I\Omega_0$, and Ω_0 is the preferred angular frequency of the rotating dust grains, Mahmoodi et $al.$ [16] derived the dielectric tensor for the dusty plasma following the standard method. For the longitudinal waves ($\omega \ll kc$) the modified dispersion relation for the DAW reads [16]

$$1 + \frac{1}{k^2 \lambda_D^2} - \frac{\omega_{pd}^2}{\omega^2} - \frac{k_\perp^2}{k^2} \frac{\Omega_r^2}{(\omega - \Omega_0)^2} = 0, \qquad (18)$$

where $\Omega_r = (4\pi d^2 n_{d0}/2I)^{1/2}$. From (18) it follows that the dust grain rotation gives the contribution only for waves with $k_\perp^2 \neq 0$. For $\Omega_0 = 0$ that contribution is expressed in the change of the dust acoustic frequency

$$\omega = \omega_{DA}\left(1 + k_\perp^2 \Omega_r^2/k^2 \Omega_{pd}^2\right)^{1/2}, \qquad (19)$$

where $\omega_{DA} = k\lambda_D \omega_{pd}/(1 + k^2 \lambda_D^2)^{1/2}$ is the DAW frequency [6] including dispersive effects. However, the presence of the dust grain rotation, (19) admits complex solutions for any rotation frequency Ω_0, satisfying the condition

$$\Omega_0 < \omega_{DA}\left[1 + \left(\frac{k_\perp^2}{k^2}\frac{\Omega_r^2}{\omega_{pd}^2}\right)^{1/3}\right]^{3/2}. \qquad (20)$$

The equality of Ω_0 on the right-hand side of (20) defines the boundary of the stability of the dust acoustic wave. Letting $\omega = \Omega_0 + i\gamma$, where $\gamma \ll \Omega_0$, we obtain from (18) the growth rate for $\omega_{pd}/\Omega_0 \approx (1 + 1/k^2\lambda_D^2)^{1/2}$,

$$\gamma = 3^{1/2} 2^{-4/3}\left(\frac{k_\perp^2}{k^2}\frac{\Omega_r^2}{\omega_{pd}^2}\right)^{1/3}\Omega_0. \qquad (21)$$

It turns out that the energy of the dust rotation can flow into plasma oscillations, driving them at nonthermal level. The scattering of transverse electromagnetic waves off these enhanced fluctuations can help to determine the existence of a preferred frequency of the dust grain rotation.

9

DISCUSSION

In this paper, we have presented two novel instabilities in an unmagnetized dusty plasma. First, we have considered the instability of a nonuniform dusty plasma sheath in the presence of a dc electric field and the equilibrium dust charge gradient. The latter is maintained by the equilibrium electron and ion currents that reach the dust grain surface, as well as by a finite equilibrium dust flow that is driven by the dc electric field. It is found that when the forces associated with the dc electric field and the dust charge gradient oppose each other, a dusty plasma is subjected to an absolute instability whose growth rate is given by (15). This instability has been observed by Nunomura [29] near the sheath boundary in a dusty plasma. Physically, instabilities arise because the dc sheath electric field does work on the dust grains to create dust charge fluctuations which cannot keep in phase with the potential of electrostatic disturbances in a nonuniform dusty plasma with a dust charge gradient. Thus, free energy stored in the latter is coupled to unstable electrostatic waves when the dc electric field (in association with the dust charge fluctuation) produces a charge imbalance in the dusty plasma. Second, we have discussed the dispersion properties of a dusty plasma containing elongated and rotating dust grains [16]. It is found that the rotational energy of the dust grains can be coupled to plasma oscillations when the wave frequency is close to the rotational angular frequency of the dust grain. The instability of the longitudinal waves occurs only in the case when the wave vector lies in the plane of the dust grains. In this case, there exists a coupling between the longitudinal electric field and charges that are placed onto the dust grain surfaces and that rotate together with the dust grains. The dust grain rotation induced electrostatic waves can account for fluctuations in astrophysical objects as well as in experiments that are presently under construction for understanding the physics of needle-shape dust grains in a controlled environment.

Acknowledgments

The author is grateful to Prof. Nodar Tsintsadze and Prof. Davy Tskhakaya for a valuable collaboration. This work was partially supported by the Deutsche Forschungsgemeinschaft through the Sonderforschungsbereich 191 "Physikalische Grundlagen der Niedertemperatur Plasmen", the Swedish Natural Science Research Council, as well as by the NATO project entitled "Studies of Collective Processes in Dusty Plasmas" through the grant SA(PST.CGL974733)5066, and by the European Union through the Human Potential- Research and Training Network (RTN) Program for carrying out the project entitled "Complex Plasmas: The science of Laboratory Colloidal Plasmas and Mesospheric Charged Dust Aerosols". The author also acknowledges the support of the International Space Science Institute (ISSI), Bern (Switzerland) for the international team "Dust Plasma Interaction in Space".

REFERENCES

1. Mendis, D. A., in *Advances in Dusty Plasmas*, Editors: P. K. Shukla, D. A. Mendis, and T. Desai, 1997, World Scientific, Singapore, pp. 3–19.
2. Verheest, F., *Waves in Dusty Space Plasmas*, 2000, Kluwer Academic Publishers, Dordrecht, The Netherlands.
3. Havnes, O., Troim, J., Blix, T. *et al.*, *J. Geophys. Res.* **101**, 10839 (1996).
4. Havnes, O, Naesheim, L. I., Hartquist, T. *et al.*, *Planet. Space Sci.* **44**, 1191 (1996).
5. Shukla, P. K., "Nonlinear Effects in Dusty Plasmas", in *Proceedings of the First Capri Workshop on Dusty Plasmas*, 1999, edited by Nappi C (Arco Felice, Napoli, Italy: Consiglio Nazionale delle Ricerche, Instituto di Cibernetica), pp. 38–39.
6. Rao, N. N., Shukla, P. K., and Yu, M. Y., *Planet. Space Sci.* **38**, 345 (1990).
7. Shukla, P. K. and Silin, V. P., *Physica Scripta* **45**, 508 (1992); Shukla, P. K., *ibid* **45**, 504; Nakamura, Y., Bailung, H., and Shukla, P. K., *Phys. Rev. Lett.* **83**, 1602 (1999).
8. Chu, J. H. and I, Lin, *Phys. Rev. Lett.* **72**, 4009 (1994); Chu, J. H., Du, J. B., and I, Lin, *J. Phys. D: Appl. Phys.* **27**, 296 (1994).
9. Thomas H., Morfill G. E., Demmel V., Goree J., Feuerbacher B., and Möhlmann, D., *Phys. Rev. Lett.* **73** 652 (1994).
10. Ikezi, H., *Phys. Fluids* **29** 1765 (1986).
11. Shukla, P. K., *Phys. Plasmas* **1**, 1362 (1994).
12. Varma, R. K., Shukla, P. K., and Krishan, V., *Phys. Rev. E* **47**, 3612 (1993); Shukla, P. K. 1996 in *The Physics of Dusty Plasmas*, edited by Shukla, P. K., Mendis, D. A., and Chow, V. W. (Singapore: World Scientific), pp. 107–121.
13. Shukla, P. K., *Phys. Lett. A* **268**, 100 (2000).
14. de Angelis, U. and Shukla, P. K., *Phys. Lett. A* **244**, 557 (1998); de Angelis, U. and Shukla, P. K., *Physica Scripta* **60**, 69 (1999).
15. Rosenberg, M. and Kalman, G., *Phys. Rev. E* **56**, 7166 (1997); Murillo, M. S., *Phys. Plasmas* **5**, 3116 (1998); Mamun, A. A., Shukla, P. K., and Farid, T., *ibid.* **7**, 2329 (2000).
16. Mahmoodi, J., Shukla, P. K., Tsintsadze, N. L., and Tskhakaya, D. D., *Phys. Rev. Lett.* **84**, 2626 (2000).
17. Shukla, P. K. and Rosenberg, M., *Phys. Plasmas* **6**, 1038 (1999).
18. Melandsø , F. , *Phys. Plasmas* **3**, 3890 (1996).
19. Morfill, G. E., Thomas, H. M., and Zuzic, M. 1996 in *Advances in Dusty Plasmas*, edited by Shukla, P. K., Mendis, D. A., and Desai, T. (Singapore: World Scientific) pp. 99–142
20. Shukla, P. K., *Phys. Rev. Lett.* **84**, 5328 (2000).
21. Nambu, M., Vladimirov, S. V., and Shukla, P. K. *Phys. Lett. A* **203**, 40 (1995); Shukla, P. K. and Rao, N. N., *Phys. Plasmas* **3**, 1760 (1996).
22. Mohideen, U., Rahman, H. U., Smith, M. A., Rosenberg, M., and Mendis, D. A., *Phys. Rev. Lett.* **81**, 349 (1998).
23. Avinash, K. and Shukla, P. K., *Phys. Lett. A* **225**, 82 (1999); *ibid.* **258** 195 (1999).
24. Rosenberg, M., *Planet. Space Sci.* **41**, 229 (1993).
25. Barkan, A., Merlino, R. L., and D'Angelo, N., *Phys. Plasmas* **2**, 3563 (1995); Prab-

hakara, H. R., and Tanna, V. L. , *ibid* **3**, 1212 (1996); Thomas Jr., E. and Watson, M., *ibid.* **6**, 4111 (1999).

26. J. B. Pieper and J. Goree, *Phys. Rev. Lett.* **77**, 3137 (1996).

27. Molotkov, V. I., Nefedov, A. P., Torchinskii, V. M., Fortov, V. E., and Kharpak, A. G., *Zh. Eksp. Teor. Fiz.* **116**, 902 (1999) [*JETP* **89**, 477 (1999)].

28. Fortov, V. E., Kharpak, A. G., Kharpak, S. A., Molotkov, V. I., Nefedov, A. P., Petrov, O. F., and Torchinsky, V. M. ,*Phys. Plasmas* **7**, 1374 (2000).

29. Nunomura, S., Misawa, T., Ohno, N., and Takamura, S., *Phys. Rev. Lett.* **83**, 1970 (1999).

Waves in Dusty Plasmas and the Concept of Fugacity

Nagesha N. Rao[1]

Theoretical Physics Division
Physical Research Laboratory
Navrangpura, Ahmedabad–380009
INDIA

Abstract. The propagation of ultra low–frequency electrostatic modes in dusty plasmas has been reviewed in the light of the concept of dust fugacity (f), which is defined by $f \equiv 4\pi n_{do}\lambda_D^2 R$ where n_{do}, λ_D and R are, respectively, the dust number density, the plasma Debye length and the grain size (radius). Dusty plasmas are defined to be tenuous, dilute or dense according as $f \ll 1$, ~ 1, or $\gg 1$, respectively. By using the fluid as well as the kinetic (Vlasov) theories, attention is focused on the "Dust–Acoustic Waves" (DAWs) and the "Dust–Coulomb Waves" (DCWs) which exist in the tenuous and the dense regimes, respectively. Unlike the DAWs which exist even for constant grain charge, the DCWs are the <u>normal modes</u> associated with grain charge fluctuations, and are driven by an effective pressure called "Coulomb Pressure". They can be considered as the electrostatic analogue of the hydromagnetic (Alfvén or magnetoacoustic) modes which are driven by the magnetic field pressure. In the dilute regime, the two modes merge into a single mode, which may be called the "Dust Charge–Density Wave" (DCDW). When the grains are closest, the DCW dispersion relation is identical with that of the "Dust–Lattice Waves" (DLWs). Dense dusty plasmas are shown to be governed by a <u>new scale–length</u> defined by $\lambda_R \equiv 1/\sqrt{4\pi n_{do}R\delta}$, where δ is a parameter related to the <u>charging frequencies</u>. The scale–length λ_R characterizes the effective shielding length due to the collective grain interactions, and plays a fundamental role in dense dusty plasmas, which is very similar to that of the Debye length (λ_D) of the tenuous regime. The frequency spectrum as well as the damping rates for the various dust modes have been analytically obtained, and compared with the numerical results.

I. INTRODUCTION

Dusty plasmas are characterized by the presence of finite–sized, highly charged grains which lead to the existence of a new, ultra low–frequency regime for waves and instabilities [1]. Among the host of dusty modes discussed in the literature during the past decade, two kinds of waves [2,3] have received wide attention [4–11]

---

[1] E–mail : raonn@prl.ernet.in

CP537, *Waves in Dusty, Solar, and Space Plasmas*, edited by F. Verheest, et al.
© 2000 American Institute of Physics 1-56396-962-9/00/$17.00

as well as experimental confirmation [12–15]. First, in the weak coupling regime when dust collective interactions are dominant, dusty plasmas support an acoustic mode called the "Dust–Acoustic Wave" (DAW) [2], which is driven by the plasma (electron and ion) thermal pressure while the inertia arises from dust mass density. Second, in the strong coupling regime when plasma crystals are formed due to the interactions between the Debye–shielded near–neighbors, dusty plasmas support a lattice mode called the "Dust–Lattice Wave " (DLW) [3]. These two wave modes exist in different frequency regimes, and have been derived from entirely different theoretical approaches. For typical dusty plasmas in the laboratory [12–15], the DLW frequency regime is lower than that of the DAWs.

The concept of "fugacity" plays a central role in Statistical Physics [16], and is most useful in characterizing "denseness" of a statistical system. The operational definition of fugacity is given by $z = \exp(\mu/k_B T)$, where μ is the chemical potential and T is the temperature. For a classical system of non–extended particles, the fugacity parameter is given by $z \approx n_0 \lambda_{dB}^3$, where n_0 is particle number density and λ_{dB} is the thermal de Broglie wavelength. Thus, the classical limit of a Fermi system requires that $z \ll 1$, which expresses the low–density, high–temperature limit of quantum systems.

While most of the dusty plasma systems that occur in the laboratory as well as in space situations are classical and non–degenerate, they do admit a dimensionless parameter f, which is similar to the fugacity parameter of Statistical Physics. The dust fugacity parameter (f) arises naturally in all problems dealing with collective processes such as waves and instabilities when dust charging effects are self–consistently included by using the charging equation. In dusty plasmas, fugacity is measure of the grain packing, and is useful in characterizing the different regimes of dusty plasmas. While different limiting forms [6,7] of this parameter has been around in the literature, the physical significance as well as the practical usefulness of dust fugacity has been clearly brought out only recently in a series of papers [17–19].

Dust fugacity is defined by $f \equiv 4\pi n_{do} \lambda_D^2 R \sim N_D R/\lambda_D$ where $N_D = n_{do} \lambda_D^3$ is the plasma parameter, and n_{do}, λ_D and R are, respectively, the dust number density, the plasma Debye length and the grain size (radius). It is then appropriate to classify [17–19] dusty plasmas as tenuous (low fugacity), dilute (medium fugacity), or dense (high fugacity) when f satisfies, $f \ll 1$, ~ 1, or $\gg 1$, respectively. A detailed fluid, kinetic as well as numerical analysis of electrostatic wave propagation in the ultra low–frequency regime reveals [17–19] that the usual DAWs are indeed applicable for the tenuous regime, while in the dense regime dusty plasmas support an entirely different and new kind of wave mode. The latter has been called the "Dust–Coulomb Wave" (DCW), and is accompanied by dust charge as well as number density perturbations which are directly proportional to each other. This is in contrast to the DAWs which exist even when the dust charge fluctuations are neglected. Furthermore, the DCWs have a simple physical interpretation in terms of an effective pressure called "Coulomb Pressure" defined [19] by $P_c = n_{do} q_{do}^2/R$, where q_{do} is the grain surface charge. Thus, the restoring force for the DCWs is provided by the Coulomb pressure and, accordingly, they can be considered as the electrostatic analogue of the hydromagnetic (Alfvén or magnetoacoustic) waves which are driven by the magnetic field pressure. Dense dusty plasmas are characterized by a new scale–length defined by $\lambda_R = 1/\sqrt{4\pi n_{do} R \delta}$ where δ is the ratio of charging frequencies. The scale–length λ_R essentially arises due to the grain collective interactions, and plays a fundamental role in dense dusty plasmas like that of the Debye length (λ_D) in the tenuous regime. In fact, the ratio of the two length scales is measure of dust fugacity, and is given through $f\delta = \lambda_D^2/\lambda_R^2$. While the existence of DAWs is experimentally well confirmed, it may be noted that the recent observation [20] on an instability associated with grain charge fluctuations may be an indication of the existence of DCWs even in the strong coupling regime.

In this review, I aim to highlight the role played by the concept of fugacity in the physics of dusty plasmas, and point out its usefulness by considering the example of electrostatic wave propagation in the ultra low–frequency regime which is dominated by dust collective dynamics.

II.   FLUID THEORY

For dust modes having phase speeds much smaller than the electron and the ion thermal speeds, it is sufficient to describe the respective number densities (n_e and n_i) by Boltzmann distributions, while the dust collective dynamics is governed by the full set of fluid equations together with the current balance equation for the grain surface charging. These are given by [2,5]

$$n_e = n_{eo} \exp\left(\frac{e\phi}{T_e}\right), \qquad n_i = n_{io} \exp\left(-\frac{e\phi}{T_i}\right), \qquad (1)$$

$$\frac{\partial n_d}{\partial t} + \frac{\partial}{\partial x}(n_d v_d) = 0, \qquad \frac{\partial v_d}{\partial t} + v_d \frac{\partial v_d}{\partial x} = -\frac{q_d}{m_d}\frac{\partial \phi}{\partial x} - \frac{\gamma_d T_d}{m_d n_d}\frac{\partial n_d}{\partial x}, \qquad (2)$$

$$\frac{\partial^2 \phi}{\partial x^2} = -4\pi\left(q_d n_d + e n_i - e n_e\right), \qquad \frac{\partial q_d}{\partial t} + v_d \frac{\partial q_d}{\partial x} = I_e + I_i, \qquad (3)$$

where n_{eo} (n_{io}) is the electron (ion) equilibrium number density, and n_d, q_d and γ_d denote, respectively, the number density, the charge, and the adiabatic index for the dust component. We shall use the standard probe theory to describe the grain surface charging. Accordingly, the electron (I_e) and the ion (I_i) currents are given by

$$I_e = -\pi e R^2 \left(\frac{8T_e}{\pi m_e}\right)^{1/2} n_e \exp\left(\frac{e\psi}{T_e}\right), \qquad I_i = \pi e R^2 \left(\frac{8T_i}{\pi m_i}\right)^{1/2} n_i \left(1 - \frac{e\psi}{T_i}\right), \qquad (4)$$

where m_e (m_i) is the electron (ion) mass, and $\psi = q_d/R$ is the dust grain surface potential relative to the plasma potential. Typically, in equilibrium, the net effect of these currents is to negatively charge the grains when secondary electron emissions are neglected.

For wave perturbations of the form $\exp[i(kx - \omega t)]$, Eqs. (1)–(4) yield the dispersion relation [17–19]

$$\frac{\omega^2}{k^2} = \frac{C_{DA}^2}{1 + k^2 \lambda_D^2 + f\Delta} + \frac{\gamma_d V_{td}^2}{2}, \tag{5}$$

where $V_{td} = \sqrt{2T_d/m_d}$, $\Delta(\omega) = \omega_2/(\omega_1 - i\omega)$, and the characteristic phase speed (C_{DA}) of the DAWs is given [2] by $C_{DA} = \lambda_D \omega_{pd}$ where $\omega_{pd} = \sqrt{4\pi n_{do} q_{do}^2/m_d}$. The charging frequencies ω_1 and ω_2 are given by [5]

$$\omega_1 = \chi + n_{eo} R e^2 \left(\frac{8\pi}{T_e m_e}\right)^{1/2} \exp\left(\frac{e\psi_0}{T_e}\right), \quad \omega_2 = \omega_1 - \frac{e\psi_0}{T_i}\chi, \quad \chi = \frac{1}{\sqrt{2\pi}} \frac{R\omega_{pi}}{\lambda_{Di}}. \tag{6}$$

where $\omega_{pi} = (4\pi n_{io} e^2/m_i)^{1/2}$ and $\lambda_{Di} = (T_i/4\pi n_{io} e^2)^{1/2}$. It may be noted that for typical dusty plasmas, $\delta \equiv \omega_2/\omega_1 \sim 1$ over a wide range of dust fugacity.

It should be mentioned that the concept of fugacity defined above is sufficient for most dusty plasmas of current interest. However, in general, it may be needed to define a "generalized fugacity" by $F = f\delta$, which incorporates the physics of charging processes through the parameter δ. We shall not, however, carry through this distinction, but use both the definitions interchangeably.

Electrostatic dust modes

The dispersion relation (5) governs the propagation of electrostatic dust modes in the entire range of dust fugacity (f) as well as the charging frequency (ω_1). We now consider below the different cases explicitly and summarize the results. More details can be found elsewhere [17–19].

(a) _Low–frequency_ : $|\omega| \ll \omega_1$

In this frequency range, dusty plasmas support different types of wave modes depending on the fugacity regime. There are three cases :
(1) _Tenuous regime_ $(f\delta \ll 1)$: In the tenuous regime, Eq. (5) yields

$$\frac{\omega_r^2}{k^2} = \frac{C_{DA}^2}{(1 + k^2\lambda_D^2)} + \frac{\gamma_d V_{td}^2}{2}, \tag{7}$$

which is the well–known DAW dispersion relation [2]. The charge fluctuation damping rate is given by

$$\frac{\omega_i}{\omega_r} = -\frac{\omega_r f\delta}{2\omega_1(1 + k^2\lambda_D^2)}, \tag{8}$$

which shows weak damping ($|\omega_i| \ll \omega_r$). Here, and in the following, we will always consider positive values of the real frequency ($\omega_r > 0$).

(2) *Dense regime ($f\delta \gg 1$) :* For dense dusty plasmas, Eq. (5) gives

$$\frac{\omega_r^2}{k^2} = \frac{C_{DC}^2}{\delta\left(1 + k^2\lambda_R^2\right)} + \frac{\gamma_d V_{td}^2}{2}, \tag{9}$$

where the characteristic phase speed C_{DC} is given by $C_{DC} = q_{do}/\sqrt{m_d R}$, and the new scale–length λ_R by $\lambda_R = 1/\sqrt{4\pi n_{do} R\delta}$. Modes represented by Eq. (9) have been called the DCWs [17–19]. The damping coefficient for the DCWs is given by

$$\frac{\omega_i}{\omega_r} = -\frac{\omega_r}{2\omega_1\left(1 + k^2\lambda_R^2\right)}, \tag{10}$$

which is weak since $|\omega_i| \ll \omega_r$ is satisfied.

(3) *Dilute regime ($f\delta \sim 1$) :*

For the general case (arbitrary $f\delta$) including the dilute regime ($f\delta \sim 1$), Eq. (5) yields [17–19] the dispersion relation for the so–called "Dust Charge–Density Waves" (DCDWs), namely,

$$\frac{\omega_r^2}{k^2} = \frac{C_{DC}^2 C_{DA}^2}{[C_{DC}^2 + C_{DA}^2\delta\left(1 + k^2\lambda_R^2\right)]} + \frac{\gamma_d V_{td}^2}{2}, \tag{11}$$

while the damping rate is determined from

$$\frac{\omega_i}{\omega_r} = -\frac{\omega_r}{2\omega_1} \frac{C_{DA}^2\delta}{[C_{DC}^2 + C_{DA}^2\delta\left(1 + k^2\lambda_R^2\right)]}. \tag{12}$$

It should be remarked that Eq. (11) yields the DAW and DCW dispersion relations, namely, Eqs. (7) and (9), as limiting cases for the tenuous and dense regimes, respectively [17–19].

(b) *High–frequency :* $|\omega| \gg \omega_1$

In this frequency regime, the grain charge is practically a constant over one wave period. Since the DCWs arise due essentially to the grain charge fluctuations, one expects *a priori* the absence of such modes in this frequency regime, while only the DAWs should exist. This follows directly also from the dispersion relation (5), as indicated below. We consider two relevant cases separately.

(1) *Case $f\delta \ll (\omega/\omega_1)$:* Here, both tenuous ($f\delta \ll 1$) as well as dilute ($f\delta \sim 1$) regimes are accessible. The relevant dispersion relation obtained from Eq. (5) is

$$\frac{\omega_r^2}{k^2} = \frac{C_{DA}^2}{\left(1 + k^2\lambda_D^2\right)} + \frac{\gamma_d V_{td}^2}{2}, \tag{13}$$

which is now applicable even in the dilute regime. Clearly, Eq. (13) shows that only the DAWs exist, as anticipated. The damping rate for $\omega_r^2 \gg k^2 V_{td}^2$ is given by

$$\frac{\omega_i}{\omega_r} = -\frac{(\omega_1/\omega_r)f\delta}{2(1+k^2\lambda_D^2)}. \tag{14}$$

Thus, for $\omega_r \gg \omega_1$, it follows that the DAWs are weakly damped both in the tenuous ($f\delta \ll 1$) as well as in the dilute ($f\delta \sim 1$) regimes, while the DCWs are totally absent.

(2) *Case $f\delta \gg (\omega/\omega_1)$* : The accessible fugacity range in this case corresponds to the (super) dense regime, and the modes are governed by

$$\omega_r^2 = \frac{\gamma_d k^2 V_{td}^2}{2}, \qquad \frac{\omega_i}{\omega_r} = -\left(\frac{\omega_r}{\omega_1}\right)\frac{C_{DA}^2}{\gamma_d V_{td}^2}\frac{1}{f\delta}. \tag{15}$$

Equations (15) represent a dust thermal wave (DTW), which is weakly damped provided $f\delta \gg (\omega_r/\omega_1)/(V_{td}^2/C_{DA}^2)$. For typical dusty plasmas of current interest ($V_{td}^2 \ll C_{DA}^2$), this corresponds to super-dense regime.

III. KINETIC THEORY

The kinetic analysis of the dust wave modes is carried out by using the Vlasov equations for the three components of the plasma, and the Poisson equation,

$$\frac{\partial f_j}{\partial t} + (\vec{v} \cdot \nabla_{\vec{x}}) f_j - \frac{q_j}{m_j} (\nabla\phi \cdot \nabla_{\vec{v}}) f_j = 0, \qquad \nabla^2\phi = -4\pi \sum_{j\equiv e,i,d}\int q_j f_j d^3v. \tag{16}$$

Equations (16) are coupled to an equation for the charge variable (q_d) which, in the kinetic theory, can be described by a Vlasov–like equation [7]

$$\frac{\partial q_d}{\partial t} + (\vec{v} \cdot \nabla_{\vec{x}}) q_d - \frac{q_d}{m_d} (\nabla\phi \cdot \nabla_{\vec{v}}) q_d = I_e + I_i, \tag{17}$$

which has the source terms arising due to the electron and the ion currents (I_e and I_i). Equation (17) is nonlinear in q_d. In order to keep the analysis tractable as well as discuss the linear modes, it is reasonable to assume the grain charge (q_d) to be velocity independent, and hence reduce Eq. (17) to the form,

$$\frac{\partial q_d}{\partial t} + (\vec{v} \cdot \nabla_{\vec{x}}) q_d = I_e + I_i, \tag{18}$$

which is similar to the fluid–like equation (3). In equilibrium, the particle species are assumed to be governed by the respective Maxwellian distributions

$$f_{jo} = n_{jo} \left(\frac{1}{\pi V_{tj}^2}\right)^{3/2} \exp\left(-\frac{v^2}{V_{tj}^2}\right); \qquad j \equiv (e, i, d), \tag{19}$$

where $V_{tj} = \sqrt{2T_j/m_j}$ is the particle thermal speed, and n_{jo} denotes the respective particle (equilibrium) number densities. For one–dimensional propagation of perturbations assumed to vary as $\sim \exp[i(kx - \omega t)]$, Eqs. (16) and (18) can be linearized and Fourier analyzed to yield the dispersion relation [7]

$$D(\omega, k) \equiv 1 + \sum_{j \equiv e,i,d} \frac{2\omega_{pj}^2}{k^2 V_{tj}^2} [1 + \xi_j Z(\xi_j)] - i \frac{4\pi R n_{do}\omega_2}{k^3 V_{td}} Z(\eta_d) = 0, \qquad (20)$$

where $\eta_d = (\omega + i\omega_1)/kV_{td}$, and $\xi_j = \omega/kV_{tj}$, $(j \equiv e, i, d)$. In Eq. (20), $Z(\zeta)$ is the plasma dispersion function defined by

$$Z(\zeta) = \frac{1}{\sqrt{\pi}} \int_{-\infty}^{+\infty} \frac{\exp(-\eta^2)}{(\eta - \zeta)} \, d\eta. \qquad (21)$$

For electrostatic modes, the dispersion relations are obtained from Eq. (20).

Electrostatic dust modes

To obtain the wave spectrum from the dispersion relation (20), we consider the case when the Landau damping effects are small. Accordingly, for wave phase speeds satisfying $V_{td} \ll (\omega/k) \ll V_{ti}, V_{te}$, we find

$$|\xi_e| \ll 1, \quad |\xi_i| \ll 1, \quad |\xi_d| \gg 1, \quad |\eta_d| \gg 1, \qquad (22)$$

and hence the dielectric response function $D(\omega, k)$ from Eq. (20) yields

$$D(\omega, k) \equiv \frac{(1 + k^2\lambda_D^2 + A_2 f \, \Delta)}{k^2\lambda_D^2} - \frac{\omega_{pd}^2 A_3}{\omega^2} + i\sqrt{\pi} \, A_1 = 0, \qquad (23)$$

where

$$A_1 = \sum_{j \equiv e,i,d} \frac{\xi_j \exp(-\xi_j^2)}{k^2\lambda_{Dj}^2}, \quad A_2 = 1 - \frac{k^2 V_{td}^2}{2(\omega_1 - i\omega)^2}, \quad A_3 = 1 + \frac{3}{2}\frac{k^2 V_{td}^2}{\omega^2}. \qquad (24)$$

The coefficient A_1 contains the contributions arising from the Landau damping, while A_2 and A_3 contain dust thermal contributions as well as the charge fluctuation damping effects. The wave spectrum is obtained by writing, $D(\omega, k) \equiv D_r(\omega, k) + i D_i(\omega, k)$ where

$$D_r(\omega, k) = \left[\frac{1 + k^2\lambda_D^2}{k^2\lambda_D^2} - \frac{\omega_{pd}^2}{\omega^2}\left(1 + \frac{3}{2}\frac{k^2 V_{td}^2}{\omega^2}\right)\right] + \frac{f\delta}{k^2\lambda_D^2} \frac{\omega_1^2}{(\omega_1^2 + \omega^2)}\left[1 - \kappa(\omega_1^2 - 3\omega^2)\right],$$

$$(25)$$

$$D_i(\omega, k) = \frac{f\delta}{k^2\lambda_D^2} \frac{\omega_1\omega}{(\omega_1^2 + \omega^2)}\left[1 - \kappa(3\omega_1^2 - \omega^2)\right] + \sqrt{\pi}A_1, \qquad (26)$$

where $\kappa = k^2 V_{td}^2/\{2(\omega_1^2 + \omega^2)^2\}$. Thus, ω_r and ω_i are calculated respectively from $D_r(\omega_r, k) = 0$ and $\omega_i = -D_i(\omega_r, k)/(\partial D_r/\partial\omega_r)$.

(a) *Low-frequency* : $|\omega| \ll \omega_1$

For frequencies much smaller than the charging frequency, we derive

$$\frac{\omega_r^2}{k^2 C_{DA}^2} = \frac{1}{(1 + k^2\lambda_D^2 + f\delta)} + \frac{3\beta_d}{2}, \qquad \frac{\omega_i}{\omega_r} = -\frac{\omega_r}{2\omega_1}\frac{\omega_r^2}{k^2 C_{DA}^2}\frac{F_1}{F_2}, \tag{27}$$

where

$$F_1 = f\delta\left(1 - \frac{3}{2}\frac{k^2 C_{DA}^2 \beta_d}{\omega_1^2}\right) + \Gamma_L, \qquad \Gamma_L \approx \sqrt{\pi}\frac{\Lambda_{ei}}{(1 + \Lambda_{ei})}\frac{\omega_1}{kC_{DA}}\frac{(1 + \epsilon)}{\sqrt{\beta_i}}, \tag{28}$$

$$F_2 = \left(1 + \frac{3k^2 C_{DA}^2 \beta_d}{\omega_r^2}\right) - \frac{\omega_r^2 f\delta}{k^2 C_{DA}^2}\frac{\omega_r^2}{\omega_1^2}\left(1 - \frac{3k^2 C_{DA}^2 \beta_d}{\omega_1^2}\right), \qquad \epsilon = \left(\frac{\beta_i}{\beta_e}\right)^{1/2}\frac{1}{\Lambda_{ei}}. \tag{29}$$

Here, we have defined the dimensionless parameters

$$\beta_d = \frac{V_{td}^2}{C_{DA}^2}, \quad \beta_e = \frac{V_{te}^2}{C_{DA}^2}, \quad \beta_i = \frac{V_{ti}^2}{C_{DA}^2}, \quad \Lambda_{ei} = \frac{\lambda_{De}^2}{\lambda_{Di}^2}, \quad \Lambda_{ed} = \frac{\lambda_{De}^2}{\lambda_{Dd}^2}. \tag{30}$$

As in the fluid case, there are different fugacity regimes.

(1) Tenuous regime (f$\delta \ll$ 1) : In this limit, we obtain the expressions

$$\frac{\omega_r^2}{k^2 C_{DA}^2} = \frac{1}{(1 + k^2\lambda_D^2)} + \frac{3\beta_d}{2}, \qquad \frac{\omega_i}{\omega_r} \approx -\frac{\omega_r}{2\omega_1}\frac{(f\delta + \Gamma_L)(1 + k^2\lambda_D^2)^{-1}}{[1 + 3\beta_d(1 + k^2\lambda_D^2)]}. \tag{31}$$

The first of Eqs. (31) is just the DAW fluid dispersion relation (7) for $\gamma_d = 3$, while the second contains the Landau damping term in addition to the charge fluctuation damping given by Eq. (8). The relative magnitude of their contributions depends on various plasma parameters.

(2) Dense regime (f$\delta \ll$ 1) : In the dense limit, Eqs. (25) and (26) yield

$$\frac{\omega_r^2}{k^2 C_{DC}^2} = \frac{1}{\delta(1 + k^2\lambda_R^2)} + \frac{3V_{td}^2}{2C_{DC}^2}, \qquad \frac{\omega_i}{\omega_r} \approx -\frac{\omega_r}{2\omega_1}\frac{[1 + (\Gamma_L/f\delta)](1 + k^2\lambda_R^2)^{-1}}{[1 + 3\beta_d f\delta(1 + k^2\lambda_R^2)]}. \tag{32}$$

The first of Eqs. (32) is the same as that obtained from the fluid theory (with $\gamma_d = 3$) for DCWs [cf. Eq. (9)], while second generalizes Eq. (10) to include the contributions due to the Landau damping.

It should be pointed out that an analysis of Eqs. (25) and (26) for the case of arbitrary fugacity (including the dilute regime, $f\delta \sim 1$) is not quite straightforward, while the resulting expressions are not simple.

(b) _High–frequency_ : $|\omega| \gg \omega_1$

In this frequency regime, the dispersion relation $D_r(\omega_r, k) = 0$ from Eq. (25) yields the bi–quadratic

$$\omega_r^2(1 + k^2\lambda_D^2) = \left(k^2 C_{DA}^2 - \omega_1^2 f\delta\right)\left(1 + \frac{3k^2 V_{td}^2}{2\omega_r^2}\right),\tag{33}$$

which sets an upper limit on the permissible fugacity values, namely, $f\delta \lesssim k^2 C_{DA}^2/\omega_1^2$ since $(k^2 C_{DA}^2 - \omega_1^2 f\delta) > 0$ is required. For weak dust thermal effects, we have expressions

$$\frac{\omega_r^2}{k^2 C_{DA}^2} \approx \left(1 - \frac{\omega_1^2 f\delta}{k^2 C_{DA}^2}\right)\Theta + \frac{3\beta_d}{2}, \qquad \frac{\omega_i}{\omega_r} \approx -\frac{\omega_1}{2\omega_r}\left(f\delta + \Gamma_L \frac{\omega_r^2}{\omega_1^2}\right)\Theta,\tag{34}$$

where $\Theta \equiv (1 + k^2\lambda_D^2)^{-1}$. Thus, as in the fluid theory [cf. Eqs. (13) and (14)], for $|\omega| \gg \omega_1$ the DAWs exist in the tenuous as well as dilute regimes, while the damping is weak when the conditions $f\delta \lesssim 1$ and $\Gamma_L \lesssim \omega_1^2/\omega_r^2$ are satisfied.

We conclude this section by pointing out a similarity [17–19] between the DCWs and the DLWs. For the case of a linear dust chain with 3–particle near–neighbor interactions, the DLW dispersion relation, in the lowest order, is given [3] by $(\omega/k)_{DLW} \sim \sqrt{2q_{do}^2/m_d d}$ where d is the inter–grain separation. This is very similar to the DCW dispersion relation [cf. Eqs. (9) and (32)] which for cold dust and for $\delta \sim 1$ as well as $k^2\lambda_R^2 \ll 1$ becomes $(\omega/k)_{DCW} \sim \sqrt{q_{do}^2/m_d R}$. In fact, for the case when the grains are closest, we have, $d \approx 2R$ and the two modes are identical even though they have been derived from entirely different approaches, and exist in different parameter regimes ! This coincidence is somewhat surprising as well as puzzling, and is possibly an indication that the DCWs or DCW–like modes may exist even in the strong coupling regime when dusty plasma crystals are formed.

IV. NUMERICAL RESULTS

The discussion in the previous two sections brings out the various normal modes that exist in different fugacity regimes. In each case, the dispersion relation has been obtained in an appropriate limit with respect to the fugacity and charging frequency. For a general case, it is not possible to obtain simple expressions, and one needs to resort to numerical work on the fluid or the kinetic dispersion relations. A detailed numerical analysis of the general dispersion relations has recently been presented elsewhere [19]. We summarize below the main results.

The numerical work has been carried out in three different ways : First, we solve the real (kinetic) dispersion relation $D_r(\omega_r, k) = 0$ from Eq. (25) for real roots by using the simple bisection method, which is sufficient. This yields the dispersion curves (ω_r vs. k), while the wave damping rates (ω_i vs. k) are analytically evaluated. Second, when Landau damping contributions are small, it is appropriate to treat the parameter A_1 from Eqs. (24) as a given numerical parameter with a small magnitude, and accordingly the general dispersion relation (23) becomes a seventh–order complex algebraic equation for the complex frequency. This has been solved for complex roots by using the standard Muller's method. Third, the full complex (transcendental) kinetic dispersion relation (23) together with Eqs. (24) has also been solved for complex roots thereby retaining self–consistently the entire contributions due to the Landau damping terms. A detailed comparison between the results obtained from the numerical as well as the analytical work shows that the dispersion curves (ω_r vs. k) obtained from the numerical solutions agree, both qualitatively as well as quantitatively, with those that follow from the analytical expressions derived from the fluid as well as the Vlasov theories. On the other hand, for the wave damping rates (ω_i vs. k), there is good agreement between the numerical and the analytical results at small wavenumbers ($k^2 \lambda_D^2 \ll 1$) in the tenuous as well as the dilute regimes. At larger wavenumbers, the net damping rate predicted by the theories is typically much lower than that found from the numerical solution of the kinetic dispersion relation.

REFERENCES

1. Verheest, F., *Space Sci. Rev.* **77**, 267 (1996).
2. Rao, N.N., Shukla, P.K., and Yu, M.Y., *Planet. Space Sci.* **38**, 543 (1990).
3. Melandso, F., *Phys. Plasmas* **3**, 3890 (1996).
4. Varma, R.K., Shukla, P.K. and Krishan, V., *Phys. Rev.* **E47**, 3612 (1992).
5. Rao, N.N., and Shukla, P.K., *Planet. Space Sci* **42**, 221 (1994).
6. Melandsø, F., Aslaksen, T.K., and Havnes, O., *Planet. Space Sci.* **41**, 321 (1993).
7. Melandsø, F., Aslaksen, T.K., and Havnes, O., *J. Geophys. Res.* **98**, 13315 (1993).
8. Rosenberg, M. and Kalman, G., *Phys. Rev. E* **56**, 7166 (1997).
9. Winske, D. and Murillo, M.S., *Phys. Rev. E* **59**, 2263 (1999).
10. Rao, N.N., *Phys. Plasmas* **6**, 2349 (1999).
11. Singh, S.V. and Rao, N.N., *Phys. Plasmas* **6**, 3157 (1999).
12. Barkan, A., Merlino, R.L., and D'Angelo, N., *Phys. Plasmas* **2**, 3563 (1995).
13. Prabhakara, H.R. and Tanna, V.L., *Phys. Plasmas* **3**, 3176 (1996).
14. Thompson, C., Barkan, A., D'Angelo, N., and Merlino, R.L., *Phys. Plasmas* **4**, 2331 (1997).
15. Homann, A., Melzer, A., Peters, A., and Piel, A., *Phys. Rev. E* **56**, 7138 (1997).
16. Huang, K., *Statistical Mechanics* (Wiley, New York, 1963), Ch. 11, p. 226.
17. Rao, N.N., *Phys. Plasmas* **6**, 4414 (1999).
18. Rao, N.N., *Phys. Plasmas* **7**, 795 (2000).
19. Rao, N.N., *Phys. Plasmas* 2000 (in Press).
20. Nunomura, S., Misawa, T., Ohno, N., and Takamura, S., *Phys. Rev. Lett.*, **83**, 1970 (1999).

The Alfvén resonance and surface waves in dusty space and astrophysical plasmas

N. F. Cramer and S. V. Vladimirov

Department of Theoretical Physics and Research Centre for Theoretical Astrophysics, School of Physics, The University of Sydney, N.S.W. 2006, Australia

Abstract. Charged dust grains can have a large effect on the dispersion characteristics of hydromagnetic Alfvén and magnetoacoustic waves propagating at frequencies well below the ion–cyclotron frequency, even if the proportion of the total charge on the dust (usually negative) is quite small in space and astrophysical plasmas, such as cometary atmospheres and interstellar molecular clouds. The dust introduces a cutoff–resonance pair. A number of effects of the dust are considered here. Wave energy propagating at oblique angles to the magnetic field in an increasing density gradient can be very efficiently damped by resonance absorption processes in a dusty plasma. It is shown that, as well as the usual Alfvén resonance, a low frequency dust–ion hybrid resonance also occurs. The effects of dust on the dispersion of surface waves in highly structured plasmas and damping due to dust charging may also play roles in such plasmas.

INTRODUCTION

Magnetoacoustic waves that propagate obliquely to the magnetic field in a nonuniform plasma can encounter the Alfvén resonance, where the wavenumber perpendicular to the magnetic field direction becomes infinite (in the limit of zero electron temperature and no resistivity) and wave energy can be absorbed there. This process has been postulated to be responsible for the heating of the solar corona [1], and has been used as an auxiliary heating mechanism for laboratory plasmas [2]. The Alfvén resonance process is modified by ion–cyclotron effects, and by the presence of multiple charged ionic species. One example is a two–ion species plasma, where a second resonance, the ion–ion hybrid resonance, arises. This resonance may play an important role in the heating of fusion plasmas with minority ion species [3]. Another example is that of a plasma with embedded charged dust grains. Dust can be a major component of space and astrophysical plasmas, such as cometary atmospheres and interstellar molecular clouds, and it may be charged due to electron attachment from the surrounding plasma or due to photoionization. For interstellar molecular clouds with electron attachment the grain charge is negative. For HII regions there can be several hundred electrons per grain, while

CP537, *Waves in Dusty, Solar, and Space Plasmas,* edited by F. Verheest, et al.
© 2000 American Institute of Physics 1-56396-962-9/00/$17.00

for HI regions there are only a few per grain [4]. The charged dust introduces an analogous second resonance, the "dust–ion hybrid" resonance.

The Alfvén resonance process is strongly modified by the presence of dust [5,6], even if the proportion of charge residing on the dust is quite small (typically the proportion is $\approx 10^{-4}$ in interstellar clouds and in cometary plasmas). The resonance absorption of waves in dusty interstellar molecular clouds could play a role in their energy balance and in the magnetic braking of protostellar clouds. Alfvén wave propagation parallel to the magnetic field in a dusty interstellar cloud was first investigated in [7]. Oblique propagation in interstellar clouds was considered in [6]. The basic properties of circularly polarized electromagnetic waves propagating parallel to the magnetic field in a plasma with static dust grains, in particular the case of frequency much less than the ion–cyclotron frequency were first studied in [8,9]. The dust can have a large effect on hydromagnetic waves propagating at frequencies well below the ion–cyclotron frequency. The right hand circularly polarized mode experiences a cutoff due to the presence of the dust. With a negligibly small charge on the dust grains, the waves have the usual shear and compressional Alfvén wave properties, while for a non–zero charge on the grains the waves are better described as circularly polarized whistler or helicon waves extending to low frequencies [8].

In this paper we review recent results in the theory of hydromagnetic waves in dusty plasmas, in particular the resonance absorption processes in such plasmas and the properties of surface waves. Some features unique to dust are discussed, such as a spectrum of dust grain sizes, and the effect of the dust grain charging process on the resonance absorption process.

THE FLUID EQUATIONS

A 3-fluid model of the cold dusty plasma is first used, which consists of the fluid momentum equations for plasma ions (singly charged), electrons, and dust grains all with the same mass and with constant (negative) charge. The uniform background magnetic field $\mathbf{B}_0$ is assumed to be in the z-direction, and the electron, ion and dust grain number densities in the uniform dusty plasma are n_{e0}, n_{i0} and n_{d0} respectively. The parameter $\delta = Z_d n_{d0}/n_{i0}$ measures the proportion of the negative charge in the plasma residing on the dust, where Z_d is the charge number of the grain. The plasma density is assumed to vary in the x–direction. Collisions between the particles and with background neutral particles are important in astrophysical plasmas, but have been ignored here to allow us to focus on the processes of interest; their effects have been discussed in [6].

The equations are linearized, so the wave fields can be taken to vary as $f(x)\exp(ik_y y + ik_z z - i\omega t)$, where k_z and ω are assumed positive. The species momentum equations may be used directly, or the usual dielectric tensor components for a three–component plasma may be used. We neglect collisional and grain charge [10] damping in the momentum equations, and assume the charge on the dust grains is not affected by the wave, although we later briefly discuss the role of the charging process. We define the local Alfvén speed using the plasma ion mass density, $v_A(x) = \sqrt{B_0^2/\mu_0\rho_{i0}}$, where $\rho_{i0}(x) = m_i n_{i0}(x)$, and the normalized frequency $f = \omega/\Omega_i$, where Ω_i is the plasma ion–cyclotron frequency.

We finally obtain the following two differential equations in x for the wave field components E_y and B_z,

$$\frac{dE_y}{dx} - \frac{k_y D}{A} E_y = i\omega \frac{A - k_y^2}{A} B_z, \qquad \frac{dB_z}{dx} + \frac{k_y D}{A} B_z = \frac{i}{\omega} \frac{A^2 - D^2}{A} E_y. \qquad (1)$$

If all the dust grains are spherical with a single size, the orbit limited motion (OLM) theory [4] tells us that, for given plasma conditions, there is the same charge on each grain. In that case, A and D are given by

$$A = \frac{\omega^2}{v_A^2} \left(\frac{1}{1 - f^2} + \frac{b}{1 - f^2/g^2} \right) - k_z^2, \qquad D = \frac{\omega \Omega_i}{v_A^2} \left(\frac{1}{1 - f^2} - \frac{b/g}{1 - f^2/g^2} \right). \qquad (2)$$

Here the ratio of charged dust mass density to plasma ion mass density is $b = m_d n_{d0}/m_i n_{i0}$, and the ratio of (magnitude of) dust grain cyclotron frequency $\Omega_d = Z_d e B_0/m_d$ to plasma ion cyclotron frequency is $g = \Omega_d/\Omega_i$.

A and D are functions of x through their dependence on v_A and b. However, if there is a continuous spectrum of dust grain sizes, the functions A and D involve integrals over the dust size (e.g., [11–13]). For a single dust grain size, A and D are real functions (provided $f \neq 1$ and $f/g \neq 1$), whereas for a size spectrum, the wave frequency may become equal to a dust–cyclotron frequency within the integrals, resulting in an imaginary part to A and D and dust–cyclotron damping of the wave.

A singularity of the equations (1) occurs where $A = 0$. For a wave of fixed frequency, the singularity occurs where the following resonance condition is satisfied:

$$\alpha^2 = f^2 \left(\frac{1}{1 - f^2} + \frac{bg^2}{g^2 - f^2} \right), \qquad (3)$$

where $\alpha = v_A k_z/\Omega_i$. For a given α, (3) has two solutions for f^2, so there are two resonance frequencies. For a cold single ion species plasma ($b = 0$), there is a single resonance frequency from (3), namely the generalized Alfvén resonance given by $\omega^2 = v_A^2 k_z^2/(1 + \alpha^2)$.

WAVES IN THE UNIFORM PLASMA

Before we proceed to a further discussion of resonance absorption and surface waves in a nonuniform plasma, it is useful to discuss the solution of Eqs. (1) in the uniform plasma. In a uniform plasma the solutions are of the form $E_y, B_z \propto \exp(ik_x x)$, where the wavenumber k_x is given by

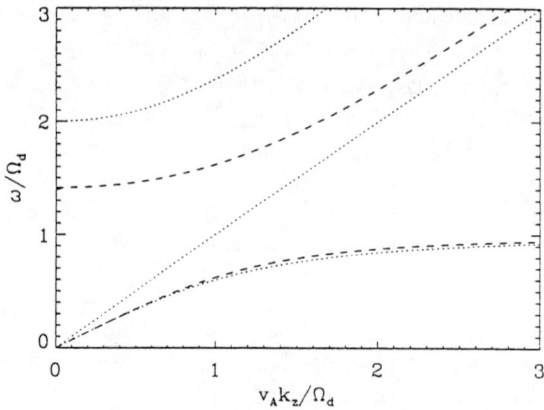

FIGURE 1. The frequency of waves in a uniform dusty plasma, normalized to the dust grain cyclotron frequency, plotted against the normalized wavenumber k_z along the magnetic field, for cutoff in k_x (dotted curves) and resonance in k_x (dashed curves). $\delta = 10^{-6}$ and $g = \Omega_d/\Omega_i = 10^{-6}$.

$$k_x^2 = -k_y^2 + \left(\frac{A^2 - D^2}{A}\right).$$ (4)

Cutoffs occur where $k_x^2 = 0$. If $k_y = 0$ also, the cutoffs correspond to the case of propagation of the wave purely parallel to the magnetic field, and are given by $A \pm D = 0$, which reduces to

$$\alpha^2 = f^2 \left[\frac{1 + g - (1 - \delta)(1 \pm f)}{(1 \pm f)(g \mp f)}\right].$$ (5)

There are three positive frequency solutions of Eq. (5) for the parallel propagating modes for given α, as shown by the dotted lines in Fig. 1. The lowest frequency mode is the dust–cyclotron mode, which is right–hand circularly polarized, has a resonance in k_z at $\omega = \Omega_d$, and has the dispersion relation at low k_z: $\omega \approx v_A k_z/(1 + b)^{1/2} = v_{AT}k_z$, where v_{AT} is the Alfvén speed based on the total mass density of plasma ions and dust grains.

The next highest frequency mode is the ion–cyclotron mode, left–hand circularly polarized, with a resonance in k_z at $\omega = \Omega_i$, and with the same dispersion relation at very low k_z, i.e. for $\omega \ll \Omega_d$, as the right–hand mode; the two modes together can form the linearly polarized Alfvén wave with phase velocity v_{AT}. For the frequency range $\Omega_d \ll \omega \ll \Omega_i$ the left–hand mode has a whistler–type dispersion relation $\omega = k_z^2 v_A^2/\delta\Omega_i$ [14,15].

The highest frequency mode is a right–hand circularly polarized mode, which is the fast Alfvén wave at high frequencies, with a k_z cutoff at $\Omega_m = (1+b)\Omega_d/(1-\delta)$. The linear and nonlinear properties of this mode for stationary dust have been discussed in detail in Ref. [15].

A resonance in k_x occurs, from Eq. (4), when $A = 0$, i.e. when Eq. (3) is satisfied. For given α this defines two resonance frequencies. The resonance frequencies are plotted in Fig. 1 (the dashed lines) against $v_A k_z / \Omega_d = \alpha / g$ for the case $s = |k_z / k_y| = 1$, $g = 10^{-6}$ and $b = 1$. The highest cutoff frequency occurs at $\omega = \Omega_m$ ($= 2\Omega_d$ in this case) for $k_z = 0$. It is useful to consider two ranges of α for a discussion of the character of the resonances.

In the limit of long wavelength along the magnetic field, such that $\alpha < g \ll 1$, i.e. $v_A k_z \ll \Omega_d$ (the left part of Fig. 1), the lower of the two resonance frequencies is given by $\omega_A = v_{AT} k_z$, i.e. the Alfvén resonance frequency for the plasma with mass density contributed by the plasma ions and the dust grains. The second (higher) resonance frequency is called the dust–ion hybrid resonance frequency here, and in this limit it is given by $\omega_H = ((1 - \delta)\Omega_d \Omega_m)^{1/2} = (1 + b)^{1/2}\Omega_d$. The dust–ion hybrid resonance frequency is the analogue of the ion–ion hybrid resonance frequency considered previously [3] in connection with resonance heating of fusion plasmas with minority ion species, usually positively charged.

In the shorter wavelength range $\Omega_d < v_A k_z \ll \Omega_i$ ($g < \alpha \ll 1$) (to the right in Fig. 1) the higher resonant frequency is given by

$$\omega_1 = (v_A^2 k_z^2 + b\Omega_d^2)^{1/2}, \tag{6}$$

and the lower by

$$\omega_2 = v_A k_z \Omega_d / (v_A^2 k_z^2 + b\Omega_d^2)^{1/2}. \tag{7}$$

Which of the two resonance frequencies is to be interpreted as the Alfvén resonance frequency now clearly depends on the size of b, or δ. If the dust mass density is small enough that $b < \alpha^2 / g^2$, we have $\omega_1 \approx v_A k_z$, which is the Alfvén resonance frequency based only on the plasma ions, while the other (dust–ion hybrid) resonance frequency is $\approx \Omega_d$. For any value of b there is always a value of α high enough that the dust motion is frozen out of the Alfvén resonance. If α becomes comparable to 1, ion–cyclotron effects modify the Alfvén resonance frequency. Even though the resonant frequency is not dependent on δ in this case, it has been shown in Ref. [5] that the two cutoff frequencies on either side of the resonant frequency are highly sensitive to the value of δ, and that this can lead to enhanced Alfvén resonance absorption in a dusty plasma. This is also the case considered in earlier work on surface waves in the dusty plasma [16], where it was shown that the presence of dust strongly modifies the dispersion relation and damping of the waves, even though the resonance condition is independent of δ.

In the case $b > \alpha^2 / g^2$, $\omega_2 \approx \alpha / b^{1/2}$, which is the Alfvén resonance frequency based on the dust mass density, and the hybrid frequency is $\omega_1 \approx b^{1/2}\Omega_d$.

Consider a wave of fixed frequency and k_z propagating in the x–direction. Around the Alfvén resonance, a wave initially at a frequency just below the local resonant frequency and propagating into an increasing plasma density, will encounter the resonance at the point in the density gradient where Eqs. (1) have a singularity.

Thus wave energy will be absorbed at the resonance position where the wave frequency satisfies Eq. (3). In the collisionless case the resonance absorption in such a nonuniform plasma can be considerably enhanced by the presence of the dust, because the wave may be cutoff downstream of the resonance [5]. Around the hybrid resonance, the wave can propagate at a frequency just below the resonant frequency, and again encounter the resonance in an increasing density profile.

Considering a spectrum of dust sizes and charges, if the resonant frequency is either much greater or much smaller than the typical dust cyclotron frequency, the dust respectively is frozen out of or into the plasma motion. In either case the resonant frequency depends simply on the effective plasma mass density. If however the resonant frequency is close to the typical Ω_d, the resonance position will depend on the mass and charge spectrum of the dust, and the resonance will be smeared out. In addition, dust cyclotron damping will contribute to the total dissipation.

SURFACE WAVES

We consider now the solutions of Eqs. (1), localized about a narrow surface of width a separating the dusty plasma and a vacuum. The dispersion relation is found by requiring that the tangential components of the electric and magnetic field, E_y and B_z, are both continuous across the boundaries $x = 0$ and $x = -a$, i.e. that the homogeneous dusty plasma and vacuum solutions be matched with the solution inside the non–zero width surface [16,17]. The surface wave solutions must have $k_x^2 < 0$ in the uniform plasma region $x < -a$, i.e., the wave fields there vary as $\exp(k_p x)$, where $k_p = |k_x|$. In the vacuum ($x > 0$) they vary as $\exp(-k_v x)$ where, since the phase velocity is assumed $\ll c$, we have to a good approximation

$$k_v^2 = k_y^2 + k_z^2. \tag{8}$$

The solution of Eqs. (1) is obtained by a perturbation technique. The wavelength in the plane of the surface is assumed much larger than the width of the surface transition, so that the parameter $\epsilon = k_z a$ is small, and the dispersion equation may be written:

$$\mathcal{D}(\tilde{\omega}) = \mathcal{D}_0(\tilde{\omega}) + \epsilon \mathcal{D}_1(\tilde{\omega}) = 0, \tag{9}$$

with $\tilde{\omega} = \omega - i\gamma$, the damping rate γ being of order $\epsilon\omega$. Here

$$\mathcal{D}_0(\tilde{\omega}) = \frac{k_p A_p - k_y D_p}{A_p - k_y^2} + \frac{k_z^2}{k_v}, \tag{10}$$

and $\mathcal{D}_1$ is the correction:

$$\mathcal{D}_1(\tilde{\omega}) = \frac{k_z}{a} \int_0^{-a} \frac{1}{A} \left[-2\frac{k_y}{k_v} D + \frac{A^2 - D^2}{k_z^2} + \frac{k_z^2}{k_v^2}(A - k_y^2) \right] dx. \tag{11}$$

28

Here, A_p and D_p are A and D evaluated in the plasma at $x = -a$.

The dispersion relation for the surface wave on a sharp dusty plasma– vacuum boundary, as $\epsilon \to 0$, is obtained by setting $\mathcal{D}_0(\omega) = 0$ in Eq. (10). This leads to the following quartic equation for f:

$$(1 - \delta)^2 f^4 - \frac{2\sigma(1 - \delta)}{(1 + s^2)^{1/2}} \alpha^2 f^3 - \left(\frac{2 + s^2}{1 + s^2}(1 + g\delta)\alpha^2 + (\delta + g)^2 \right) f^2$$

$$- \frac{2\sigma}{(1 + s^2)^{1/2}}(\delta - g^2)\alpha^2 f + \frac{2 + s^2}{1 + s^2}g(\delta + g)\alpha^2 = 0, \qquad (12)$$

where $\sigma = \text{sign}(k_y)$.

In the dust–free case ($\delta = 0$ and $g = 0$) there is one possible positive frequency solution for each sign of the wavenumber component in the $\mathbf{B}_0 \times \mathbf{n}$ or y direction, a fast wave (f^+) for positive k_y, and a slow wave (f^-) for negative k_y. For mobile dust grains there can be two surface wave solutions for each sign of k_y. We consider each sign of k_y separately.

Fig. 2 shows the frequencies of the surface waves (the solid curves) obtained from (12) for positive k_y and $s = 1$, plotted as a function of $v_A k_z / \Omega_d$ and for two values of δ. The frequencies of the cutoffs and resonances of k_x are also shown. There are two surface modes over a range of wavenumber, and the larger the proportion δ of negative charge on the dust, the greater is the separation in frequency between the two modes. The lower frequency surface wave stops at an upper frequency close to the dust– cyclotron frequency, where a resonance in k_x occurs, and so is referred to here as the dust–cyclotron surface wave. For small α/g, the dust–cyclotron surface wave has the dispersion relation

$$\omega = \left(\frac{2 + s^2}{1 + s^2} \right)^{1/2} v_{AT} k_z - \frac{b}{(1 + b)^2 (1 + s^2)^{1/2}} \frac{(v_A k_z)^2}{\Omega_d}. \qquad (13)$$

In the limit $k_z \to 0$ where the quadratic term in k_z can be neglected, this mode is the usual non-dispersive Alfvén surface wave, e.g. [18], but in the plasma of combined ion and dust mass density.

The higher frequency (fast) surface wave has the following dispersion relation for small wavenumber:

$$\omega = \Omega_m \left[1 + \frac{b}{2(1 + b)^3} \frac{(1 + (1 + s^2)^{1/2})^2}{1 + s^2} \left(\frac{v_A k_z}{\Omega_d} \right)^2 \right]. \qquad (14)$$

Thus this mode has a frequency cutoff at $\omega = \Omega_m$ as $k_z \to 0$. The fast surface mode has no upper cutoff for the range of wavenumber shown, but at much higher frequency (close to the ion–cyclotron frequency) it stops where a cutoff in k_p (the upper dotted line) is encountered [16]. A second surface mode exists generally in any two–ion species plasmas [3], and may strongly influence the Alfvén wave heating of plasmas with minority positive species. Fig. 2(c) and (d) shows the

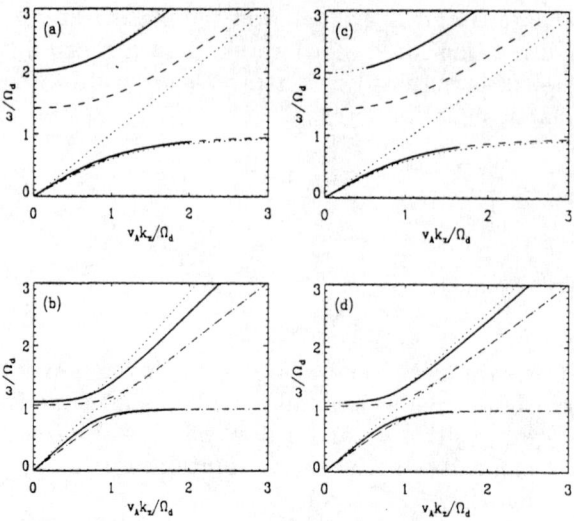

FIGURE 2. The normalized frequency of surface waves, for positive k_y, as a function of normalized wavenumber k_z (solid curves). Cutoffs (dotted curves) and resonances (dashed curves) in the dusty plasma are also shown. $g = 10^{-6}$. (a) $s = 1$, $\delta = 10^{-6}$, (b) $s = 1$, $\delta = 10^{-7}$, (c) $s = 1.3$, $\delta = 10^{-6}$, (d) $s = 1.3$, $\delta = 10^{-7}$.

same plots as in Fig. 2(a) and (b), but for $s = 1.3$. The major change is that the fast surface wave has a lower cutoff at a non–zero wavenumber k_z, where the cutoff in k_p is encountered.

For negative k_y, again there are two distinct modes, however there is only a small overlap range of wavenumber of the modes. The lower frequency mode has the same dispersion relation, for $k_z \to 0$, as the corresponding positive k_y mode. However the negative k_y mode stops at the second cutoff of k_p, rather than at a resonance. The higher frequency surface wave stops at a lower wavenumber k_z where it encounters the higher resonance. The higher frequency surface wave, at much higher frequency (close to the ion–cyclotron frequency), stops where a resonance (the Alfvén resonance) is again encountered. As s decreases, the two modes are practically continuous, except for a narrow stop-band delineated by the second cutoff and the higher resonance in k_p. The width of the stop–band between the two modes increases as the proportion of negative charge on the dust increases.

DAMPING OF THE SURFACE WAVES

A surface wave of fixed frequency on a dusty plasma–vacuum interface can experience damping in the transition region between the plasma and the vacuum at either the Alfvén resonance or the hybrid dust–ion resonance, but not both. In

30

the transition region the plasma ion density decreases from its value at $x = -a$ to zero at $x = 0$, and the local Alfvén speed correspondingly increases, so that the local value of α increases from its value in the dusty plasma at $x = -a$ to ∞ in the vacuum. Thus a surface wave of given frequency shown in Fig. 2 experiences resonance damping in the surface only if a resonance curve lies to the right of the surface wave solution curve. Whether the resonance encountered is to be interpreted as the Alfvén resonance or hybrid resonance depends on the k_z dependence of the local resonance frequency. Resonance cannot occur if the frequency lies in the range separating the lower and upper resonance curves: $\Omega_d < \omega < (1+b)^{1/2}\Omega_d$. It is evident from Fig. 2 that the positive k_y surface waves always encounter a resonance in the surface layer. Similarly, the higher frequency surface mode for negative k_y always encounters a resonance. However, the lower frequency mode for negative k_y extends into the above frequency range and so does not encounter resonance in that range, and indeed beyond it. If the resonance is not encountered, no Alfvén resonance damping occurs, and the only damping will be due to global resistive, viscous or other collisional processes, or collisionless Landau damping.

The damping rate of the surface wave due to the non-zero width of the surface is proportional to the imaginary contribution to $\mathcal{D}_1$, which comes from the pole of the integrand in Eq. (11), i.e. at that point in the surface transition where $A = 0$, i.e. the Alfvén resonance or dust–hybrid resonance condition. The resonant damping rate is independent of the dissipative mechanism in the plasma. It is well known that the actual dissipative mechanism depends on the plasma parameters, and can be resistive or viscous for collisional plasmas, or it can be mode conversion into a short wavelength mode, which is subsequently Landau damped, in a collisionless plasma. All of these processes may occur in a dusty plasma as well, but an additional unique dissipative mechanism in the presence of dust grains is the dust charging process [10]. We have shown [19] that dust charging induces the existence of an additional short wavelength highly damped mode, into which a magnetoacoustic wave (such as the surface waves considered here) will mode convert as it encounters a resonance. In a sense, this short wavelength mode is the shear Alfvén wave, modified by the charge perturbations in the plasma during the grain charging process. Another effect of the dust charging is that, even in the case of a sharp surface and thus with no resonant damping, the short wavelength mode will couple to the usual surface mode because of the boundary conditions at the surface and the existence of the electric field component E_z, with a resulting modification of the dispersion relation and a global damping due to the dust charging. Provided the dust charging frequency is small compared with the wave frequency, these effects will not change the above results on surface waves significantly.

CONCLUSIONS

We have shown that the presence of dust particles which acquire a proportion of the negative charge in a plasma can strongly modify the process of Alfvén resonance absorption, and induce an additional, dust–ion hybrid resonance. The effects of dust, including the effects of the Alfvén and the hybrid resonances, on surface waves and their damping have also been considered, and it has been shown that a second surface mode can arise. If dust charging is included, a short wavelength dissipative mode will couple to the surface wave and provide a damping mechanism, even when the surface is very sharp and no resonance damping can occur.

The presence of charged dust can also influence the excitation of hydromagnetic waves by instabilities caused by pickup ions in cometary plasmas [20]. These waves may undergo resonant absorption in the structured plasma near the cometary bow shock, and the presence of dust may enhance this absorption. The presence of charged dust in interstellar molecular clouds may influence the heating and momentum balance due to resonance absorption of waves in the clouds, and modify the properties of surface waves in such highly structured environments.

ACKNOWLEDGMENTS

Support for this work has been provided by the Australian Research Council.

REFERENCES

1. Ionson, J. A., *Astrophys. J.* **226**, 650 (1978).
2. Hasegawa, A., and Chen, L., *Phys. Fluids* **19**, 1924 (1976).
3. Cramer, N. F., and Yung, C.-M., *Plasma Phys. Controlled Fusion* **28**, 1043 (1986).
4. Spitzer, L., Jr., "Physical Processes in the Interstellar Medium", (John Wiley, New York) (1978).
5. Cramer, N. F., and Vladimirov, S. V., *Phys. Scripta* **53**, 586 (1996).
6. Cramer, N. F., and Vladimirov, S. V., *Publ. Astron. Soc. Australia* **14**, 170 (1997).
7. Pilipp, W., Hartquist, T. W., Havnes, O., and Morfill, G. E., *Astrophys. J.* **314**, 341 (1987).
8. Mendis, D. A., and Rosenberg, M., *IEEE Trans. Plasma Sci.* **20**, 929 (1992).
9. Shukla, P. K., *Phys. Scripta* **45**, 504 (1992).
10. Vladimirov, S. V., *Phys. Plasmas* **1**, 2762 (1994).
11. Bliokh, P., Sinitsin, V., and Yaroshenko, V., "Dusty and self-gravitational plasmas in space", (Kluwer Academic Publishers, Dordrecht) (1995).
12. Tripathi, K. D., and Sharma, S. K., *Phys. Plasmas* **3**, 4380 (1996).
13. Wardle, M., and Ng, C., *Mon. Not. R. Astron. Soc* **303**, 239 (1999).
14. Verheest, F., and Meuris, P., *Phys. Lett. A* **198**, 228 (1995).
15. Vladimirov, S. V., and Cramer, N. F., *Phys. Rev. E* **49**, 6762 (1996).
16. Cramer, N. F., and Vladimirov, S. V., *Phys. Plasmas* **3**, 4740 (1996).
17. Cramer, N. F., Yeung, L. K., and Vladimirov, S. V., *Phys. Plasmas* **5**, 3126 (1998).
18. Wentzel, D. G., *Astrophys. J.* **233**, 756 (1979).
19. Cramer, N.F., and Vladimirov, S.V., *Physica Scripta* **T75**, (1998).
20. Cramer, N. F., Verheest, F., and Vladimirov, S. V., *Phys. Plasmas* **6**, 36 (1999).

Non-Ideal Magnetized Dusty Plasma : Self-Similar Expansion

R Bharuthram[1] , NN Rao[2] and SR Pillay[3]

[1] *ML Sultan Technikon, P O Box 1334, Durban 4000, South Africa,*
[2] *Theoretical Physics Division, Physical Research Laboratory, Ahmedabad 380009, India,*
[3] *Department of Physics, University of Durban-Westville, Private Bag X54001, Durban 4000, South Africa*

Abstract. The self–similar expansion of a warm non–ideal magnetized dusty plasma filling semi–infinite half space has been investigated by incorporating the van der Waals equation of state for the dust species. The analysis has been carried out by using an MHD model wherein the plasma is frozen to the magnetic field lines and quasi charge–neutrality is maintained. The expansion dynamics is described by a self–similar set of nonlinear equations, which are numerically solved for the dust density, velocity and pressure profiles.

I INTRODUCTION

A common feature of recent studies on thermal dusty plasmas is the assumption of an ideal equation of state for the dust component. Such models are valid for sub-micron and micron size dust grains and dilute plasmas where the characteristic Coulomb potential energy is much less than the mean thermal energy. However, for larger grains in the super-micron range and higher grain densities, the interaction between neighbouring dust grains is enhanced by the reduced inter-particle distance. Hence, such dusty plasmas are essentially strongly coupled and the effects due to the non-ideal nature of the plasma cannot be neglected [1,2].

This paper considers the self-similar expansion into vacuum of a non-ideal magnetised dusty plasma filling semi-infinite half-space. To simplify the complex set of equations governing the dynamics of the system, we restrict our analysis to a one dimensional transverse expansion of a collisionless plasma in which the dust grains have constant charge. The effect of plasma parameters, such as temperatures and densities of the charged species, on the expansion profile is examined by solving the derived set of self-similar equations. Our results are compared with those obtained previously for an ideal dusty plasma [3].

The basic theory is given in the next section and the self-similar equations are derived in section III. Finally, in section IV we present the numerical solutions and

CP537, *Waves in Dusty, Solar, and Space Plasmas,* edited by F. Verheest, et al.
© 2000 American Institute of Physics 1-56396-962-9/00/$17.00

discussion.

II BASIC THEORY

We consider the expansion of a non-ideal plasma consisting of electrons, ions and dust grains in the presence of an ambient magnetic field $\mathbf{B} = B_0\hat{\mathbf{z}}$. The dynamical evolution of the plasma during the expansion is described by means of the MHD model developed by Rao [4,5]. This approach was also used by [3] in their study of the self-similar expansion of an ideal plasma.

The MHD model of dusty plasmas has been derived [4,5] based on the three fluid equations coupled to the Maxwell equations. Since the plasma expansion subsequent to its onset is dominated by the dynamics of the heavier dust particles, the expansion proceeds relatively slowly, reaching self-similar state asymptotically [6]. In view of this, one can neglect the displacement current in Ampere's law and replace the Poisson equation by the quasi-neutrality condition between the three species. Furthermore, since the electron mass is very small compared with the ion as well as the dust particle mass, we neglect the electron inertia effects. It is then possible to eliminate the electron fluid velocity and the electric field from the basic fluid equations to obtain a two–fluid MHD model for dusty plasmas. Omitting the relevant details which can be found elsewhere [4,5], we have

$$m_d n_d D_d \mathbf{v}_d = \frac{Z n_d}{4\pi n_e} (\nabla \times \mathbf{B}) \times \mathbf{B} + \frac{Z_d e}{c} \frac{n_i}{n_e} n_d (\mathbf{v}_d - \mathbf{v}_i) \times \mathbf{B} - Z_d \gamma_e T_e \frac{n_d}{n_e} \nabla n_e - \nabla p_d ,$$

$$(1)$$

$$m_i n_i D_i \mathbf{v}_i = \frac{n_i}{4\pi n_e} (\nabla \times \mathbf{B}) \times \mathbf{B} - \frac{Z_d e}{c} \frac{n_i}{n_e} n_d (\mathbf{v}_d - \mathbf{v}_i) \times \mathbf{B} - \gamma_e T_e \frac{n_i}{n_e} \nabla n_e - \gamma_i T_i \nabla n_i ,$$

$$(2)$$

where n_d, $\mathbf{v}_d$ and m_d denote, respectively, the number density, the fluid velocity and the mass of the dust particles, the subscript "i" denotes the corresponding quantities for the ions, Z_d is the charge number of the dust particles, γ_i and γ_e are, respectively the adiabatic indices for the ion and the electron fluids, and p_d is the dust fluid pressure. The evolution of the magnetic field $\mathbf{B}$ is governed by the induction equation,

$$\frac{\partial \mathbf{B}}{\partial t} + \frac{m_d c}{Z_d e} \nabla \times (D_d \mathbf{v}_d) - \nabla \times (\mathbf{v}_d \times \mathbf{B}) = 0 \qquad (3)$$

which is obtained from Faraday's law and the equation of motion for the dust fluid. In equations (1)–(3), $D_d \equiv \partial/\partial t + \mathbf{v}_d \cdot \nabla$ denotes the convective derivative for the dust fluid flow, and $D_i \equiv \partial/\partial t + \mathbf{v} \cdot \nabla$ for the ion fluid. The number densities are determined by the continuity equations,

$$\frac{\partial n_j}{\partial t} + \nabla \cdot (n_j \mathbf{v}_j) = 0, \tag{4}$$

where $j = d(i)$ for the dust(ion) fluid.

The non–ideal contributions due to the dust fluid have to be incorporated by means of a suitable equation of state which is derived by microscopic considerations. While the exact nature of the latter is still not known, to keep the analysis simple, as well to obtain the effect of such contributions, the well–known van der Waals equation of state is used to model the non-ideal effects due to the dust fluid. Accordingly, the dust fluid pressure is governed by the van der Waals equation expressed in terms of the dust number density (n_d) , namely,

$$(p_d + An_d^2)(1 - Bn_d) = n_d k_d T_d \tag{5}$$

where A and B have their usual definitions, namely, $A = 9K_d T_c/8n_c$ and $B = 1/3n_c$. The quasi-neutrality condition $n_e = n_i + Z_d n_d$ closes the system of governing equations.

It may be noted that the governing equations (1) and (2) have new forcing terms which are proportional to the relative velocity between the dust and ion fluid velocity. Furthermore, they are equal in magnitude but have opposite signs. This is due to the fact that in the absence of any external forces, the total momentum of the dust and ion fluid should be conserved. Further discussion on this peculiar nature of the two–fluid MHD model can be found in [3–5]. Furthermore, in the special case when ion inertial effects are neglected, (1) and (2) may be combined to yield [3–5]

$$n_d m_d D_d \mathbf{v}_d = \frac{1}{4\pi} (\nabla \times \mathbf{B}) \times \mathbf{B} - \nabla p_d - (\gamma_e T_e + \gamma_i T_i) \nabla n_i - Z_d \gamma_e T_e \nabla n_d \tag{6}$$

In the limit $m_i \to 0$, (2) and (4) may be combined to yield

$$\frac{\partial n_i}{\partial t} + \nabla_\perp \cdot \left\{ \frac{n_i}{B^2} \mathbf{B} \times (\mathbf{v}_d \times \mathbf{B}) - \frac{c}{Z_d e} \frac{n_i}{4\pi n_d B^2} \mathbf{B} \times [(\nabla \times \mathbf{B}) \times \mathbf{B}] \right.$$

$$\left. + \frac{c}{Z_d e} \frac{n_i}{n_d B^2} \left[\gamma_e T_e (\mathbf{B} \times \nabla n_e) + \frac{n_e}{n_i} \gamma_i T_i (\mathbf{B} \times \nabla n_i) \right] \right\} = 0 \tag{7}$$

where, for plasma expansion perpendicular to the magnetic field, the term $\nabla_\parallel v_{i\parallel}$ has been dropped.

The set of equations (3)–(7) describe the transverse expansion of a quasi-neutral, magnetised non-ideal dusty plasma in the frozen-in-field limit. In the next section, we consider the self–similar expansion as described by these equations.

III   SELF-SIMILAR EQUATIONS

The non-ideal dusty plasma embedded in an external field is assumed to occupy at the initial time the semi-infinite half-space $x < 0$. At later times, we consider its expansion transverse to **B** into a vacuum. Treating all variables as a function of x and t only, the y-component of (8) yields $v_{dy} = constant$, which without loss of generality can be set to zero. Defining $u_d \equiv v_{dx}$ (3), (4) and (7) may then be written in the form

$$\frac{\partial f}{\partial t} + \frac{\partial}{\partial x}(u_d f) = 0 \tag{8}$$

where $f \equiv B$, n_d and n_i.

At this point, we introduce the self-similar variable $\xi = x/\lambda t$, where the velocity scaling factor λ will be chosen *a posteriori*. Thus, following Rao and Bharuthram [3], the quasineutrality condition and (8) can be combined to yield the frozen-in-field condition

$$\frac{n_d}{n_{d0}} = \frac{n_i}{n_{i0}} = \frac{n_e}{n_{e0}} = \frac{B}{B_0} \tag{9}$$

where the subscript " 0 " indicates the corresponding quantities at some initial value of ξ, say $\xi = 0$. From (5), we have

$$\frac{dp_d}{d\xi} = \left[\frac{k_d T_d}{(1 - Bn_d)^2} - 2An_d \right] \frac{dn_d}{d\xi} \tag{10}$$

which when combined with (6) gives us

$$(u_d - \lambda\xi)\frac{du_d}{d\xi} = -\left[\frac{V_A^2}{n_{d0}} + \frac{C_s^2}{n_d} + \frac{1}{m_d n_d} \left\{ \frac{k_d T_d}{(1 - Bn_d)^2} - 2An_d \right\} \right] \frac{dn_d}{d\xi} \tag{11}$$

where $V_A = (B_0^2/4\pi n_{d0} m_d)^{1/2}$ is the dust–Alfvén speed, and we have defined the acoustic speed by $C_s = \{(n_{e0}\gamma_e k_B T_e + n_{i0}\gamma_i k_B T_i)/n_{d0}m_d\}^{1/2}$.

The dust continuity equation (4) in terms of the self-similar variable reduces to

$$(u_d - \lambda\xi)\frac{dn_d}{d\xi} = -n_d \frac{du_d}{d\xi} \tag{12}$$

For a non-trivial solution describing the non-steady expansion of the dusty plasma, we require the conditions $dn_d/d\xi \neq 0$ and $du_d/d\xi \neq 0$ to be satisfied. Accordingly, Eqs. (11) and (12) yield the secularity condition

$$u_d - \lambda\xi = \pm\sqrt{H} \tag{13}$$

where

$$H = V_A^2 \frac{n_d}{n_{d0}} + C_s^2 + \frac{1}{m_d} \left[\frac{K_d T_d}{(1 - \frac{n_d}{3n_c})^2} - \frac{9}{4} \frac{n_d K_d T_c}{n_c} \right] \tag{14}$$

Equation (13) self–consistently governs the flow velocity of the dust fluid during the self–similar expansion of the plasma.

We now select $\lambda = V_A$ so that $\xi = x/V_A t$ becomes dimensionless.
Defining the dimensionless quantities by $\tilde{n}_d = n_d/n_{d0}$, $\tilde{u}_d = u_d/V_A$, $\tilde{p}_d = p_d/n_{d0}k_dT_d$, $\beta = C_s^2/V_A^2$, $\alpha = n_{d0}K_dT_d/(B_0^2/4\pi)$, $\theta = n_{d0}/n_c$, $\tau = T_c/T_d$, the set of equations (11)—(14) are suitably combined to yield the following equation for the normalized dust number density $(\tilde{n}_d)$:

$$\frac{d\tilde{n}_d}{d\xi} = \pm \frac{\tilde{n}_d}{\sqrt{\tilde{H}}}\left[1 + \frac{1}{2}\frac{\tilde{n}_d\tilde{F}}{\tilde{H}}\right]^{-1} \tag{15}$$

where $\tilde{H} = \tilde{n}_d + \beta + \alpha(1 - \frac{\theta}{3}\tilde{n}_d)^{-2} - \frac{9}{4}\alpha\tau\tilde{n}_d$ and $\tilde{F} = 1 + \frac{2}{3}\alpha\theta(1 - \frac{\theta}{3}\tilde{n}_d)^{-3} - \frac{9}{4}\theta\alpha\tau$.
From (5) and (13) we then obtain, respectively,

$$\tilde{p}_d = \frac{\tilde{n}_d}{(1 - \frac{\theta}{3}\tilde{n}_d)} - \frac{9}{8}\tau\theta\tilde{n}_d^2 \tag{16}$$

and

$$\tilde{u}_d = \xi \pm \sqrt{\tilde{H}}. \tag{17}$$

We now briefly comment on the case of ideal dusty plasmas which has been discussed elsewhere [3]. For this case, $\tau \to 0$ and $\theta \to 0$, and hence the expression for $\tilde{H}$ reduces to $\tilde{H} = \tilde{n}_d + \beta + \alpha$ which is the same as Eq. (22) of [3] corresponding to the value of $\gamma = 1$ assumed in the Van der Waals equation (5). Then, Eq. (15) simplifies to

$$\frac{d\tilde{n}_d}{d\xi} = \pm \frac{2\tilde{n}_d[\tilde{n}_d + \beta + \alpha]}{[2\beta + 3\tilde{n}_d + 2\alpha]}$$

which is the Eq. (23) of [3] for $\gamma = 1$.

IV   NUMERICAL RESULTS AND DISCUSSION

Equation (15) governs the dust density profile during the self–similar expansion of the magnetized dusty plasma, while the pressure and velocity profiles are given by Eqs. (16) and (17), respectively. For arbitrary values of the plasma parameters, the density profiles are obtained by numerically solving Eq. (15) for specified initial/boundary conditions.

In Figures 1–2, the numerical solutions of Eqs. (15)–(17) are illustrated. In Figures 1(a)–(c) we set $\alpha = 1.0$, $\beta = 1.0$ and $\theta = 1.0$. The variation of the dust pressure with the expansion parameter ξ (figure 1(c)) for different value of τ show that as τ decreases, the non-ideal behaviour diminishes and the pressure approaches that of ideal gas in behaviour. This is not surprising since decreasing τ (T_c/T_d) implies an increase in the temperature T_d and therefore an increase in the thermal energy of the dust grains. As a result the cohesive forces are reduced and the plasma becomes less strongly coupled. From figure 1(a) it is observed that the more non-ideal the plasma becomes (larger values of τ), the less freely does it expand due to the stronger inter-dust grain coupling as a result of the cohesive forces. This effect is alo observed in figure 1(b), where in the initial stages the dust expansion velocity $\tilde{v}_d$ decreases as the non-ideal nature of the plasma increases (τ increasing).

Figures 2(a)–(c) are plotted for different values of θ (n_{do}/n_c). Increasing θ implies an increase in the number density of the dust grains which reduces the volume available to dust grains for free travel. Hence the volume reduction effects are increased and the plasma becomes more non-ideal in nature. This is seen to be most significant for the dust pressure $\tilde{p}_d$ during the expansion.

It is also found that the effect of increasing β (increasing magnetic field strength B_o) has the same effect as observed for the ideal plasma expansion by [3]. As B_o increases, the dust grains are more strongly tied to the field lines. Therefore their transverse (to $\mathbf{B}_o$) expansion is restricted. The plasma is found to expand slowly over a large distance ξ as β increases. It also retains its non-ideal character for a large distance during the expansion process.

Acknowledgement

One of us (N.N.R.) would like to thank the NRF (South Africa) for supporting his visit to the University of Durban–Westville.

REFERENCES

1. Fortov, V.E. and Iakubov, I.T., *Physics of Nonideal Plasmas* (Hemisphere, New York, 1990).
2. Rao, N.N. (1998a), *J. Plasma Phys.* **59**, 561–574.
3. Rao, N.N and Bharuthram, R. (1995), Planet. Space. Sci. **43**, 1087.
4. Rao, N.N., (1993), J. Plasma Phys. **49**, 375.
5. Rao, N.N., (1993), Planet. Space Sci. **41**, 21.
6. Gurevich, A.V., Pariiskaya, L.V. and Pitaevski, L.P., Sov. Phys. JETP, **22**, 449.

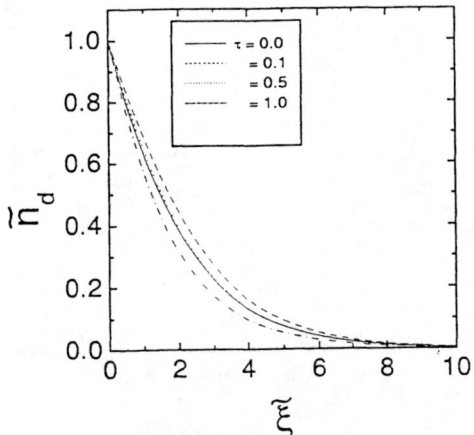

Fig 1a : Graph of dust density $\tilde{n}_d$ versus
expansion parameter $\tilde{\xi}$ for fixed
$\alpha = 1.0$, $\beta = 2.0$, $\theta = 1.0$

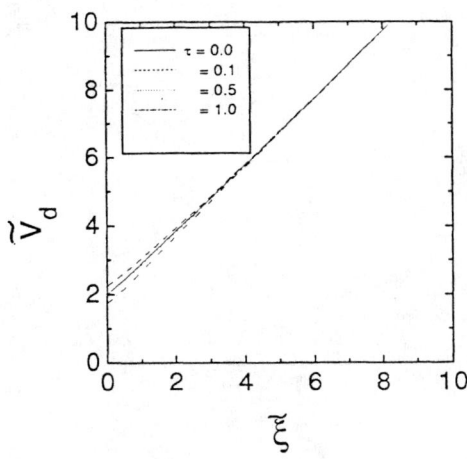

Fig 1b : Graph of dust density $\tilde{v}_d$ versus
expansion parameter $\tilde{\xi}$ for fixed
$\alpha = 1.0$, $\beta = 2.0$, $\theta = 1.0$

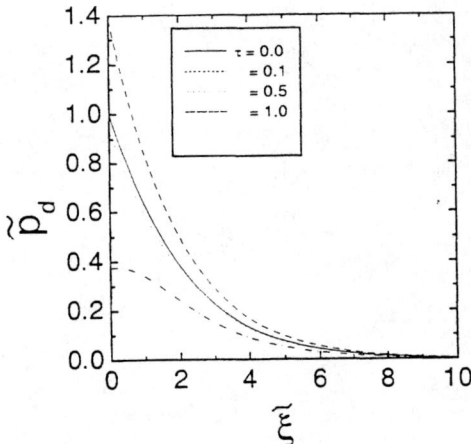

Fig 1c : Graph of dust density $\tilde{p}_d$ versus
expansion parameter $\tilde{\xi}$ for fixed
$\alpha = 1.0$, $\beta = 2.0$, $\theta = 1.0$

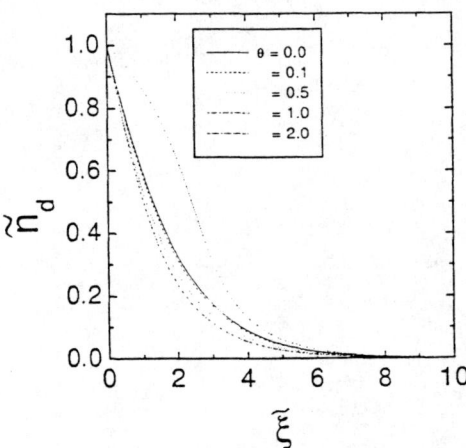

Fig 2a : Graph of dust density $\tilde{n}_d$ versus
expansion parameter $\tilde{\xi}$ for fixed
$\alpha = 1.0$, $\beta = 1.0$, $\tau = 1.0$

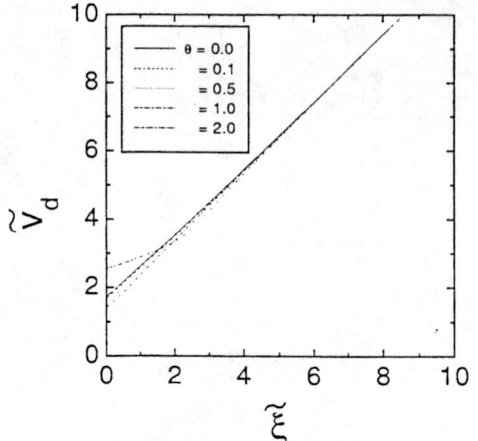

Fig 2b : Graph of dust density $\widetilde{v}_d$ versus
expansion parameter $\widetilde{\xi}$ for fixed
$\alpha = 1.0$, $\beta = 1.0$, $\tau = 1.0$

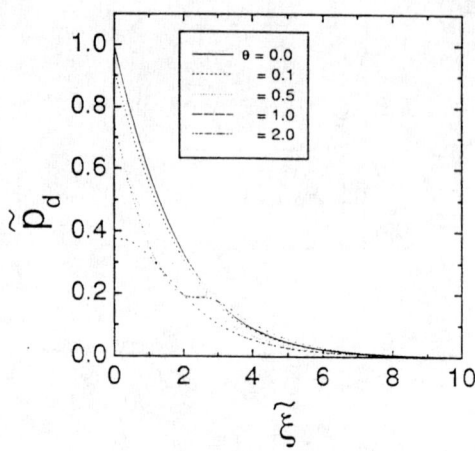

Fig 2c : Graph of dust density $\widetilde{p}_d$ versus
expansion parameter $\widetilde{\xi}$ for fixed
$\alpha = 1.0$, $\beta = 1.0$, $\tau = 1.0$

The effect of dust charge inhomogeneity on low-frequency modes in a strongly coupled plasma

T. Farid, A. A. Mamun, and P.K. Shukla

Institut fuer Theoretische Physik IV, Ruhr Universitaet Bochum, 44780-Bochum, Germany

Abstract. An analysis of low-frequency modes accounting for dust grain charge fluctuation and equilibrium grain charge inhomogeneity in a strongly coupled dusty plasma is presented. The existence of an extremely low frequency mode, which is due to the inhomogeneity in the equilibrium dust grain charge, is reported. Besides, the equilibrium dust grain charge inhomogeneity makes the dust-acoustic mode unstable. The strong correlations in the dust fluid significantly drive a new mode as well as the existing dust-acoustic mode. The applications of these results to recent experimental and to some space and astrophysical situations are discussed.

INTRODUCTION

The prediction of the phase transition [1] of a classical Coulomb plasma in the presence of a micron-sized dust particles in the strongly coupled regime (i.e., the coupling constant $\Gamma \; [=(q_d^2/a_d k_B T_d) \; exp(-a_d/\lambda_D)$, where q_d is dust charge, a_d is the inter grain distance, λ_D is the dusty plasma Debye radius, k_B is the Boltzmann's constant, and T_d is the dust temperature] for micron sized dust grains can easily be order of unity or larger) is a milestone for the study of complex plasma systems. It is noteworthy that an increase in the dust temperature melts the dust crystals, which vaporize afterwards (i.e., the weakly coupled ideal Coulomb plasma can be recovered). This prediction is verified by a number of experiments [2–4] and theoretical models [5–8]. Thus, the study of transitions from the strongly coupled to weakly coupled regimes in the dusty plasmas are confirmed by the laboratory experiments.

There are two well recognised techniques to measure the fluctuation of charges on the dust grain: first, it varies randomly due to the collection of individual electrons and ions at random times; second, in the presence of perturbations it varies when a particle moves in an inhomogeneous plasma or a plasma that varies with time; this is supported by the experiment [9], which observed a self-excited vertical oscillation of dust particles, and which is supposed to arise from the charge deviation from its equilibrium value, due to the delay in charging.

CP537, *Waves in Dusty, Solar, and Space Plasmas,* edited by F. Verheest, et al.

In this paper, we are considering a dusty plasma system consisting of the Boltzman distributed electrons and ions, and strongly correlated dust grains where the charge on a dust particle fluctuates and its equilibrium value also varies with space. The numerical analysis of the effects of the dust grain charge inhomogeneity or strong correlations in the dust fluid is assumed.

MODEL EQUATIONS

Let us consider a three-component dusty plasma; the dust grains are strongly coupled because of their lower temperature and larger electric charge (charge varies due to the dust charge gradient and fluctuating currents flowing onto the dust grain surface), whereas electrons and ions are very weakly coupled due to their higher temperatures and smaller electric charges. Hence, on the time scale of the dust plasma period, the electrons and ions are in local thermodynamic equilibrium, and their number densities, N_e and N_i, obey the Boltzmann distributions [10,11], $N_e = n_{e0} \exp(e\Phi/k_B T_e)$, and $N_i = n_{i0} \exp(-e\Phi/k_B T_i)$, (with different parameters; n_{e0} (n_{i0}) and T_e (T_i) are the unperturbed number density and the temperature of electrons (ions), respectively, Φ is the electrostatic wave potential, and e is the magnitude of the electron charge).

The spherical symmetric dust particles are assumed with the radius r_d and the surface charge q_d. If the dust particles are charged due to the collection of the electrons and ions from the background plasma, we have for the electron (ion) current I_e (I_i) [12–16]

$$I_e(q_d, \Phi) = -\pi r_d^2 e \left(\frac{8k_B T_e}{\pi m_e}\right)^{1/2} N_e \exp\left(\frac{eq_d}{r_d k_B T_e}\right), \tag{1}$$

and

$$I_i(q_d, \Phi) = \pi r_d^2 e \left(\frac{8k_B T_i}{\pi m_i}\right)^{1/2} N_i \left(1 - \frac{eq_d}{r_d k_B T_i}\right), \tag{2}$$

where m_e (m_i) is the mass of the electron (the ion). To study low-frequency electrostatic dust-modes in the strongly coupled dusty plasma system under consideration, we consider normal mode analysis; express our dependent variables N_d, U_d, Φ, and q_d in terms of their equilibrium and perturbed parts as $N_d = n_{d0} + n_d$, $\mathbf{U_d} = 0 + \mathbf{u_d}$, $q_d = -eZ_{d0}(x) - eZ_d$, and $\Phi = 0 + \phi$ (for $Z_{d0}(x)$ is number of electrons residing on the dust grain at equilibrium, which is not constant, but varies with x). The dynamics of charge fluctuating dust oscillations is governed by the well known generalized hydrodynamic (GH) equations [17–19], and the charging equation [12–16], which in the linearized form are :

$$\frac{\partial n_d}{\partial t} + n_{d0}(\nabla \cdot \mathbf{u_d}) = 0, \tag{3}$$

$$\left(1 + \tau_m \frac{\partial}{\partial t}\right)\left(\frac{\partial u_d}{\partial t} - \frac{Z_{d0}e}{m_d}\nabla\Phi + \frac{\mu_d v_{td}^2}{n_{d0}}\nabla n_d\right) = \eta'\nabla^2 \mathbf{u_d} + \left(\zeta' + \frac{1}{3}\eta'\right)\nabla(\nabla\cdot\mathbf{u_d}), \quad (4)$$

$$\left(\frac{\partial}{\partial t} + \nu_1\right)Z_d + u_{dx}\left(\frac{\partial Z_{d0}}{\partial x}\right) = \nu_2\left(\frac{n_e}{n_{e0}} - \frac{n_i}{n_{i0}}\right), \quad (5)$$

and

$$\nabla^2\phi = 4\pi e\left(n_e - n_i + Z_{d0}n_d + Z_d n_{d0}\right), \quad (6)$$

where $v_{td} = (k_B T_d/m_d)^{1/2}$, $\nu_1 = |I_{e0}|e[1/(r_d k_B T_e) + 1/(r_d k_B T_i + Z_{d0}e^2)]$, $\nu_2 = |I_{e0}|/e$, $I_{e0} = I_e(q_d = -Z_{d0}e, \Phi = 0)$, $D_t = \partial/\partial t + \mathbf{U_d}\cdot\nabla$, N_d is the dust particle number density, $\mathbf{U_d}$ is the dust fluid velocity, m_d is the dust particle mass, N_d and P_d are the dust fluid density and the pressure, τ_m is the viscoelastic relaxation time, η and ζ are transport coefficients of shear and bulk viscosities. There are various models to calculate these transport coefficients [17–24]. The viscoelastic relaxation time τ_m is given by [18,19] $[\tau_m = \frac{m_d \eta_l}{k_B T_d}\left(1 - \mu_d + \frac{4}{15}u(\Gamma)\right)^{-1}$ with μ_d the compressibility [18], and $u(\Gamma)$ [23] the measure of the excess internal energy of the system].

The set of equations (1)-(6), yield the dispersion relation [24]

$$1 + \left(1 + \frac{i\nu_2\frac{n_{d0}}{n_{e0}}}{\omega + i\nu_1}\right)\frac{1}{k^2\lambda_{De}^2} + \left(1 + \frac{i\nu_2\frac{n_{d0}}{n_{i0}}}{\omega + i\nu_1}\right)\frac{1}{k^2\lambda_{Di}^2}$$

$$- \omega_{pd}^2\left(1 - i\frac{k_d}{k}\frac{\omega}{\omega + i\nu_1}\right)\left(\omega^2 - \mu_d k^2 v_{td}^2 + i\frac{\eta_l\omega k^2}{(1 - i\omega\tau_m)}\right)^{-1} = 0, \quad (7)$$

where $k_d = (1/Z_{d0})(\partial Z_{d0}/\partial x)$ is the inverse of the dust grain charge inhomogeneity scale length. It is noted here that $k_d = 0$ for homogeneous dust grain charge (i.e., for Z_{d0}=constant). It can be shown that if we consider the dust grains with constant and homogeneous charge (i.e. , $\nu_{1,2} = 0$ and $k_d = 0$) and weakly coupled dusty plasma (i.e., $\Gamma << 1$ or $\tau_m = 0$, $\mu_d = 0$, and $\eta_l = 0$) , equation (7) stands for the dispersion relation for the dust-acoustic mode studied by *Rao et al.* [11]. Again, if we consider a weakly coupled dusty plasma with the dust grain charge fluctuation (but the dust grain charge at equilibrium being constant, viz. $k_d = 0$), (7) becomes the dispersion relation for the damped dust-acoustic waves [12–16].

NUMERICAL ANALYSIS

We have numerically analyzed our general dispersion relation (7) for different values of k_d (such as, $k_d = 0$, $k_d = 0.2$, $k_d = 0.4$, and $k_d = 0.8$) with typical plasma parameters of interest: $T_e = 10^4 \,°K$, $T_i = 10^3 \,°K$, $T_d = 300 \,°K$, $m_d = 10^{-12}gm$, $n_{e0} = 10^7 cm^{-3}$, $n_{d0} = 10^4 cm^{-3}$, and $Z_{d0} \approx 10^2$. The results are displayed in Figures 1-2.

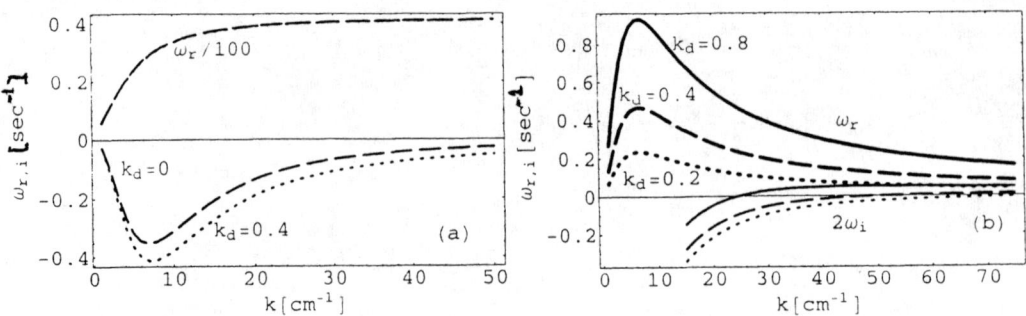

FIGURE 1. a) The real frequency ω_r and the growth rate ω_i of the existing dust-acostic mode for different values of k_d (b) The real and immaginary parts of a low- frequency mode.

Figure 1 is a plot of equation (7) for different values of $k_d = 0, 0.4$ and $\tau_m = 0$, $\mu_d = 0$, and $\eta_l = 0$; figure 1a shows a damped dust-acoustic mode: the variation of ω_r with k (postive region of figure 1a) and the damping rate (ω_i) verses k (the negative region of plot). It is observed that this agrees with the main charecteristics of the dust-acoustic waves in a dusty plasma with the dust grain charge fluctuation: the dust-acoustic mode is damped due to the dust grain charge fluctuation [11–14]. The real frequency is found to be unaffected by the dust grain charge inhomogeneity. On the other hand, figure 1b represents the variation of the real and immaginary parts of other root of (7) for $k_d = 0.8$ (solid curve), $k_d = 0.4$ (dashed curve), and $k_d = 0.2$ (dotted curve). This figure shows the properties of an extremely low-frequency stable mode which only exists in the presence of equilibrium dust grain charge inhomogeneity. It is evident from figure 1b that the growth rate increases with increasing values of $k_d = 0.2 - 0.8$. Hence, it is concluded that the equilibrium dust grain charge inhomogeneity introduces a completely new extremely low frequency stable mode and causes an unstable branch as well.

In Figure 2, the effects of the correlations of the dust fluid (for different values of the transport coefficients, such as, μ_d, η_l, and τ_m are shown for parameters used before, and for $Z_{d0} = 10^2$, we have $\Gamma = 1$. If, we replace Z_{d0} by 3×10^2 and 10^3, but use the other parameters as before, we find that Γ becomes 10 and 160, repectively. The typical values [18] of η_l are $1.04a_d^2\omega_{pd}$ for $\Gamma = 1$, $0.08a_d^2\omega_{pd}$ for $\Gamma = 10$, and $0.3a_d^2\omega_{pd}$ for $\Gamma = 160$.

Figure 2a shows the dispersion curves (ω_r vs. k) of the dust-acoustic mode (thick lines) and our new mode (thin lines) are modified by strong correlations in the dust fluid (in both cases, the solid curves represent for $\Gamma = 1$ and the dashed curves for $\Gamma = 10$). Figure 2b shows the change in the damping rate; it is deduced that the growth rate of the dust-acoustic mode (thick lines) is decreased by the strong

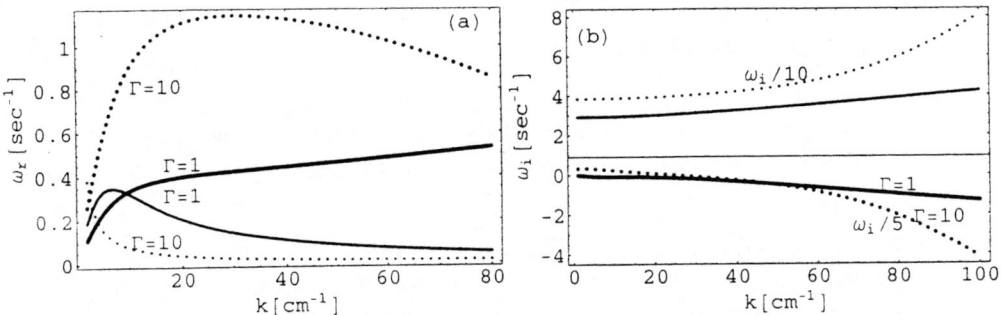

FIGURE 2. (a) The real frequency ω_r dust-acoustic mode (thick lines) and a low-frequency mode (thin lines) for different values of $\Gamma = 1$ (dotted line) and $\Gamma = 10$ (solid line), respectively. (b) The corresponding growth rate ω_i of the existing dust-acostic mode for different values of Γ.

correlations in the dust fluid. On the other hand, the other new mode (thin lines) is more unstable due to strong correlations in the dust fluid.

CONCLUSIONS

We have investigated low-frequency electrostatic dust-modes in a strongly coupled dusty plasma, in the presence of the dust grain charge fluctuation and the equilibrium dust grain charge inhomogeneity. We have assumed a three component unmagnetized dusty plasma comprising the Boltzmann electrons and ions, and strongly coupled dust fluid where the dust grain charge at equilibrium is not constant, but varies with x. We first derived the general dispersion realtion for the low-frequency electrostatic dust-mode and then numerically analyzed different roots of this general dispersion relation. The description of the modes is as follows:

- A stable low-frequency mode, which is only in the presence of the equilibrium dust grain charge inhomogeneity (cf. lower plot of Fig. 1b). It disappears in the absence of the equilibrium dust grain charge inhomogeneity and one recovers the results of earlier published works [12–14] (cf. Fig. 1a).

- The nature of the existing dust-acoustic mode is unaffected by the dust grain charge inhomogeneity (cf. Fig. 1a).

- The other branch of the dust-acoustic mode has been found to be unstable due to the effect of the equilibrium dust grain charge inhomogeneity. Also the increase in the dust grain charge inhomgeneity (i.e., k_d) stimulates the unstable wave mode for lower values of k , i.e., the wave mode of larger wavelength becomes unstable (cf. Fig. 1b).

- The effects of strong correlations in the dust fluid significantly modify the dispersion properties of this new mode as well as of the dust-acoustic mode. This modification irregularly changes with the wavelength of the mode considered (cf. Fig. 2a).

- The nature of damping of the dust-acoustic mode and of the growth rate of the instability of the dust-acoustic mode are also found to be modified significantly by strong correlations in the dust fluid (cf. Fig. 2b).

Acknowledgments

T. Farid gratefully acknowledges the support of the Deutscher Akademischer Austauschdienst (DAAD). This research was partially supported by the International Space Science Institute at Bern (Switzerland) through its international team " Dust Plasma Interaction in Space". One of the authors (A. A. Mamun) gratefully acknowledges the financial support of the Alexander von Humboldt-Stiftung (Bonn, Germany) and the study leave granted by the authority of Jahangirnagar University (Dhaka, Bangladesh).

REFERENCES

1. H. Ikezi, *Phys. Fluids* **29**, 1764-1766 (1986).
2. Y. Hayashi and K. Tachibana, *Jpn. J. Appl. Phys.***33**, 804 (1994).
3. J. H. Chu and Lin I, *Phys. Rev. Lett.* **72**, 4009 (1994).
4. H. Thomas, G. E. Morfill, V. Demmel, J. Goree, B. Feuerbacher, and D. Mohlmann, *Phys. Rev. Lett.* **73**, 652 (1994).
5. M. Rosenberg and G. Kalman, *Phys. Rev. E.* **56**, 7166 (1997).
6. H. Totsuji, T. Kishimoto, and C. Totsuji, *Phys. Rev. Lett.* **78**, 3113 (1997).
7. S. Hamaguchi, R. T. Farouki, and D. H. E. Dubin, *Phys. Rev. E.* **56**, 4671 (1997).
8. H. C. Lee and D. Y. Chen,*Phys. Rev. E*, **56** 4596 (1997).
9. S. Nunomura, T. Misawa, N. Ohno, and S. Takamura, *Phys. Rev. Lett.* **83**, 1970 (1999).
10. C. K. Goertz, *Rev. Geophys.* **27**, 271 (1989).
11. N. N. Rao, P. K. Shukla, and M. Y. Yu, *Planet. Space Sci.* **38**, 543 (1990).
12. R. K. Varma, P. K. Shukla, and V. Krishan, *Phys. Rev. E.* **47**, 3612 (1993).
13. F. Melandsø, T. K Aslaksen, and O. Havnes, *Planet. Space Sci.* **41**, 321 (1993).
14. N. N. Rao and P. K. Shukla, *Planet. Spacee Sci.* **42**, 221 (1994).
15. J. X. Ma and P. K. Shukla, *Phys. Plasmas.* **1**, 1506 (1995).
16. P. K. Shukla, in *Physics of Dusty Plasmas*, edited by P. K. Shukla, D. A. Mendis, V. W. Chow , World Scientific, Singapure, 1996, pp. 107-121.
17. S. Ichimaru and S. Tanaka, *Phys. Rev. Lett.* **56**, 2815 (1986).
18. S. Ichimaru, H. Iyetomi, and S. Tanaka, *Phys. Rep.* **149**, 91 (1987).

19. M. A. Berkovsky, *Phys. Lett. A* **166**, 365 (1992).

20. P. K. Kaw and A. Sen, *Phys. Plasmas*, **5**, 3552 (1998).

21. J. P. Boon and S. Yip, *Molecular Hydrodynamics*, McGraw-Hill, New York, 1980.

22. R. Abe, *Prog. Theor. Phys.* **21**, 475 (1959).

23. W. L. Slattery, G. D. Doolen, and H. E. DeWitt, *Phys. Rev. A* **21**, 2087 (1980).

24. A. A. Mamun, P.K. Shukla , and T. Farid, Phys. Plasma. **7**, 2329 (2000).

Auroral emissions due to a dusty plasma instability

L J Gelinas and M C Kelley

Dept. of Electrical Engineering,
Cornell University, Ithaca, NY

M.F. Larsen

Dept. of Physics,
Clemson University, Clemson, SC

Abstract. A TMA (tri-methyl aluminum) release rocket launched during the 1998 Coqui II sounding rocket campaign in Puerto Rico produced what appeared to be an artificially-generated aurora. This phenomenon has been has been observed on several other occasions, but has not yet been fully investigated. Charged particulates in the TMA trail may excite plasma or dusty plasma instabilities resulting in particle acceleration and auroral-like atmospheric emissions. We discuss several possible scenarios here, though determination of the appropriate particle acceleration mechanism will only be possible after more experimental and theoretical work.

INTRODUCTION

Anomalous light emissions generated by TMA releases have been observed on several occasions, but the phenomena has heretofore not been reported, in part because of the anecdotal nature of the observations. During the recent Coqui II sounding rocket campaign in Puerto Rico, the phenomena was again observed, though this time ground observations have characterized the artificial aurora to the extent that we feel confident that it is demonstrably a real, physical process. In this paper we describe the the anecdotal reports of artificial aurora and show that observations during the Puerto Rico campaign fit the profile. We then discuss possible artificial auroral generation mechanisms, including the possibility of a dusty plasma instability involving charged particulates in the TMA trail. We also relate the observations to similar phenomena in planetary atmospheres.

CP537, *Waves in Dusty, Solar, and Space Plasmas*, edited by F. Verheest, et al.
© 2000 American Institute of Physics 1-56396-962-9/00/$17.00

FIGURE 1. Artificial aurora location with respect to rocket trajectory.

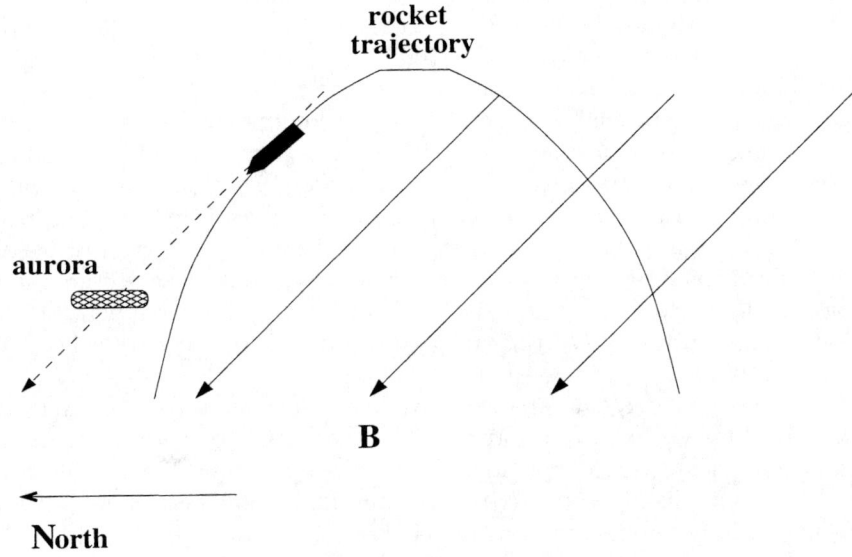

DESCRIPTION OF PHENOMENON

Observations of artificial aurora from rocket-borne TMA releases appear to have begun several years ago, when for the first time a small amount of kerosene was added to the TMA liquid. Since TMA reacts violently in air, nitrogen was originally used to purge the filling lines during loading of the TMA payloads. However, the excess nitrogen introduced bubbles into the TMA, resulting in gaps during the release. Instead, a small amount of kerosene (about 100 g), rather than nitrogen, was used to purge the fill lines; these payloads apparently produce artificial aurora.

There are several other characteristics common to all TMA related artificial aurora besides the presence of a kerosene contaminant. The aurora occur during the downleg TMA release, and only when there has also been an upleg release. The aurora do not occur during the downleg release when there has been no upleg release. The aurora also appear to be a projection of some source down along the magnetic field line which the rocket is crossing. At high latitudes where the magnetic field lines are nearly vertical, the aurora appears below the rocket, and follows it along the downleg trajectory [1]. This effect also appears, but the apparent motion is more difficult to observe, at mid latitudes where the magnetic field dip angle is near 45 degrees from the vertical (see Fig. 1).

ANALYSIS

The general description of the artificial aurora phenomena above leads us to a few assumptions about the generation mechanism. First, that the TMA byproducts themselves (CO_2, H_2O, AlO_2) are not sufficient to cause the auroral display, since "pure" TMA payloads do not produce aurora, therefore we assume some other contaminant (eg. kerosene) or process (coagulation of byproducts into macroscopic particles) must be responsible. Given that the aurora are only produced during downleg release after an upleg release, the energy source for the auroral light is likely coming from material evaporated onto the rocket from an upleg release. If the TMA itself is not the culprit, then perhaps kerosene/ice/Al particles attached to the payload could be the cause. As these particles are released (evaporated) from the payload they acquire a charge, either from the payload itself, which is probably at about -1V with respect to the ambient plasma, or from charge collection in the surrounding plasma.

Approximately 1 erg/cm^2/s is required to produce an aurora of intensity 1 kR, the threshold for visible natural aurora [2]. The dimensions of the artificial aurora patch generated by the Puerto Rico TMA rocket and captured by the white light camera with a 3 second exposure indicates an illuminated area roughly 4 km by 20 km. The total energy involved in this is about $2.4 \times 10^5 J$. To obtain this energy, the total mass of material that must be released at the rocket velocity near 140 km altitude is:

$$M = \frac{2E}{v_R^2} = 0.48kg \tag{1}$$

where the rocket velocity v_R^2 is 1 km/s. This is a sizeable fraction of the total mass released on both the upleg and downleg of the TMA releases ($\approx$3 kg), that must be evaporated from the payload over the short time scale in which the rocket is near the source region. Since we know of no other source of free energy to tap into, it seems reasonable to assume that the kinetic energy associated with the moving rocket is converted by some process, possibly via a dusty plasma interaction. We thus tentatively conclude that the disturbance must draw energy from the rocket kinetic energy and not just the material released from the rocket; we discuss possible scenarios for this process in the following section.

DISCUSSION

Natural (discrete) aurora can be produced via electrons accelerated along magnetic field lines by lower hybrid waves generated by streaming ions at the plasma sheet boundary [3]. Lower hybrid waves have also been observed in spacecraft wakes, including Galileo's wake during its 1990 Earth fly-by [4] and in the Space Shuttle wake [5]. The lower hybrid waves observed by Galileo's plasma wave instrument were in the near-wake region, 10.6 m from the spacecraft body. The near wake region extends out a few spacecraft radii, proportional to the Mach number [6]:

$$L_{wake} = R_0 \frac{V_{flow}}{(2kT_E/M_+)^{\frac{1}{2}}} \tag{2}$$

where R_0 is the body radius, T_e is the electron temperature and M_+ is the ionic mass. Lower hybrid waves in the Shuttle wake have been observed in both the near and far wake regions, with wave intensification in both regions at times of thruster firings parallel to $\mathbf{B}$. Generation of lower hybrid waves in these spacecraft wakes is presumed to be due to excitation of a drift instability in the spacecraft wake, either by plasma density gradients in the near wake region or expansion of the Shuttle exhaust plume into the surrounding plasma [4,5].

For the present case of artificial aurora generation, the wave energy source must come primarily from the rocket motion, rather than dust motion through the plasma. Most sounding rockets do not produce artificial auroras solely from their supersonic motion, therefore, if lower hybrid waves are generated in the wake, they must be quickly damped by the thermal plasma in the absence of a charged dust population. Growth of lower hybrid drift waves occurs for parallel wavelengths longer than a critical length, about 10 times the perpendicular density gradient scale length [7]. If the scale length of the wake is approximately equal to the radius of the rocket, then the then the required parallel wavelength for growth of the instability must be larger than 10 rocket radii (about 5 m in length for a typical sounding rocket). For a sounding rocket with velocity of 1 km/s (Mach number 3) the rocket wake barely extends to 10 radii, according to Eqn. 2; waves longer than the critical wavelength cannot be excited and wave growth is suppressed. Note that the near wake regions of the Galileo spacecraft and the Shuttle Orbiter described above extend further due to their higher velocities (6-7 km/s) and Mach numbers (near 6), and therefore wave growth is possible in the near wake region.

However, if the sounding rocket wake were to extend further along $\mathbf{B}$ than normally allowed by the thermal plasma, wave growth may occur. If heavy, charged particles are mingled with the thermal plasma wake, the dust particle inertia may slow thermal plasma migration into the depleted wake region. This effective lengthening of the rocket wake parallel to $\mathbf{B}$ by heavy charged particles may therefore make lower hybrid wave growth possible. Motion of the (falling) charged dust through the plasma wake may also contribute to wave growth, although as mentioned previously, the energetics of artificial aurora creation make it unlikely that this is the sole energy source. The delay between the time the Puerto Rico rocket crossed the source region at 140 km and the occurrence of the aurora 20 seconds later is likely related to the growth rate of the instability.

For charged dust particles to influence the plasma wake dynamics, the dust particles in the wake region must be charged very quickly to have any effect on the length of the wake, as thermal plasma motion would ordinarily neutralize the wake within a few milliseconds. Therefore, the particles must be released from the rocket carrying a charge. Uncharged TMA particles should not produce an artificial aurora since the charging time for dust in the thermal plasma is on the order of a few seconds [8]. If kerosene-based ice/dust particles are attached to the payload, which is presumably negatively charged, they could be released carrying a charge, depending on the payload potential and the work function of the dust/ice material.

CONCLUSION

Particle acceleration processes are very common in natural plasmas but are as yet not very well understood. In most acceleration processes magnetic fields play a role, and of course, some source of free energy must exist. Understanding the variety of ways nature finds to support acceleration mechanisms is a major goal of space physics. We have described a recently observed phenomenon, the creation of an artificial aurora by a TMA release rocket. The artificial aurora was observed at mid-latitudes, far from the auroral oval, so that the energy must clearly come from the rocket-TMA system, resulting in field-aligned particle acceleration. The particle acceleration process in the case of artificial aurora generation is not easily explained without considering a dusty plasma interaction. We have considered here the possible explanation of excitation of the lower hybrid drift instability in the sounding rocket wake. The feasibility of this process for auroral particle acceleration depends on the presence of charged dust which can effectively lengthen the rocket wake, allowing growth of the instability.

REFERENCES

1. Hecht, J., pers. comm. (1999).
2. Arnoldy, R. , pers. comm. (2000).
3. Bingham, R., D.A. Bryant, D.S. Hall, *Annales Geophysicae*, **6**, 159-168 (1988).
4. Keller, A.E., D.A. Gurnett, W.S. Kurth, Y.Yuan, A. Bhattacharjee, *Planet. Space Sci.* **45**, 201-219 (1997).
5. Feng, W., D.A. Gurnett, I.H. Cairns, *JGR* **98**, 21,571-21,580 (1993).
6. Oran, W.A., N.H. Stone, U. Samir, *JGR* **80**, 207-209 (1975).
7. Ichimaru, S, *Statistical Plasma Physics, Vol. 1* (1992).
8. Goertz C. K., *Rev. Geophys.* **27**, 271 (1989).

Jeans modes in dusty plasmas with dust mass distributions

G. Jacobs[1], F. Verheest[1], M.A. Hellberg[2], S.R. Pillay[3]
and V.V. Yaroshenko[1]

[1] *Sterrenkundig Observatorium, Universiteit Gent, Krijgslaan 281, B-9000 Gent, Belgium*
[2] *School of Pure and Applied Physics, University of Natal, Durban 4041, South Africa*
[3] *Department of Physics, University of Durban-Westville, Durban 4000, South Africa*

Abstract. Self-gravitational modes in dusty plasmas have been extensively studied for both extreme cases of wave propagation: waves propagating parallel to an external magnetic field and perpendicularly propagating waves. Provided the presence of neutrals is insignificant, the presence of charged dust species stabilizes plasmas against gravitational collapse, the mechanisms being induced by plasma thermal effects and/or magnetic field pressure. We present here the modifications of magnetosonic modes and electrostatic, parallel modes due to the possible distributions of the dust mass. The necessary adaptations of the corresponding critical Jeans lengths are discussed and extended to intermediate angles of propagation.

INTRODUCTION

Plasmas and dust are basic building blocks of our universe and the synergy of the two components displays interesting new phenomena. Among the playgrounds of these phenomena one has noctilucent clouds, circumsolar dust rings and cometary comae and tails. Dusty plasmas contain charged dust grains that can be treated as additional ionic species with very distinct and unusual characteristics, like high and variable charges, and even higher masses. For a general discussion of dusty plasma waves and instabilities we can refer to recent overviews [1–4]. Due to the very heavy dust grain masses, self-gravitation can become important in very large systems, even comparable to electromagnetic force magnitudes. Self-gravitational effects alter wave dispersion and introduce the possibility of gravitational collapse [5–7]. Low-frequency modes, as found in dusty plasmas, can act as an antagonist, they oppose gravitational instability when neutrals are almost absent. This is translated mathematically into a drastic modification of the Jeans length, the criterion for gravitational instability. For perpendicular propagation the Alfvén and magnetosonic velocities play a role whereas for parallel propagation there are

CP537, *Waves in Dusty, Solar, and Space Plasmas*, edited by F. Verheest, et al.
© 2000 American Institute of Physics 1-56396-962-9/00/$17.00

purely thermal effects, expressed via a dust-acoustic velocity [8]. Here we review the modifications of magnetosonic modes and electrostatic, parallel modes due to the possible distributions of the dust mass. For intermediate angles of propagation, the corresponding critical Jeans lengths are briefly discussed.

DISPERSION TENSOR

We start from a multispecies description: in addition to the fluid and Maxwell equations, there is a gravitational Poisson's equation in order to describe the self-gravitating effects. A barotropic law for the scalar pressures completes the model. At arbitrary angles of propagation, after linearization and Fourier transforming, one can deduce a 4×4 dispersion tensor D by eliminating the three components of the electric field and the scalar gravitational potential. Details of the derivation and the expressions for the elements of the dispersion tensor can be found in earlier papers [9–11]. The dispersion law becomes

$$\det [D_{ij}] = 0. \tag{1}$$

All dispersion laws in the following sections can be deduced from (1).

ELECTROSTATIC, PARALLEL MODES

At propagation parallel to an external magnetic field, self-gravitational effects modify only the longitudinal modes. The dispersion law for these longitudinal, electrostatic modes is [7,12,13]

$$\left(1 - \sum_\alpha \frac{\omega_{p\alpha}^2}{\mathcal{L}_\alpha}\right)\left(1 + \sum_\alpha \frac{\omega_{J\alpha}^2}{\mathcal{L}_\alpha}\right) + \left(\sum_\alpha \frac{\omega_{p\alpha}\omega_{J\alpha}}{\mathcal{L}_\alpha}\right)^2 = 0, \tag{2}$$

with

$$\mathcal{L}_\alpha = (\omega - kU_\alpha)^2 - k^2 c_{s\alpha}^2 \tag{3}$$

and introducing the Jeans frequencies $\omega_{J\alpha}^2 = 4\pi G N_\alpha m_\alpha$ and the speeds of sound per species $c_{s\alpha}$. Other notations are standard. Capital letters denote equilibrium values and neutral dust species are included, in principle explicit expressions for the neutrals are obtained by setting the appropriate charges equal to zero. Equation (2) highlights the coupling between self-gravitational effects and electrostatic plasma oscillations, and these two phenomena can be isolated by taking the low and high frequency limit, respectively . The high frequency limit of (2) is the dispersion law for electrostatic plasma oscillations, including the Buneman modes when streaming between the species is present:

$$\sum_\alpha \frac{\omega_{p\alpha}^2}{\mathcal{L}_\alpha} \approx 1. \tag{4}$$

The low frequency limit of (2), when plasma effects are unimportant,

$$1 + \sum_\alpha \frac{\omega_{J\alpha}^2}{\mathcal{L}_\alpha} \approx 0, \tag{5}$$

describes a multispecies Jeans instability. Only low frequencies allow the heavy dust grains to respond to the perturbations and these low frequency modes are the ones under consideration. Therefore, in the remainder of this section we shall treat the ions and a fortiori the electrons as being Boltzmann distributed.

Monodisperse, nonstreaming dust

If only one charged dust species is present and there is no streaming between species, the dispersion law (2) simplifies to

$$\omega^2 = k^2(c_{sd}^2 + c_{da}^2) - \omega_{Jd}^2. \tag{6}$$

Here $c_{da\alpha} = \lambda_D \omega_{pd\alpha}$ is the dust-acoustic speed. For cold dust, (6) reduces to a single Jeans dust mode [5,6]. If, in addition to ignoring the dust thermal speed, self-gravitational effects are excluded, the dust-acoustic mode is recovered [8]. Equation (6) is the archetypal dispersion law for "Jeans"-modes. Setting $\omega^2 = 0$ determines the critical Jeans length:

$$k_J^2 = \frac{\omega_{Jd}^2}{c_{sd}^2 + c_{da}^2}. \tag{7}$$

Polydisperse, nonstreaming but cold dust species

If the possibility of multiple charged dust species is retained, under the restriction that the dust species are cold, the dispersion law turns into

$$\omega^4 + \left(\sum_d \omega_{Jd}^2 - k^2\lambda_D^2 \sum_d \omega_{pd}^2\right)\omega^2 - \frac{k^2\lambda_D^2}{2}\sum_d\sum_{d'}(\omega_{pd}\omega_{Jd'} - \omega_{pd'}\omega_{Jd})^2 = 0, \tag{8}$$

as discussed by Meuris et al [13]. The last term clearly represents the modifications due to the presence of dust species of different characteristics. It can only vanish if all dust species have the same chargo-to-mass ratio.

Two dust species

For two charged dust species, equation (2) reduces to [14]:

$$\left[\omega^2 - k^2(c_{sd1}^2 + c_{da1}^2) + \omega_{Jd1}^2\right]\left[\omega^2 - k^2(c_{sd2}^2 + c_{da2}^2) + \omega_{Jd2}^2\right]$$
$$= (\omega_{Jd1}\omega_{Jd2} - k^2 c_{da1}c_{da2})^2, \tag{9}$$

clearly showing the coupling between the two Jeans dust modes. This is a quadratic dispersion law in ω^2 with positive discriminant, implying that both roots in ω^2 are real. Unstable modes require $\omega^2 < 0$, but this can only happen for one of the two roots, the other then being positive. A necessary and sufficient condition for instability is thus a negative constant term in (9), synonymous with a negative product of both real roots. This condition is always fulfilled for sufficiently small wavenumbers. This yields the following expression for the Jeans wave number:

$$k_J^2 = \frac{\omega_{Jd1}^2 c_{sd2}^2 + \omega_{Jd2}^2 c_{sd1}^2 + (\omega_{Jd1}c_{da2} - \omega_{Jd2}c_{da1})^2}{c_{sd1}^2 c_{da2}^2 + c_{sd2}^2 c_{da1}^2 + c_{sd1}^2 c_{sd2}^2} \tag{10}$$

$$\simeq \frac{(\omega_{Jd1}c_{da2} - \omega_{Jd2}c_{da1})^2}{c_{sd1}^2 c_{da2}^2 + c_{sd2}^2 c_{da1}^2}, \tag{11}$$

assuming that the pure plasma pressures are much larger than the dust pressures, so that for the dust species $c_{sd\alpha} \ll c_{da\alpha}$. For one charged (with subscript d) and one neutral dust or gas component (with subscript n), equation (11) simplifies to

$$k_J^2 = \frac{\omega_{Jcd}^2}{c_{scd}^2 + c_{da}^2} + \frac{\omega_{Jnd}^2}{c_{snd}^2}. \tag{12}$$

MAGNETOSONIC MODES

Average dust dispersion

If the charged dust is represented by a single species with some average properties and there is no neutral gas at all, the dispersion law for waves propagating perpendicularly to the static magnetic field is [9,10]

$$\omega^2 = k^2(V_{ms}^2 + c_{sd}^2) - \omega_{Jd}^2, \tag{13}$$

where the ions and electrons are considered quasi-inertialess. The magnetosonic velocity V_{ms} is defined in general through

$$V_{ms}^2 = \frac{B_0^2/\mu_0 + P_e + P_i}{\sum_d N_d m_d}, \tag{14}$$

where the electron and ion pressures are given as $P_{e,i} = N_{e,i}m_{e,i}c_{se,si}^2$. It is reasonable to suppose that the effective dust pressure is smaller than the electron and ion pressures rendering $c_{sd} \ll V_{ms}$. The critical Jeans wavenumber obtained from (13) then becomes

$$k_J \approx \frac{\omega_{Jd}}{V_{ms}}. \qquad (15)$$

Thus the magnetosonic velocity here comes into play, expressing the dominant influence of the magnetic field, combined with the plasma pressures, on the critical Jeans lengths for perpendicular propagation. On the other hand, in (6) only plasma thermal effects, represented by c_{da}, determine k_J.

Two charged dust species

For two charged dust species, the dispersion law transforms into a biquadratic equation with positive discriminant. The constant term simplifies to [15]

$$C = \Omega_1^2 \Omega_2^2 \omega_{J,total}^4 \left[k^2 V_{ms}^2 (1 - \mathcal{K}) - \omega_{J,total}^2 \right], \qquad (16)$$

when thermal speeds are neglected. This is a valid approximation when $\mathcal{K} \leq \mathcal{O}\left(\frac{1}{3}\right)$, and $\mathcal{K}$ and the total Jeans frequency $\omega_{J,total}$ have been introduced as

$$\mathcal{K} = \frac{\omega_{J1}^2 \omega_{J2}^2}{\omega_{J,total}^2} \left(\frac{1}{\Omega_1} - \frac{1}{\Omega_2} \right)^2, \qquad (17)$$

$$\omega_{J,total}^2 = \omega_{J1}^2 + \omega_{J2}^2. \qquad (18)$$

If $C > 0$, the biquadratic dispersion law has two positive roots for ω^2 whereas $C < 0$ gives one positive and one negative root for ω^2. The latter root clearly corresponds to the Jeans unstable case. The transition occurs at $C = 0$ or equivalently $k = k_J$ given through

$$k_J^2 = \frac{\omega_{J,total}^2}{(1 - \mathcal{K}) V_{ms}^2}. \qquad (19)$$

Several charged dust species

In order to determine the critical Jeans length, one has to consider the dispersion law as a polynomial in ω^2 with parameter k^2. Below the threshold value k_J at least one of the polynomial roots leaves the positive real axis. As noted from the previous paragraphs, the Jeans criterion involves passing from $\omega^2 > 0$ to $\omega^2 < 0$ through the origin. It follows that Taylor expansion up to first order in ω^2 will contain all the useful information and as a consequence that

$$k^2 = k_J^2 = \frac{\omega_{J,total}^2}{V_{ms}^2 (1 - \mathcal{K})} \qquad \text{at} \qquad \omega^2 = 0. \qquad (20)$$

Under the assumptions $\omega^2 \ll c^2 k^2$ and $k\lambda_D \ll 1$, the dispersion law (1) is already linear in k^2 and after the first order Taylor expansion simplifies to [15]

$$\omega^2 = \frac{1 - \mathcal{K}}{1 + \mathcal{L}} \left[(1 - \mathcal{K}) k^2 V_{ms}^2 - \omega_{J,total}^2 \right], \tag{21}$$

where the definitions of $\omega_{J,total}^2$ and $\mathcal{K}$ have been expanded to

$$\omega_{J,total}^2 = \sum_d \omega_{Jd}^2, \tag{22}$$

$$\mathcal{K} = \frac{1}{2\omega_{J,total}^2} \sum_d \sum_{d'} \omega_{Jd}^2 \omega_{Jd'}^2 \left(\frac{1}{\Omega_d} - \frac{1}{\Omega_{d'}} \right)^2. \tag{23}$$

$\mathcal{L}$ is a small positive parameter, its precise form does not furnish us with valuable physical insight and is not mentioned here. The different dust species are indicated by taking d and d' as dummy indices in the summations. If the Jeans frequencies are in some sense small compared to the gyrofrequencies, as is typically the case in astrophysical plasmas, then $0 \le \mathcal{L} \ll \mathcal{K} \ll 1$.

Presence of neutral gas

If neutrals are included, the derivation starting from the dispersion tensor (1) becomes even more complicated. Therefore we focus on the determination of the critical Jeans lengths, instead of dealing with the formidable algebraic expressions for the dispersion laws. Previous sections indicated a shortcut for obtaining the critical Jeans lengths; putting $\omega = 0$ in the elements of the dispersion tensor conserves all the necessary information required for determining the critical Jeans wave number. Following this strategy, the critical wave number can be expressed as

$$k_J^2 = \frac{1}{1 - \mathcal{K}} \left[\frac{\omega_{J,total}^2}{V_{ms}^2} + \sum_g \frac{\omega_{Jnd}^2}{c_{snd}^2} \right]. \tag{24}$$

Comparing this with (20) shows that the stabilizing effects of the charged dust components are superseded by the presence of neutral gas components. Seeing that $c_{sg} \sim c_{sd} \ll V_{ms}$, the neutral terms dominate (24), provided that the Jeans frequencies of all species are comparable.

INTERMEDIATE ANGLES OF PROPAGATION

For intermediate angles of propagation between strictly parallel and perpendicular we have to revert to the full dispersion law (1). For simplicity, we first delve into the "classic" dusty plasma. This model deals with only one effective dust species, and electrons and ions are treated as being inertialess ($\omega^2 \ll k^2 c_{se,si}^2$). We also suppose that $\omega^2 \ll c^2 k^2$ and $k^2 \lambda_D^2 \ll 1$. All modes are coupled and incorporated in the following dispersion law [11]:

$$\omega^6 - \left[\Omega_d^2\kappa^2\left((1+\kappa^2)\cos^2\vartheta+1\right)+\Delta\right]\omega^4$$
$$+\left[(\Omega_d^2+\Delta)\kappa^2+2\Delta\right](\Omega_d^2\kappa^2\cos^2\vartheta)\omega^2-(\Omega_d^4\kappa^4\cos^4\vartheta)\Delta=0, \qquad (25)$$

a cubic polynomial in ω^2. Here $\kappa = ck/\omega_{pd}$, and we note that all thermal and self-gravitational effects are contained in

$$\Delta = k^2(c_{da}^2 + c_{sd}^2) - \omega_{Jd}^2. \qquad (26)$$

For $\Delta > 0$ but small, all three roots of (25) are real and positive. When Δ shifts, for small absolute values, towards negative values, one of the roots ω^2 goes simultaneously with Δ through zero. The only negative root, obtained when Δ turns negative but remains small, corresponds to the purely growing Jeans instability. The latter is clearly activated through a slight change of Δ from positive to negative values. At Δ very small, the smallest root becomes $\omega^2 \simeq \Delta\cos^2\vartheta$ and of course $\Delta = 0$ gives the familiar result

$$k_{cr}^2 = \frac{\omega_{Jd}^2}{c_{da}^2 + c_{sd}^2}. \qquad (27)$$

When adding more dust species to the dusty plasma model, the full dispersion law becomes unwieldy and would require careful disentangling. It is preferable to cut this Gordian algebraic knot and go straight ahead to the critical Jeans lengths. Previous sections marked out the strategy of putting $\omega = 0$ in the elements of the dispersion tensor (1). This strategy does not allow that ϑ approaches 90^0, after having set $\omega = 0$ in the elements of the dispersion tensor. This is imputable to the fact that the limits $\omega \to 0$ and $\vartheta \to 90°$ are not interchangeable, in the model used where electrons and ions are treated as massless the limit $\omega \to 0$ & $\vartheta \to 90^0$ does not exist mathematically. Taking $k_{cr}\lambda_D \ll 1$, and using in addition the following reasonable assumptions, $Z_dm_e \ll Z_dm_i \ll m_d$, $c_{sd} \ll c_{si}, c_{se}$ and $\omega_{Je} \ll \omega_{Ji} \ll \omega_{Jd}$, then gives for the critical Jeans length that

$$k_{cr}^2\left(1+\lambda_D^2\sum_d\frac{\omega_{pd}^2}{c_{sd}^2}\right) = \sum_d\frac{\omega_{Jd}^2}{c_{sd}^2} + \frac{\lambda_D^2}{2}\sum_d\sum_{d'}\frac{(\omega_{pd}\omega_{Jd'}-\omega_{pd'}\omega_{Jd})^2}{c_{sd}^2c_{sd'}^2}. \qquad (28)$$

This result is independent of the angle of wave propagation ϑ, showing that the longitudinal effects determine the Jeans instability. For one dust species (28) matches (27) and a combination of charged and neutral dust gives

$$k_{cr}^2 = \frac{\omega_{Jcd}^2}{c_{da}^2 + c_{scd}^2} + \frac{\omega_{Jnd}^2}{c_{snd}^2}. \qquad (29)$$

Again, because of $c_{sn} \sim c_{sd} \ll c_{da}$, such critical wavenumbers are close to what is obtained for neutral dust alone [7]. Finally, we emphasize again that it is imperative to use a separate treatment for $\vartheta = 90°$. The appropriate treatment of the extraordinary mode at strictly perpendicular propagation [7,9] uncovers an abrupt transition of the critical lengths.

CONCLUSIONS

We have mapped out previous studies of self-gravitational effects in dusty plasmas that dealt with dust mass distributions. We focussed on the lowest order modes because the ultimate objective was to determine the critical Jeans lengths for all angles of propagation. It is interesting to note that the critical Jeans lengths for intermediate angles of propagation are the same as for parallel propagation, except close to or at perpendicular propagation where an abrupt transition occurs. Results point out that there is little gain in including several charged dust species; a monodisperse description, where the only dust species has some average properties, is almost as accurate. As usual, the stabilizing effects of Jeans modes against gravitational collapse are dominated by the presence of neutral gas components.

ACKNOWLEDGMENTS

This work was supported by the Flemish Government (Department of Science and Technology) and the (South African) National Research Foundation in the framework of the Flemish–South African Bilateral Scientific and Technological Co-operation on the Physics of Waves in Dusty, Solar and Space Plasmas. The Bijzonder Onderzoeksfonds of the Universiteit Gent is thanked for research (GJ and FV) and foreign visitor (VVY) grants.

REFERENCES

1. Shukla, P.K., *Physica Scripta* **45**, 504–507 (1992).
2. Mendis, D.A. and Rosenberg, M., *Annu. Rev. Astron. Astrophys.* **32**, 419–463 (1994).
3. Horányi, M., *Annu. Rev. Astron. Astrophys.* **34**, 383–418 (1996).
4. Verheest, F., *Space Sci. Rev.* **77**, 267–302 (1996).
5. Avinash, K. and Shukla, P.K., *Phys. Lett. A* **189**, 470–472 (1994).
6. Pandey, B.P., Avinash, K. and Dwivedi, C.B., *Phys. Rev. E* **49**, 5599–5606 (1994).
7. Bliokh, P., Sinitsin, V. and Yaroshenko, V., *Dusty and Self-gravitational Plasmas in Space*, Kluwer, Dordrecht, 1995.
8. Rao, N.N., Shukla, P.K. and Yu, M.Y., *Planet. Space Sci.* **38**, 543–546 (1990).
9. Verheest, F., Meuris, P., Mace, R.L. and Hellberg, M.A., *Astrophys. Space Sci.* **254**, 253–267 (1997).
10. Verheest, F., Hellberg, M.A. and Mace, R.L., *Phys. Plasmas* **6**, 279–285 (1999).
11. Verheest, F., Jacobs, G. and Hellberg, M.A., *Physica Scripta* **T84**, 171-174 (2000).
12. Bliokh, P.V. and Yaroshenko, V.V., *Sov. Astron.* **29**, 330–336, 1985.
13. Meuris, P., Verheest, F. and Lakhina, G.S., *Planet. Space Sci.* **45**, 449–454 (1997).
14. Verheest, F., Jacobs, G. and Yaroshenko, V.V., *Phys. Plasmas* **7**, *in press* (2000).
15. Jacobs, G., Verheest, F., Hellberg, M.A. and Pillay, S.R., "Magnetosonic modes with dust mass distributions", *Phys. Plasmas, submitted* (2000).

Crossfield driven electrostatic instabilities in a two-dust component plasma

S. K. Maharaj[1] and R. Bharuthram[2]

[1]Department of Physics, University of Durban-Westville, Private Bag X54001, Durban, 4000, South Africa

[2]M. L. Sultan Technikon, P.O. Box 1334, Durban 4000, South Africa

Abstract. Using a kinetic approach, a parametric investigation of electrostatic instabilities is undertaken in a two-dust component plasma comprising of ions, electrons and two dust species (of different mass and charge) in the low dust plasma frequency regime. These instabilities are driven by an equal $\mathbf{E} \times \mathbf{B}$ drift of the ions and electrons relative to the stationary, unmagnetized dust grains. The effects of the variation of plasma parameters such as particle drift speeds, densities and temperature on the instability real frequency and growth rate are examined. Numerical solutions of the full dispersion relation are compared with approximate solutions.

I INTRODUCTION

The motivation behind the study of linear instabilities in dusty plasmas [1-6], is largely due to the fact that relative drifts between charged particle species, provides the necessary free energy to drive such instabilities. Using a standard Vlasov analysis, Rosenberg [7] investigated the conditions for the excitation of the dust ion-acoustic (DIA) and dust-acoustic (DA) instabilities driven by weak ion and/or electron drifts in an unmagnetized dusty plasma. Approximate expressions for the real frequency and growth rate for [8] studied the electrostatic ion cyclotron instability (EICI) in a three component plasma comprising of ions, electrons and charged dust grains. The EICI is driven by a drift of the electrons in the direction of an externally applied magnetic field. Modified two stream instabilities in a dusty plasma were studied by Rosenberg and Krall [9] in both the high frequency ($\omega_{ci} << \omega << \omega_{ce}$) and the low frequency ($\omega_{cd} << \omega << \omega_{ci}$) regimes. In the former case, the instability is driven by drifting electrons. The low frequency instability is driven by an equal $\mathbf{E} \times \mathbf{B}$ drift of the magnetized ions and electrons relative to the unmagnetized dust particles. Bharuthram [5] studied low frequency electrostatic instabilities in a three-component dusty plasma driven by an equal crossfield $\mathbf{E_o} \times \mathbf{B_o}$ drift of the magnetized ions and electrons relative to the un-

CP537, *Waves in Dusty, Solar, and Space Plasmas*, edited by F. Verheest, et al.
© 2000 American Institute of Physics 1-56396-962-9/00/$17.00

magnetized dust grains. These results could be applicable to Saturn's rings where both the ions and the electrons have an azimuthal crossfield drift relative to the levitated dust grains.

The work that follows in this paper is an extension of the three-component (one dust species) model of Bharuthram [5] to a four-component (two dust species) plasma and serves to examine the effect of the presence of an additional dust species on low frequency electrostatic instabilities. Here the driving mechanism is the $\mathbf{E_o} \times \mathbf{B_o}$ drift of the magnetized electrons and ions relative to the unmagnetized dust components. Rosenberg and Krall [9] assumed angles of wave propagation for which $k_z \ll k_\perp$ (where the directions of $k_z(k_\perp)$ are relative to $\mathbf{B_o} = B_o\hat{\mathbf{z}}$) and found only the dust-modified two stream instability (DMTSI) as being the solution to the dispersion relation. We have extended their work to include much wider angles of wave propagation and find that in addition to the DMTSI, the dust-acoustic instability (DAI) could also be excited as first reported by Bharuthram [5]. In the next section the kinetic dispersion relation is presented, numerical solutions of the full dispersion relation are presented in Section III and a summary and discussion of our results are presented in Section IV.

II   THE KINETIC DISPERSION RELATION

For a four-component plasma comprising of magnetized electrons and ions, and two unmagnetized dust components, the kinetic dispersion relation obtained for electrostatic waves following the procedure outlined by Gary and Sanderson [10] and Bharuthram and Pather [4] is given by:

$$
\begin{aligned}
0 = k^2 &+ \frac{1}{\lambda_{de}^2}\left[1 + \frac{\omega - k_\perp V_E - p\Omega_e}{\sqrt{2}k_z C_e}\sum_{p=-\infty}^{\infty} Z(z_{pe})\Gamma_{pe}\right] \\
&+ \frac{1}{\lambda_{di}^2}\left[1 + \frac{\omega - k_\perp V_E - p\Omega_i}{\sqrt{2}k_z C_i}\sum_{p=-\infty}^{\infty} Z(z_{pi})\Gamma_{pi}\right] \\
&+ \frac{1}{\lambda_{ddl}^2}\left[1 + \frac{\omega}{\sqrt{2}k_z C_{dl}}\sum_{p=-\infty}^{\infty} Z(z_{pdl})\Gamma_{pdl}\right] \\
&+ \frac{1}{\lambda_{ddh}^2}\left[1 + \frac{\omega}{\sqrt{2}k_z C_{dh}}\sum_{p=-\infty}^{\infty} Z(z_{pdh})\Gamma_{pdh}\right]
\end{aligned}
\tag{1}
$$

where e, i, dl and dh represent the electron, ion, light and heavy dust components respectively. The wave frequency is given by $\omega = \omega_r + i\gamma$ and Z is the plasma dispersion function. $\Gamma_{pj} = e^{-\alpha_j}I_p(\alpha_j)$ is the gamma function where $\alpha_j = \frac{k_\perp^2 C_j^2}{\Omega_j^2}$ for $j = e, i, dl$ and dh. $C_j = \sqrt{\frac{T_j}{m_j}}$, Ω_j and $\lambda_{dj} = (T_j/4\pi n_{jo}q_j^2)^{\frac{1}{2}}$ is the thermal velocity, gyrofrequency and Debye length of species j respectively (where the charge on the species is $q_j = -e, +e, -Z_{dl}e$ and $-Z_{dh}e$ for $j = e, i, dl$ and dh respectively) while I_p is the modified Bessel function of order p. Here, $Z_{dl,dh} < 0$ (> 0) indicates that the grains are positively (negatively) charged. The components of the wave vector $\mathbf{k}$ are given by k_z $(k_\perp)$ along (perpendicular) to the external magnetic field $\mathbf{B}_o = B_o\hat{z}$. Here $z_{pe} = (\omega - k_\perp V_E - p\Omega_e)/\sqrt{2}k_z C_e$, $z_{pi} = (\omega - k_\perp V_E - p\Omega_i)/\sqrt{2}k_z C_i$, $z_{odl} = \omega/\sqrt{2}k_z C_{dl}$ and $z_{odh} = \omega/\sqrt{2}k_z C_{dh}$. $\mathbf{V}_E$ is the equal $\mathbf{E}_o \times \mathbf{B}_o$ drift of the ions and electrons in the presence of an equilibrium electric field $\mathbf{E}_o = -E_o\hat{x}$.

III   NUMERICAL RESULTS

For simplicity the analysis is restricted to the $y - z$ plane, with $\mathbf{k} = (k_y, k_z) = (k_\perp, k_z)$. Results are presented in normalized form. Time is normalized by the inverse total ion plasma frequency $\omega_{pi}^{-1} = (4\pi n_o e^2/m_i)^{-\frac{1}{2}}$. Speed is normalized by the ion sound speed $C_s = (T_e/m_i)^{\frac{1}{2}}$, spatial lengths by $\lambda_d = (T_e/4\pi n_o e^2)^{\frac{1}{2}}$, densities by the total plasma density n_o and temperatures by the electron temperature T_e; the normalizations having being chosen so as to keep the dust mass $m_{dl,h}$ and charge $Z_{dl,h}$ as variable parameters in the numerical work that follows. Quasi-neutrality at equilibrium requires:

$$n_{eo} + Z_{dl}n_{dlo} + Z_{dh}n_{dho} = n_{io} = n_o \tag{2}$$

The fixed parameters values are given by $T_e = T_i$, $T_{dl} = T_{dh} = 0.01\ T_e$, and $n_{eo} = 0.1\ n_o$. For the two dust species, recalling that $q(a)/m(a) \propto a^{-2}$ (Meuris et al. [12]), we select $(Z_{dl} = 10^2, m_{dl}/m_i = 10^6)$ and $(Z_{dh} = 10^4, m_{dh}/m_i = 10^{12})$ corresponding to the light and heavy dust species, respectively.
The ion to electron mass ratio is set to $m_i/m_e = 1836$ corresponding to a hydrogen plasma.

Figure 1 depicts the normalized instability growth rates as a function of $k_y/k = \sin\theta$ for different values of V_E. Figure 2 shows the real frequency curves corresponding to the growth rate curves of Figure 1. The high peaks in growth rate in Figure 1 for low k_y/k correspond to the dust-acoustic instability (DAI) whilst the peaks in the vicinity of $k_y/k \approx 1.0$ correspond to the dust-modified two-stream instability (DMTSI) as observed by Bharuthram [5]. This becomes clear in Figure 3 by plotting the real frequency and growth rate against normalized wave number $k\lambda_d$ for (a) $k_y/k = 0.175$ $(\theta = 10°)$ and (b) $k_y/k = 0.993$ $(\theta = 83°)$. It is evident from Figures 1 and 2 that for lower drift speeds only the DMTSI is excited but for higher drift velocities, the DAI dominates.
The numerical real frequency in Figure 3 is compared with the approximate analytical solution of the full dispersion relation (1), $viz.$

$$\omega_r^2 = k^2 C_s^2 \frac{\left\{\frac{m_i}{m_{dl}}Z_{dl}(\delta - 1 - Z_{dh}\frac{n_{dho}}{n_{eo}}) + \frac{m_i}{m_{dh}}Z_{dh}(\delta - 1 - Z_{dl}\frac{n_{dlo}}{n_{eo}})\right\}}{(1 + \delta\frac{T_e}{T_i} + \delta k^2 \lambda_d^2)} \tag{3}$$

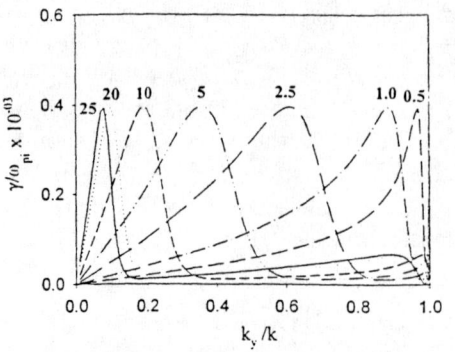

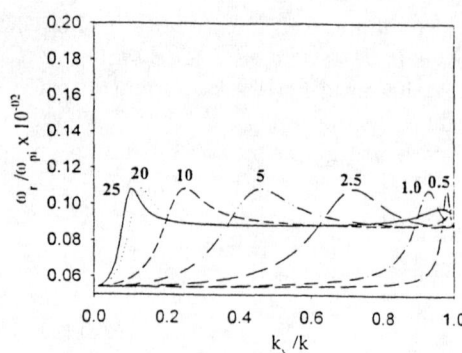

Figure 1. Plot of the normalized growth rate versus k_y/k (sinθ). The parameter labelling curves is the normalized equal ion and electron drift speed V_E/C_s. The fixed parameters are $T_i = T_e$, T_{dl} = T_{dh} = 0.01 T_e. n_{eo} = 0.1 n_o, n_{dlo} = n_{dho}, m_{dl} = 10^6 m_i, Z_{dl} = 10^2, m_{dh} = 10^{12} m_i, Z_{dh} = 10^4, $k\lambda_d$ = 0.7 and ω_{pi} = 0.01 ω_{ci}.

Figure 2: Plot of the normalized real frequencies corresponding to the growth rate curves in Figure 1. The parameter labelling the curves is V_E/C_s.

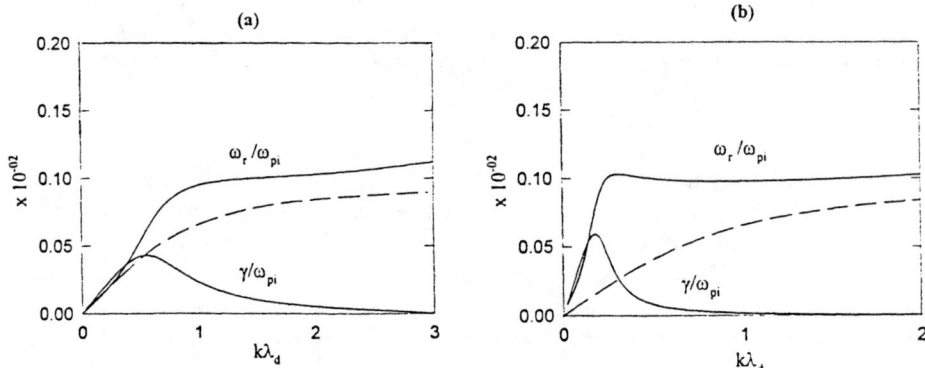

Figure 3: Plot of the normalized real frequency and growth rate versus $k\lambda_d$ for (a) k_y/k = 0.175 ($\theta = 10°$) and (b) k_y/k = 0.993 ($\theta=83°$). Here V_E/C_s = 10. The broken line is the result from the approximate solution (3). All other fixed parameters are as in Figure 1.

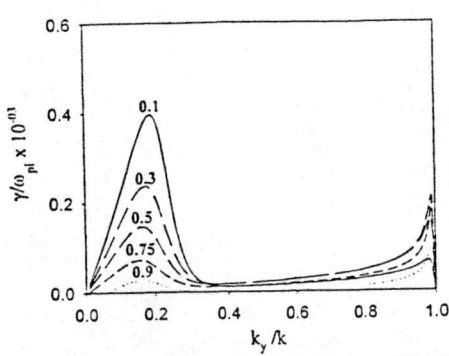
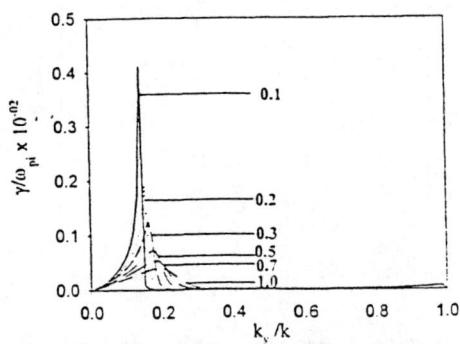

Figure 4: Plot of the normalized growth rate versus k_y/k for $V_E = 10\,C_s$. The parameter labelling the curves is the normalized equilibrium electron density n_{eo}/n_o. All other fixed parameters are as in Figure 1.

Figure 5: Plot of the normalized growth rate against k_y/k. The parameter labelling the curves is the normalized ion temperature T_i/T_e. All other fixed parameters are as in Figure 4.

which was obtained by Maharaj and Bharuthram [11] for a two-dust species plasma for a parallel drift of the ions relative to $\mathbf{B_o}$.

Figure 4 demonstrates the effect of varying the equilibrium electron density on the instability growth rate for fixed $n_{dho} = n_{dlo}$. It is seen that for the DAI, the maximum growth rate γ_{max} decreases as the equilibrium electron density n_{eo} increases. This behaviour was also observed by Bharuthram [5] for a single dust species plasma and for the ion-acoustic and EIC modes seen by D'Angelo [3].

The opposite behaviour is found to be true for the DMTSI which is consistent with the findings of Bharuthram [5]. The DAI will therefore dominate in dusty plasmas with small electron concentrations whereas large electron concentrations are needed for the DMTSI to be excited.

Figure 5 shows the effect of varying the normalized ion temperature T_i/T_e on the growth rate of the instability. Due to the kinetic nature of the dust-acoustic instability, the growth rate $\gamma \propto \frac{\partial f_{io}}{\partial v_\perp}$. The lowering of the growth rates with an increase in T_i/T_e is due to the 'flattening' of the ion velocity distribution. A wave propagating across $\mathbf{B_o}$ with phase speed $\omega_r/k_\perp$ would therefore 'see' this smaller slope, which accounts for the reduction in the growth rates.

Figure 6 shows the effect of varying the ratio of the light relative to the heavy dust densities (n_{dlo}/n_{dho}) on the instability growth rate. The equilibrium electron density is fixed at $n_{eo} = 0.1\,n_o$. The corresponding real frequencies are shown in Figure 7. It is seen that as n_{dlo}/n_{dho} decreases (n_{dho} increasing), the growth rate and real frequency of the DAI decreases due to the instability becoming dominated by the heavier species and hence the slower inertial response of the heavier particles to perturbations in potential. For $n_{dlo} < 0.01\,n_{dho}$, the growth rate becomes negligibly

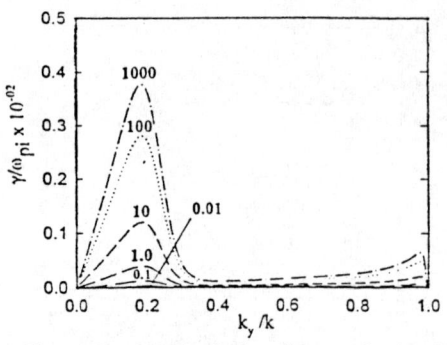

Figure 6: Plot of the normalized growth rate as a function of k_y/k. The parameter labelling the curves is n_{dlo}/n_{dho}. Here $n_{eo} = 0.1\, n_o$. All other fixed parameters are as in Figure 4.

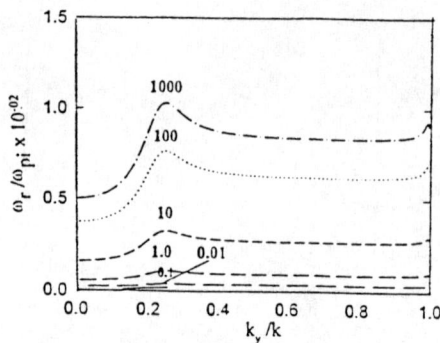

Figure 7: Plot of the normalized real frequencies corresponding to the growth rate curves in Figure 6. The parameter labelling the curves is n_{dlo}/n_{dho}.

small.

Thus when the fraction of the light dust grains is very small, the instability is present on the 'heavy dust fluid' time scale with growth rates two orders of magnitude lower than those associated with the light dust grains. We find that the growth rate of the DMTSI also decreases as the concentration of the heavy dust grains increases.

IV   CONCLUSION

In this paper we have examined low frequency electrostatic instabilities in a four-component dusty plasma. These instabilities are driven by an equal $\mathbf{E_o} \times \mathbf{B_o}$ drift of the magnetized ions and electrons relative to the unmagnetized dust grains of the two dust species. This model is relevant to the planetary rings of Saturn where the unmagnetized, levitated dust grains are stationary, while the magnetized electrons and ions have an azimuthal $\mathbf{E_o} \times \mathbf{B_o}$ drift (Rosenberg and Krall [9]). We find that the DAI dominates for small angles of wave propagation relative to $\mathbf{B_o}$, whereas for almost perpendicular propagation, the DMTSI is excited. This is consistent with the findings of Bharuthram [5]. The DAI is more easily excited in plasmas with low electron concentrations. The opposite was found to be true for the excitation of the DMTSI. An increase in T_i significantly affects the DAI due to the dependence of the growth rate on the slope of the equilibrium ion velocity distribution. A variation of T_i has no noticable effect on the DMTSI. As the concentration of the heavy dust grains starts increasing (decreasing n_{dlo}/n_{dho}), the instability becomes present on the time scale of the 'heavier dust fluid' which accounts for the reduction in the growth rate and real frequency. These results are important in understanding electrostatic phenomena in the ring plasmas of Saturn's rings where the electrons and ions have an azimuthal drift relative to the dust grains.

REFERENCES

1. Havnes, O., *Astron. Astrophys.* **90**, 106 (1980).
2. Havnes, O., *Astron. Astrophys.* **193**, 309 (1988).
3. D'Angelo, N., *Planet. Space Sci.* **38**, 1143 (1990).
4. Bharuthram, R. and Pather, T., *Planet. Space Sci.* **44**, 137 (1996).
5. Bharuthram, R., *Planet. Space Sci.* **45**, 379 (1997).
6. Rosenberg, M. and Chow, V.W., *Planet. Space Sci.* **46**, 103 (1998).
7. Rosenberg, M., *Planet. Space Sci.* **41**, 229 (1993).
8. Chow, V.W. and Rosenberg, M., *Planet. Space Sci.* **43**, 613 (1995).
9. Rosenberg, M. and Krall, N.A., *Planet. Space Sci.* **43**, 619 (1995).
10. Gary, S.P and Sanderson, J.J., *J. Plasma Phys.* **4**, 739, (1970).
11. Maharaj, S.K. and Bharuthram, R., (submitted to *J. Plasma Phys.*)
12. Meuris, P., Verheest, F. and Lakhina, G.S., *Planet. Space Sci.* **45**, 449, (1997).

Linear and non-linear dust acoustic waves in non-ideal dusty plasmas with grain charge fluctuations

S.R. Pillay[1], N.N. Rao[2] and R. Bharuthram[3]

[1] *Department of Physics, University of Durban-Westville, Durban, South Africa*
[2] *N.N. Rao Theoretical Physics Division, Physical Research Laboratory, Ahmedabad, India*
[3] *R. Bharuthram M.L. Sultan Technikon Durban, South Africa*

Abstract. The linear as well as the nonlinear propagation of dust-acoustic waves (DAWs) in non-ideal dusty plasmas comprising of electrons, ions and dust grains has been studied by self–consistently including the grain charge variations. The non-ideal effects are incorporated through the van der Waals equation of state for the dust fluid, while the charge fluctuation effects come in through the current balance equation. The real frequency and the damping rate for the linear DAWs are investigated and the reductive perturbation approach has been used to derive a K-dV-like equation governing the nonlinear evolution of the DAWs. The combined effects of the non–ideal contributions and the grain charge fluctuations in the linear as well as the nonlinear regimes of wave propagation are discussed.

I INTRODUCTION

An interesting feature of dusty plasmas relates to the finite size of the dust grains. Due to their relatively large sizes, dusty plasmas are inherently non-ideal in nature [1–2]. The ideal gas approximation used in most analyses of dusty plasmas, while being valid for grain sizes in the micrometre range and for dilute plasmas, breaks down for larger grains in the supermicrometre range. In this regime, the intergrain interactions between the neighbouring grains becomes increasingly significant. Hence, the usual equation of state used to describe ideal gas behaviour has to be replaced by one that accounts for nonideal effects.

In the present paper, we consider the regime of tenuous dusty plasmas, and extend the model of Rao [2] to study the combined effects of dust charge fluctuation and the non–ideal contributions on both the linear as well as the nonlinear propagation of the DAWs. Non-ideal effects come in through the use of the van der Waals equation of state to describe the dust grains, while dust charge variational effects are incorporated through a standard charge current balance equation. Our model and basic equations are presented in section (II). In section (III), the linear evolution

CP537, *Waves in Dusty, Solar, and Space Plasmas*, edited by F. Verheest, et al.

of the DAW is discussed, while section (IV) examines nonlinear propagation. In the latter case, we derive a KdV–like nonlinear equation to describe the DAW propagation.

II BASIC EQUATIONS

We consider a one-dimensional (along the $\hat{x}$ direction) propagation of the DAW in a collisionless dusty plasma consisting of electrons, ions and finite-sized, charged dust grains. In the very low-frequency regime of the DAW the inertia of the ions and electrons may be neglected. Thus the electron and ion number densities are governed by the Boltzmann distributions, each defined by their respective temperatures which can be assumed to be constant. Hence the electron and the ion number densities are given by

$$n_\alpha = n_{\alpha o} \exp\left(\frac{-q_\alpha \phi}{k_B T_e}\right) \quad , \tag{1}$$

where $q_\alpha = -e(+e)$ is the charge on the electron(ion), $n_{\alpha o}$ is the equilibrium density of the electrons (ions), k_B is the Boltzmann constant and ϕ is the self-consistent electric field potential.

The dust grains, on the other hand, being the heaviest component, provide the inertia for the DAW and therefore the wave dynamics are governed by the full set of dust fluid equations including the Poisson equation. Thus, we have the continuity, momentum balance and Poisson equation, respectively,

$$\frac{\partial n_d}{\partial t} + \frac{\partial}{\partial x}(n_d v_d) = 0 \quad , \tag{2}$$

$$\left(\frac{\partial v_d}{\partial t} + v_d \frac{\partial v_d}{\partial x}\right) = -\frac{q_d}{m_d}\frac{\partial \phi}{\partial x} - \frac{1}{m_d n_d}\frac{\partial p_d}{\partial x} \quad , \tag{3}$$

$$\frac{\partial^2 \phi}{\partial x^2} = -4\pi(q_d n_d + e n_i - e n_e) \quad , \tag{4}$$

where the subscript d refers to the dust component, and n, v, m and q represent the number density, fluid velocity, mass and charge, respectively.

Dust charge variation effects come in through the charge current balance equation (e.g. [3]), namely,

$$\frac{\partial q}{\partial t} + v\frac{\partial q}{\partial x} = I_e + I_i \quad , \tag{5}$$

which is valid for the case of grain charging arising from plasma currents due to electrons and ions reaching the grain surface. When the streaming velocities of the electrons and ions are much smaller than their respective thermal velocities, the expressions for the electron and ion currents for spherical grains of radius R are given by [4]

69

$$I_e = -\pi R^2 e \left(\frac{8T_e}{\pi m_e}\right)^{1/2} n_e \exp\left(\frac{e\psi}{T_e}\right) \quad , \quad \text{and} \quad I_i = \pi R^2 e \left(\frac{8T_i}{\pi m_i}\right)^{1/2} n_i \left(1 - \frac{e\psi}{T_i}\right) \quad ,$$

$$(6)$$

where $\psi \equiv q_d/R$ denotes the dust grain surface potential relative to the plasma potential.

The set of equations (1)–(6) is closed with an appropriate choice of equation of state for the dust fluid to describe the non-ideal behaviour of the finite sized grains. To keep the analysis tractable, we use the van der Waals equation of state [2] expressed in terms of the dust number density (n_d),

$$(p_d + An_d^2)(1 - Bn_d) = n_d k_d T_d \quad . \tag{7}$$

The constants A and B are as defined by Rao[2], viz., $A = 9k_d T_c/8n_c$ and $B = 1/3n_c$, with the subscript 'c' indicating the respective values at the critical point. Equations (1)–(7) constitute a complete set of equations to describe the behaviour of a non-ideal dusty plasma including grain charge fluctuation effects. In the next section, we examine the propagation of linear DAW's.

III   LINEAR DAW'S

For linear DAWs, the current balance equation (5) yields [5]:

$$\frac{\partial q_1}{\partial t} = -\omega_1 q_1 - R\omega_2 \phi_1 \quad , \tag{8}$$

where

$$\omega_1 = \lambda + (8\pi)^{1/2} R \frac{e^2 n_{eo}}{(m_e T_e)^{1/2}} \exp\left(\frac{e}{T_e}\psi_o\right) \quad ,$$

$$\omega_2 = \lambda\left(1 - \frac{q_i \psi_o}{T_i}\right) + (8\pi)^{1/2} R \frac{e^2 n_{eo}}{(m_e T_e)^{1/2}} \exp\left(\frac{e}{T_e}\psi_o\right) \quad ,$$

and $\lambda = R\omega_{pi}/(2\pi)^{1/2}\lambda_{Di}$. The equilibrium potential ψ_o has been calculated as a function of the dust density parameter [3]. For $T_e = T_i = T$, the ratio $e\psi_o/T$ is typically in the range 0 to -3, for near-earth parameters.

Applying the standard normal-mode analysis to equations (1)–(8) we obtain the dispersion relation,

$$\frac{\omega^2}{k^2} = \frac{\omega_{pd}^2 \lambda_D^2}{(1 + k^2\lambda_D^2)(1 + \mathcal{Q})} + V_{td}^2 + \frac{1}{m}\left[k_d T_d Bn_{do}\frac{(2 - Bn_{do})}{(1 - Bn_{do})^2} - 2An_{do}\right] \quad , \tag{9}$$

with $\mathcal{Q} = f\delta\omega_1/(\omega_1 - i\omega)(1 + k^2\lambda_D^2)$. Here, $f = 4\pi n_{do}\lambda_D^2 R$ is the dust fugacity parameter [6,7], $\delta = \omega_2/\omega_1$ is the ratio of the charging frequencies, $\omega_{pd} = (4\pi n_{do}q_{do}^2/m_d)^{1/2}$ is the dust plasma frequency, n_{do} is the equilibrium dust density,

$V_{td} = (k_d T_d/m_d)^{1/2}$ is the dust thermal speed and λ_D is a plasma Debye length defined through $1/\lambda_D^2 = 1/\lambda_{De}^2 + 1/\lambda_{Di}^2 = (4\pi e^2/k_B)(n_{eo}/T_e + n_{io}/T_i)$. The quantity f characterizes the dust fugacity [6,7], and plays a fundamental role in distinguishing the parameter regimes for the existence of different types of ultra low–frequency modes in dusty plasmas. In fact, the latter can be considered as tenuous, dilute or dense according as $f \ll 1$, $f \sim 1$, or $f \gg 1$. In the following analysis, we consider the case of tenuous dusty plasmas.

Substituting for the parameters A and B the dispersion relation (9) may be expressed as

$$\frac{\omega^2}{k^2 C_D^2} = \frac{(1-\beta)}{(1+k^2\lambda_D^2)}[1 - \mathcal{Q}] + \beta + \epsilon_{\nu r} + \epsilon_{cf} \quad , \tag{10}$$

where $\beta = V_{td}^2/C_D^2$, and $C_D^2 = C_{DA}^2 + V_{td}^2$. Furthermore, we define, $\eta = n_{do}/n_c$, $\alpha = T_c/T_d$. The quantities $\epsilon_{\nu r} = \beta\eta(6-\eta)/(3-\eta)^2$ and $\epsilon_{cf} = -9\beta\alpha\eta/4$ denote, respectively, the contributions due to the volume reduction coefficient and the molecular cohesive forces [2]. In the long wavelength limit we recover from (10) eq.(11) of Rao [2] for the constant dust charge case.

Setting $\omega = \omega_r + i\omega_i$ with $\omega_i \ll \omega_r$, equation (10) yields for the real frequency,

$$\frac{\omega_r^2}{k^2 C_D^2} = \frac{(1-\beta)}{(1-k^2\lambda_D^2)}\left[1 - \frac{f\delta}{(1+k^2\lambda_D^2)}\frac{\omega_1^2}{(\omega_1^2+\omega_r^2)}\right] + (\beta + \epsilon_{\nu r} + \epsilon_{cf}), \tag{11}$$

and for the damping rate,

$$\frac{\omega_i}{\omega_r} = -\frac{1}{2}\frac{(1-\beta)f\delta}{(1+k^2\lambda_D^2)(1+\epsilon_{\nu r}+\epsilon_{cf})}\frac{\omega_r\omega_1}{(\omega_1^2+\omega_r^2)} \quad . \tag{12}$$

In the limit $\omega_r^2 \ll \omega_1^2$, Eq. (11) reduces to:

$$\frac{\omega_r^2}{k^2 C_D^2} = \frac{(1-\beta)}{(1-k^2\lambda_D^2)}\left[1 - \frac{f\delta}{(1+k^2\lambda_D^2)}\right] + (\beta + \epsilon_{\nu r} + \epsilon_{cf}) \tag{13}$$

A numerical study of the response of the normalised damping rate and real frequency to changes in the non-ideal as well as charge fluctuation parameters is performed by using the respective equations (12) and (13). All frequencies are normalised by kC_D and we define $\bar{\omega}_1 = \omega_1/kC_D$. The numerical results are obtained for fixed parameters $\beta = 0.1$ and $k\lambda_D = 0.1$ and $\bar{\omega}_1 = 0.1$.

Figure 1 examines the normalised real frequency as a function of η for $f\delta = 0.1$ and $\alpha = 0.1$; 1.0 and 2.0 respectively. It is noted that there is a reduction in the real frequency compared to the constant charge case over the range of η and α values shown. An interesting feature concerning the behaviour of the real frequency is that for higher α values (corresponding to stronger cohesive force corrections to the ideal case) there is a significant drop in the real frequency over a range of low η values both for the constant and fluctuating grain charge case. In this parameter range of α and η the effects of the cohesive forces supercede those of volume reduction leading to a decreased pressure in the gas, and hence a drop in real frequency.

Figure 2 examines the behaviour of the normalised damping rate as a function of the non-ideal parameter η at $f\delta = 0.1$. We note that at high α there is a range of η values over which the cohesive forces dominate over the effects of volume reduction and add further to the damping produced by charge variation effects. Of particular interest is the fact that the damping effect resulting from charge variation is opposed in the presence of strong volume reduction effects (i.e. high η), when the cohesive forces are at their minimum (i.e. low α). This suggests that under appropriate conditions the damping effects resulting from charge variation are reduced in a non-ideal dusty plasma.

Figure 3 examines the normalised real frequency for higher dust fugacity, viz. $f\delta = 0.5$. Here we observe a wide separation between the real frequencies obtained for the constant and varying charge cases. Whilst the effects of the variation in non-ideal parameters remain very much the same, it is apparent that damping due to charge variational effects dominate, and there is a significant drop in the real frequency. Hence in the dilute or dense plasma regime charge variation effects are dominant, despite the compensating influence of the volume reduction effects at low α.

Figure 4 examines the normalised damping rate as a function of the non-ideal parameter η for $f\delta = 0.5$. Here we note an overall increase in the damping rate as is expected from the analytical expression (12).

IV   NONLINEAR ANALYSIS

In this section, we examine the non-linear evolution of small, but finite, amplitude DAWs in a non-ideal plasma by including dust charge fluctuations. Equations (1)–(4) and (7) remain as the basic set of equations to describe the nonlinear propagation. In order to use the standard reduction perturbation technique, we introduce the stretched co–ordinates ξ and τ through the expressions,

$$\xi = \epsilon^{1/2}(x - \lambda t) \qquad ; \qquad \tau = \epsilon^{3/2}t \quad , \tag{14}$$

where ϵ is the expansion parameter, and λ is the speed of the co–moving stationary frame to be determined self–consistently later. The electrostatic potential together with the dust number density, speed, pressure and the charge are expanded as follows:

$$n_d = n_{do} + \epsilon n_{d1} + \epsilon^2 n_{d2} + ... \tag{15}$$
$$v_d = \epsilon v_1 + \epsilon^2 v_2 + ... \tag{16}$$
$$\phi = \epsilon\phi_1 + \epsilon^2\phi_2 + ... \tag{17}$$
$$p_d = p_{do} + \epsilon p_{d1} + \epsilon^2 p_{d2} + ... \tag{18}$$
$$q_d = q_{do} + \epsilon^2 q_{d1} + \epsilon^3 q_{d2} + ... \quad . \tag{19}$$

It should be remarked here that the scaling chosen in the expansion for the dust charge (q_d) is appropriate as well as necessary, as discussed later, for describing DAWs since they exist even without the grain charge fluctuations (q_{d1}), and hence the latter should have an order higher than the dust number density perturbation (n_{d1}).

Equations (15)–(19) are substituted into our basic set of equations and retain terms to order ϵ^2. From the first–order equations for the perturbations, we self–consistently determine the speed λ through $\lambda^2 = CD_{DN}^2$, where the wave phase speed C_{DN} is defined by, $C_{DN}^2 = C_D^2 + V_{td}^2(\beta + \epsilon_{vr} + \epsilon_{cf})$. We transform equation (20) back to the usual (x, t) co-ordinates by using the definition for λ. For acoustic-like perturbations with phase speed near the dust acoustic speed, we have, $\partial/\partial t \sim -C_{DN}\partial/\partial x$ and hence we finally have:

$$\frac{\partial \Phi_1}{\partial t} + C_{DN}\frac{\partial \Phi_1}{\partial x}\frac{1}{2}\frac{C_{DA}^4}{\omega_{pd}^2 C_{DN}}\frac{\partial^3 \Phi_1}{\partial x^3} + \frac{q_{do}}{m_d}\frac{C_{DN}}{C_{DA}^2}\left(\frac{3}{2} + \mu\right)\frac{\partial \phi_1^2}{\partial x} = \frac{1}{2}\frac{m_d}{q_{do}^2}\frac{C_{DA}^3}{C_{DN}}\frac{\partial Q_{d1}}{\partial t} \quad (20)$$

where we have defined the scaled perturbations $\Phi_1 \to \epsilon\phi_1$ and $Q_{d1} \to \epsilon^2 q_{d1}$, while μ is given by $\mu = (V_{td}^2/C_{DN}^2)(9\eta/(3 - \eta^2) - 9/(3 - \eta)^3) + (C_{DA}^2/C_{DN}^2)(A_1 A_2^2 q_{do}/n_{do}, A_1 = (e^3 n_{io}/2k_B^2 T_i^2 - e^3 n_{eo}/2k_B^2 T_e^2)$ and $A_4 = 4\pi n_{do}\lambda_D^2$.

It may be noted that the scaling implied in the perturbed quantity Q_{d1} is consistent with the fact that the DAWs exist even when the charge fluctuations are neglected, and typically occur only the in low fugacity, tenuous ($f \ll 1$) regime [6,7] when the wave frequency is much smaller than the charging frequency ($\omega \ll \omega_1$). On the other hand, in the high fugacity, dense ($f \gg 1$) regime, the effect of the grain charge fluctuations is important. In fact, in this regime, the dust number density perturbation is directly proportional to the grain charge fluctuation and the latter propagate as a new kind of normal mode [6,7] which is driven by the Coulomb pressure. The analysis of these in the dense regime is beyond the scope of the present work, but would be considered separately elsewhere.

Equation (20) is coupled to the perturbed charging equation (8) which we now write in terms of the scaled quantities Q_{d1} and Φ_1 as

$$\frac{\partial Q_{d1}}{\partial t} = -\omega_1 Q_{d1} - R\omega_2\Phi_1 \quad . \quad (21)$$

Equations (20)–(21) thus constitute a set of equations which govern the propagation of nonlinear DAWs in a non-ideal dusty plasma. It should be remarked that these need to be solved, in general, by numerical methods. However, there exist two limiting cases wherein it is possible to analytically derive the qualitative nature of the solutions. We consider these two cases separately below.

Case (1) : $\omega \ll \omega_1$

When the wave frequency (ω) is much smaller than the grain charging frequency (ω_1), the grains can attain an average charge within one wave period, and hence the wave damping effects due to the charge fluctuations can be neglected. Accordingly, we neglect the left hand side of Eq. (21) to obtain, $Q_{d1} \approx -R\Phi_1\delta$, where $\delta = \omega_2/\omega_1$.

Substituting this expression for Q_{d1} into Eq. (20), we obtain the K–dV type of equation,

$$\alpha \frac{\partial \Phi_1}{\partial t} + C_{DN} \frac{\partial \Phi_1}{\partial x} \frac{1}{2} \frac{C_{DA}^4}{\omega_{pd}^2 C_{DN}} \frac{\partial^3 \Phi_1}{\partial x^3} + \frac{q_{do}}{m_d} \frac{C_{DN}}{C_{DA}^2} \left(\frac{3}{2} + \mu \right) \frac{\partial \Phi_1^2}{\partial x} = 0 \quad , \tag{22}$$

where $\alpha = 1 + R\delta/2(m_d/q_{do}^2)(C_{DA}^3/C_{DN})$. Equation (23) above admits the usual sech-type of localized solutions which are undamped. This is consistent with the result predicted by linear theory, where we have $\gamma \to 0$ for $\omega_1 \ll \omega_r$.

Case (2) : $\omega \gg \omega_1$

In this case, the equilibrium charge is almost constant over a wave period and, hence, the wave damping is small. For this limit, we approximate Eq. (21) by $\partial Q_{d1}/\partial t \approx -R\omega_2 \Phi_1$. Using this equation in Eq. (20), we obtain the K-dV like equation

$$\frac{\partial \Phi_1}{\partial t} + C_{DN} \frac{\partial \Phi_1}{\partial x} \frac{1}{2} \frac{C_{DA}^4}{\omega_{pd}^2 C_{DN}} \frac{\partial^3 \Phi_1}{\partial x^3} + \frac{q_{do}}{m_d} \frac{C_{DN}}{C_{DA}^2} \left(\frac{3}{2} + \mu \right) \frac{\partial \phi_1^2}{\partial x} = -\gamma \Phi_1 \quad , \tag{23}$$

where $\gamma = fR\omega_2/2(C_{DA}/C_{DN})$. Making use of the transformation, $\Phi_1 = \tilde{\Phi}_1 \exp(-\gamma t)$, we obtain

$$\frac{\partial \tilde{\Phi}_1}{\partial t} + C_{DN} \frac{\partial \tilde{\Phi}_1}{\partial x} \frac{1}{2} \frac{C_{DA}^4}{\omega_{pd}^2 C_{DN}} \frac{\partial^3 \tilde{\Phi}_1}{\partial x^3} + \frac{q_{do}}{m_d} \frac{C_{DN}}{C_{DA}^2} \left(\frac{3}{2} + \mu \right) \exp(-\gamma t) \frac{\partial \tilde{\Phi}_1^2}{\partial x} = 0, \tag{24}$$

It therefore follows that the localised solutions of equation (23) for a non-ideal dusty plasma, are also damped due to dust charge variation effects, at earlier times, but now with redefined coefficients compared to [5]. Interestingly, in the cold dust limit we have, $\gamma = 2\pi n_{do} \lambda_D^2 R\omega_2$, as obtained by Rao and Shukla [5].

REFERENCES

1. Fortov, V.E. and Iakubov, I.T. (1990), *Physics of Nonideal Plasmas*. Hemisphere, New York.
2. Rao, N.N. (1998a), *J. Plasma Phys.* **59**, 561–574.
3. Melandso, F., Aslaksen, T.K. and Havnes, O. (1993a), *Planet. Space Sci.* **41**, 312.
4. Varma, R.K., Shukla, P.K. and Krishan, V. (1993), in dusty plasmas. *Phys. Rev.* **E47**, 3612.
5. Rao, N.N. and Shukla, P.K. (1994), *Planet. Space. Sci.* **42**, 221.
6. Rao, N.N. (1999), *Physics of Plasmas* **6**, p4414–4417.
7. Rao, N.N. (2000), *Physics of Plasmas* **7**, p795–807.

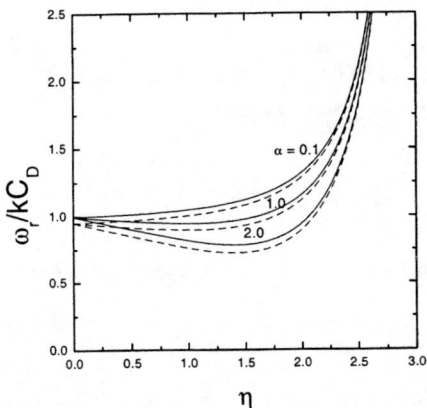

Fig 1 : ω_r/kC_D versus η for fixed
$k\lambda_D = 0.1$; $\beta = 0.1$; $f\delta = 0.1$.
Values of α are indicated on the curves. The
solid lines represent the constant charge case.

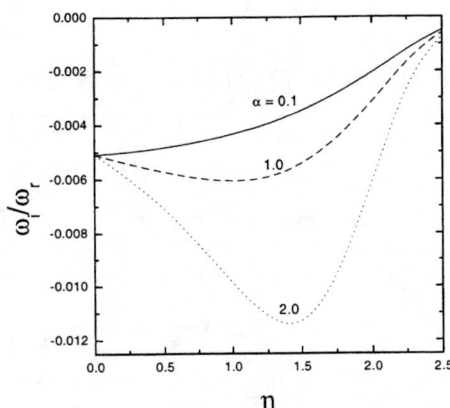

Fig 2 : ω_i/ω_r versus n for fixed
$k\lambda_D = 0.1$; $\beta = 0.1$; $f\delta = 0.1$.
Values of α are indicated on the graph.

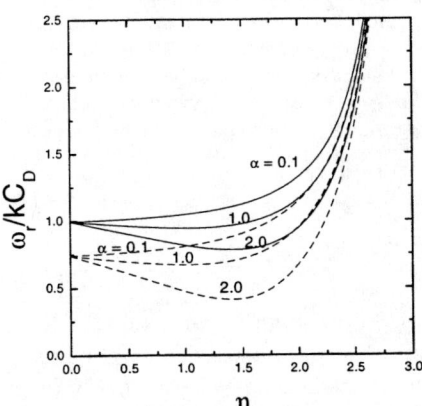

Fig 3 : ω_r/kC_D versus η for fixed
$k\lambda_D = 0.1$; $\beta = 0.1$; $f\delta = 0.5$.
Values of α are indicated on the curves. The
solid lines represent the constant charge case.

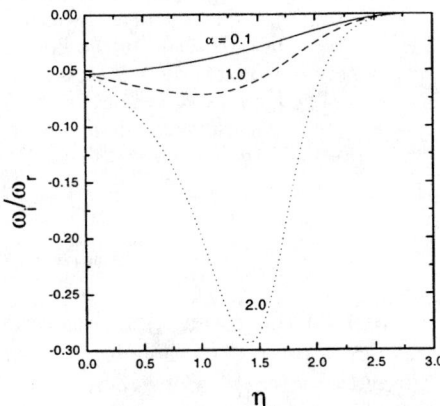

Fig 4 : ω_i/ω_r versus n for fixed
$k\lambda_D = 0.1$; $\beta = 0.1$; $f\delta = 0.5$.
Values of α are indicated on the graph.

Nonmodal phenomena in differentially rotating dusty plasmas

Stefaan Poedts[1]

Centre for Plasma-Astrophysics, K.U.Leuven, Celestijnenlaan 200B, 3001 Leuven, Belgium.

Andria D. Rogava

Department of Theoretical Astrophysics, Abastumani Astrophysical Observatory Tbilisi, Georgia and Abdus Salam International Centre for Theoretical Physics, Trieste, Italy.

Abstract. In this paper the foundation is layed for the nonmodal investigation of velocity shear induced phenomena in a differentially rotating flow of a dusty plasma. The simplest case of nonmagnetized flow is considered. It is shown that, together with the innate properties of the dusty plasma, the presence of differential rotation, Coriolis forces, and self-gravity casts a considerable richness on the nonmodal dynamics of linear perturbations in the flow. In particular: (i) dust-acoustic waves acquire the ability to extract energy from the mean flow and (ii) shear-induced, nonperiodic modes of collective plasma behaviour — *shear-dust-acoustic vortices* — are generated. The presence of self-gravity and the nonzero Coriolis parameter ("epicyclic shaking") makes these collective modes transiently unstable.

INTRODUCTION

Much of the recent progress in astrophysical fluid dynamics and plasma astrophysics is related to the efforts to properly deal with inhomogeneities. The biggest current challenge is posed by the inhomogeneous or sheared velocity fields (shear flows) which are present in a wide variety of astrophysical objects. Shear flows are known to exhibit a number of remarkable, so-called "nonmodal" phenomena, which are essentially linear and which exist due to the non-self-adjointness of the governing mathematical operators. It was found that:

- Waves sustained by the flow acquire the ability to exchange energy with the flow [1–3]. These waves can damp or amplify at the expense of the free energy in the equilibrium flow.

---

[1] Research Associate of the Flemish Fund for Scientific Research (FWO-Vlaanderen).

CP537, *Waves in Dusty, Solar, and Space Plasmas*, edited by F. Verheest, et al.

- In flows sustaining more than one type of wave motion, the shear provides a *linear* mechanism for their reciprocal transformations with corresponding energy exchange between the waves [4]- [6].

- Shear can lead to the excitation of characteristic beat waves both in hydrodynamic and hydromagnetic flows [6]- [7].

- Shear gives birth to a unique non-periodic mode of plasma collective behaviour— 'shear vortices' [8,9]. These vortices are able to extract energy from the mean flow either in a transient (2-D vortices) or in an asymptotic (3-D vortices) regime (cf. *Kelvin modes* in hydrodynamics [10]).

- In flows with moderate or high shearing rates, shear vortices eventually acquire wave-like features ("convert" into waves). This phenomenon occurs both in the hydrodynamic [11] and the plasma [9] channel.

The aim of this paper is to initiate the study of velocity shear induced nonmodal processes in *dusty space plasmas*. The presence of charged dust grains (besides the electrons and ions) in plasma flows is very common in astrophysical situations. The best known examples in the Solar System [15] include regions of the Earth's lower magnetosphere, cometary comae and tails, asteroid zones and planetary rings. The presence of the dust component may lead to remarkable observations such as the periodically appearing and disappearing dark transversal bands (so-called "spokes") in the B-ring and the "braids" in the F-ring of Saturn, closely observed by the *Voyager I and II* spacecraft cameras [16,17].

The physical model that will be developed is based on the well-known "shearing sheet approximation" [12,13], which was successfully employed by many authors in the study of galactic gaseous disks and accretion disks [18,14,7].

The next section contains the main mathematical foundation of the model. We consider *ultra-low-frequency phenomena* in a thin, differentially rotating disk flow and, therefore, assume a Boltzmann distribution for but the electrons and the ions. The dust particle dynamics is studied in the fluid approximation and the disk is considered to be nonmagnetized. The linear nonmodal analysis is performed and the problem is reduced to the solution of a single, second-order ordinary differential equation, which describes the temporal evolution of dust acoustic waves and shear dust vortices. The possible applications of shear-induced nonmodal phenomena of dusty plasma disk flows are discussed in the last section.

THEORY

Let us consider a differentially rotating, uniform and nonmagnetized plasma consisting of electrons, singly charged positive ions and negatively charged dust particles. In the present study we restrict ourselves to the study of ultra-low-frequency electrostatic waves. This allows to neglect both the electron and the ion inertia and implies the use of the Boltzmann distribution for number densities N_s ($s = e, i$) of both the electrons and the ions:

$$N_e = \mathcal{N}_e \exp[e\,\phi/k_B T_e] \approx \mathcal{N}_e(1 + e\,\phi/k_B T_e), \tag{1a}$$

$$N_i = \mathcal{N}_i \exp[-e\,\phi/k_B T_i] \approx \mathcal{N}_i(1 - e\,\phi/k_B T_i), \tag{1b}$$

where ϕ is the electrostatic potential, T_e and T_i are the electron and the ion temperatures, k_B is the Boltzmann constant, and e is the absolute charge of an electron.

The dust grains are supposed to have mass m_d, and their number density, temperature, thermal pressure and velocity are denoted by N_d, T_d, P_d, and $\mathbf{V}_d$, respectively. In reality, the charge and the mass of a dust grain can vary from one grain to the other. Moreover, in external magnetic fields dust grain charges may vary periodically in time. But for the sake of simplicity, here we assume that all dust grains have the same mass m_d and the same charge state Z ($Q_d = Ze$).

We assume that the steady thermodynamic state of the system is uniform, with number densities of electrons ($\mathcal{N}_e$), ions ($\mathcal{N}_i$) and dust particles ($\mathcal{N}_d$) satisfying the quasi-neutrality condition:

$$\mathcal{N}_i = \mathcal{N}_e + Z\mathcal{N}_d. \tag{2}$$

The equilibrium state of the system is specified by the constant vectors of the angular rotation velocity $\mathbf{\Omega} \equiv (0, 0, \Omega < 0)$ and by the locally plane-parallel steady state velocity field of the "dust fluid" $\mathbf{V}_0 \equiv (0, 2\mathcal{A}x, 0)$, with $\mathcal{A} \equiv (r/2)\partial_r\Omega > 0$ being the first Oort constant. The second Oort constant is defined as: $\mathcal{B} \equiv \mathcal{A} + \Omega$ (see Refs. [21,14,7]).

For the electric (ϕ) and gravitational (ψ) potentials of the system we have the following Poisson (field) equations [with $\Delta \equiv \partial_x^2 + \partial_y^2 + \partial_z^2$]:

$$\Delta\phi = 4\pi e \left[N_e + ZN_d - N_i\right], \tag{3}$$

$$\Delta\psi = 4\pi G m_d N_d \delta(z), \tag{4}$$

where G is the gravitational constant and $\delta(z)$ is the Dirac delta function.

For the dust component of the plasma we adopt the conventional fluid model governed by the continuity equation [with $D_t \equiv \partial_t + (\mathbf{V} \cdot \nabla)$]:

$$D_t\Sigma + \Sigma\nabla \cdot \mathbf{V} = 0, \tag{5}$$

and the equation of motion:

$$D_t\mathbf{V} = -\frac{1}{\Sigma}\nabla P + \frac{Ze}{m_d}\nabla\phi - 2\mathbf{\Omega} \times \mathbf{V} - \nabla\Psi. \tag{6}$$

where Σ and P are the vertically integrated two-dimensional surface number density and the thermal pressure of dust particles, respectively.

From this set it is rather straightforward to obtain linearized equations for the perturbations. In particular, from the continuity equation for the dust-fluid component we get [with $\mathcal{D}_t \equiv \partial_t + 2\mathcal{A}x\partial_y$]:

$$\mathcal{D}_t\sigma + \Sigma_0(\partial_x u + \partial_y v) = 0; \tag{7}$$

the two components of the equation of motion yield

$$\mathcal{D}_t u = -\partial_x S + 2\Omega v, \tag{8a}$$

$$\mathcal{D}_t v = -\partial_y S - 2Bu; \tag{8b}$$

while the Poisson equation for the electric potential takes the form:

$$\Delta\phi = 4\pi e\,(n_e + Zn_d - n_i); \tag{9}$$

and the equation for the gravitational potential reads:

$$\Delta\psi = 4\pi G\sigma\delta(z), \tag{10}$$

where

$$S \equiv \chi - (Ze/m_d)\phi, \tag{11a}$$

and

$$\chi \equiv C_i^2\sigma/\Sigma_0 + \psi. \tag{11b}$$

In these equations σ, u and v are the perturbations of the dust surface number density and the velocity radial (u) and azimuthal (v) components, respectively; while $C_i \equiv (k_B T_e/m_i)^{1/2}$ is the ion sound speed.

According to the basic principles of the "nonmodal approach" the system sustains, together with conventional exponentially evolving disturbances, also a class of solutions with nonexponential time evolution. Components of the wave number vector for this class of perturbations may depend on time [23]. For the separation of variables we use the ansatz: $F(\mathbf{r}, t) \equiv \hat{F}(\mathbf{k}(t), t)\exp[i(k_x(t)x + k_y y + k_z z)]$ with $k_x(t) = k_x(0) - 2\mathcal{A}tk_y$ [22], which reduces mathematics to the solution of an initial value problem. By introducing new dimensionless parameters: $R \equiv \mathcal{A}/C_i k_y$, $\beta_0 \equiv k_x(0)/k_y$, $\omega \equiv \Omega/C_i k_y$, $d \equiv -2\pi G\Sigma_0/C_i^2 k_y$; and new dimensionless variables $\tau \equiv tC_i k_y$, $\beta(\tau) \equiv \beta_0 - 2R\tau$, $\alpha \equiv \mathcal{N}_i/\mathcal{N}_e$, $D \equiv i\hat{\sigma}/\Sigma_0$, $\mathcal{U} \equiv \hat{u}/C_i$, $\mathcal{V} \equiv \hat{v}/C_i$, we can reduce our initial set of equations to the following system of first order ordinary differential equations:

$$D^{(1)} = \beta(\tau)\mathcal{U} + \mathcal{V}, \tag{12}$$

$$\mathcal{U}^{(1)} = -\beta(\tau)S + 2\omega\mathcal{V}, \tag{13a}$$

$$\mathcal{V}^{(1)} = -S - 2(\omega + R)\mathcal{U}, \tag{13b}$$

where the functions S and D are linked via the relation [with $\mathcal{K}(\tau) \equiv \sqrt{1 + \beta^2(\tau)}$]:

$$S = f(\tau)D, \tag{14a}$$

$$f(\tau) \equiv \epsilon\nu + \frac{\epsilon Z(\alpha - 1)}{1 + \alpha/\mu + \xi^2\mathcal{K}^2(\tau)} - \frac{d}{\mathcal{K}(\tau)}, \tag{14b}$$

with $\mu \equiv T_i/T_e$, $\nu \equiv T_d/T_e$, $\xi \equiv k_y\lambda_D$, and $\epsilon \equiv m_i/m_d$

It is easy to notice that the functions D, $\mathcal{U}$ and $\mathcal{V}$ obey the following algebraic relation:

$$-2(\omega + R)D + \mathcal{U} - \beta(\tau)\mathcal{V} \equiv C \quad (= \text{const}). \tag{15}$$

The latter relation helps to derive an explicit second order differential equation for the function $Y \equiv \mathcal{K}^{-1}D$, which may be written in the following way:

$$Y^{(2)} + \varpi^2(\tau)Y = F(\tau), \tag{16}$$

where for the "effective frequency" $\varpi(\tau)$ and the "external force" $F(\tau)$ we have:

$$\varpi^2(\tau) \equiv f(\tau)\mathcal{K}^2(\tau) + 4\omega(\omega + R) + \frac{8\omega R}{\mathcal{K}^2(\tau)} + \frac{12R^2}{\mathcal{K}^4(\tau)}, \tag{17a}$$

$$F(\tau) \equiv -\frac{2\omega C}{\mathcal{K}(\tau)} - \frac{4RC}{\mathcal{K}^3(\tau)}. \tag{17b}$$

Thus we are able to reduce the mathematical contents of the problem to the solution of this equation. All other (physical) variables are readily expressed through the functions Y and $Y^{(1)}$. The equation is of the same type as the ones appearing in various problems concerning acoustic phenomena in neutral fluids [1,2], ion-acoustic perturbations [8] and Langmuir oscillations [9] in usual two-component electrostatic plasmas. However, in this particular case the physical medium is more complex (electron-ion plasma complemented by the admixture of charged dust particles) and the effects of Coriolis forces and self-gravity are taken into account. Hence, the expressions for the effective frequency and the external force are rather complicated: they depend on a number of free parameters associated both with the plasma and with the flow characteristics. The format of this contribution does not allow us to study all possible regimes of solutions in a full extent. Below we only try to outline principal classes of solutions.

The convenient way to classify solutions is to consider separately initial perturbations with $C = 0$ and with $C \neq 0$

1. **Dust-acoustic waves:**

   When the perturbations are exactly wave-like ($C = 0$ and, hence, $F(\tau) \equiv 0$) [1,2] the system sustains shear-modified dust-acoustic waves. The flow inhomogeneity ($R \neq 0$) leads to the temporal variability of the wave-number vector $k_x(t)$, which causes a temporal variation of the effective frequency of dust-acoustic waves $\varpi(\tau)$. This means that at a certain stage of evolution the wave acquires the ability to draw (extract) energy from the background flow and to amplify [1,2,8,9]. In addition, the effect of the nonzero Coriolis force ($\omega \neq 0$) and the nonzero self-gravity ($d \neq 0$) may become substantial, because even when both these parameters are small, the low value of the dust-acoustic wave characteristic frequency still gives a chance for these physical factors to affect the evolution of the system quite substantially. In particular, the combined action of the Coriolis force ("epicyclic shaking") and the self-gravity effect (Jeans instability) may easily lead to the appearance of the *specific shear-induced transient instability of dust-acoustic waves.*

Even though this instability is transient (i.e., exists only for the limited time interval, where $\varpi^2(\tau) < 0$) it may easily succeed, while it is acting, to increase significantly the amplitude of the waves. As a result, the initial low-amplitude dust-acoustic waves may amplify considerably and may easily reach magnitudes, where nonlinear effects may switch on and make the system turbulent. This kind of "nonmodal" transition to turbulence (with initial stage of linear and transient increase of the perturbations) has already been presumed in the hydrodynamic context [22]- [24] and it is quite possible that the similar scenario may lead to the onset and maintenance of the *dust-acoustic turbulence* in self-gravitating, differentially rotating dusty plasma flows.

2. **Shear-Dust-acoustic vortices:**

There is another class of solutions, we call them *shear-dust-acoustic vortices*, which are characterized by $C \neq 0$ and which are (at least, initially) totally nonperiodic. Such solutions are known as Kelvin modes in hydrodynamics [1,2,10] and recently the analogous vortical patterns were identified and studied in electrostatic plasmas as well [8,9,19]. For low-enough shearing rates the approximate solutions for these vortices are given by the simple expression: $Y(\tau) \simeq F(\tau)/\varpi^2(\tau)$. This is a very specific class of initial perturbations, which in the coarse of time, may or may not acquire oscillating (wave-like) features. In neutral fluids and in usual electrostatic plasmas such "conversion" of the vortices into waves [11] happens only in the flows with moderate or high shearing rates [11,9]. While in dusty plasmas, where the generic characteristic frequency of dust-acoustic modes is usually very low, such transitions are expected and found to take place even in the flows with rather low shearing rates [19].

Notice that Coriolis forces result in a second term in the external force $F(\tau)$ (see Eq. (17b)) so that this term may change sign in the course of the shear-induced drift of the perturbations[2].

Like dust-acoustic waves, shear-dust-acoustic vortices may also exhibit very strong transient growths and thus, potentially, they can also give birth to large-amplitude perturbations. These finite amplitude disturbances, in turn, may lead to the development of the turbulence. Certainly it is naive to expect that real perturbations in real systems are purely of the periodic or nonperiodic type. Real systems will always sustain different mixtures of both kinds of perturbations. Therefore, the origin of the turbulence in such flows may be twofold. However, the clear distinction between these two types of perturbations, which is problematic even in hydrodynamic shear flows of neutral fluids [1,2], is even more difficult in differentially rotating flows of dusty plasmas.

---

[2] With our sign convention $R > 0$ and $\omega < 0$.

CONCLUSION

We considered acoustic phenomena in a differentially rotating, nonmagnetized and self-gravitating disk of dusty plasma. We found that the presence of the velocity shear leads to a number of notable *shear-induced nonmodal processes*. In particular, we found that:

- The dust acoustic waves become able to extract energy from the mean flow. This process may be both adiabatic and exponential. In the former case the energy of the waves may increase quasi-linearly, staying "tuned" (proportional) to the effective time-dependent frequency of the wave. In the latter case the velocity shear together with the Coriolis forces and/or self-gravity effects may ensure the onset of the specific shear-induced *transient shear instability* of the dust acoustic waves.

- The differentially rotating flow of dusty plasma sustains a class of nonperiodic perturbations, which we call *shear dust acoustic vortices*, which are characterized by a nonperiodic motion of the fluid particles.

- The actual perturbations, arising in this system are always containing both the wave and the vortical components. The distinction between these two types of perturbations is thus rather conditional and for the disks with moderate and large shearing rates the actual perturbations exhibit the features of both kinds of behaviour on certain stages of their temporal evolution.

In order to apply this simple model to the study of specific processes in planetary dusty rings, three additional physical aspects must be taken into account to make the model more complete. First, one has to take into account the probable large-scale magnetic field in the rings. If the rings are magnetized than, even in the electrostatic limit, the dusty plasma flows sustain together with acoustic waves also dust-cyclotron waves and this will make the rings even more "wavy".

Second, it is likely that the charge of dust grains is *not* constant, but periodically variable [20] due to the variation of the total velocity of the grain at the Larmor frequency[3]. This may lead to a number of remarkable physical consequences.

Third, it should be noted that the simple, plane-parallel shear profile may be an over-simplification in a number of astrophysical cases including the dusty plasma rings of Saturn. These shear flows may be kinematically more complex [25], leading to a number of highly nontrivial additional effects associated with the multidimensionality, with complex geometric and/or kinematic features of involved flows.

One of the most remarkable discoveries of Voyager I and II was the phenomenon of periodically appearing and disappearing dark transversal bands (so called "spokes") in Saturn's B-ring. Since it is known that (a) the rings are largely composed of micron and/or submicron-sized dust grains; and (b) there exists a correlation between the evolution period of spokes and that of Saturn's rotation; it is reasonable

---

[3] Caused, in turn, by the gyration of dust particles about their respective guiding centers.

to assume that this phenomenon is likely to be related with some electrodynamic processes in the ring dusty plasma flows. We argue that the role of shear-induced nonmodal processes can be rather important in this context.

Acknowledgments

This work is supported in part by the INTAS grant No. 97-0504. Andria Rogava wishes to thank Abdus Salam International Centre for Theoretical Physics for supporting him, in part, through an Associate Membership Award (ICTP).

REFERENCES

1. Chagelishvili G.D., Rogava A.D. & Segal I.N.: Phys. Rev. E, **50**, 4283 (1994).
2. Chagelishvili G.D., Khujadze G.R., Lominadze J.G. & Rogava A.D.: Phys. Fluids, **9**, 1955 (1997).
3. Rogava A.D., Mahajan S.M., & Berezhiani V.I.: Phys. Scripta, **58**, 622 (1998).
4. Chagelishvili G.D., Rogava A.D. & Tsiklauri D.G.: Phys. Rev. E, **53**, 6028 (1996).
5. Rogava A.D., Mahajan S.M., & Berezhiani V.I.: Phys. Plasmas, **3**, 3545 (1996).
6. Rogava A.D. & Mahajan, S.M.: Phys. Rev. E, 55, 1185 (1997).
7. Rogava A.D., Poedts S. & Heirman S., Mon. Not. R. Astron. Soc., **307**, L31 (1999).
8. Rogava A.D., Chagelishvili G.D., & Berezhiani V.I.: Phys. Plasmas, **4**, 4201 (1997).
9. Rogava A.D., Chagelishvili G.D., & Mahajan S.M.: Phys. Rev. E, **57**, 7103 (1998).
10. Marcus P., & Press W.H.: J. Fluid Mech., **79**, 525 (1977).
11. Chagelishvili G.D., Tevzadze A.G., Bodo G., & Moiseev, S.S., Phys. Rev. Lett., **79**, 3178 (1997).
12. Goldreich P., Lynden-Bell D., Mon. Not. R. Astron. Soc., **130**, 125 (1965).
13. Fan Z.H., Lou Y.Q., Nature, **383**, 800 (1996).
14. Fan Z.H., Lou Y.Q., Mon. Not. R. Astron. Soc., **291**, 91 (1997).
15. Verheest F., Space Sci. Rev., **77**, 267 (1996).
16. Smith B.A. et al. Science, **212**, 163 (1981).
17. Smith B. A. et al. Science, **215**, 504 (1982).
18. Balbus S.A. & Hawley J.F., Ap. J., **376**, 214 (1991).
19. Poedts S., Khujadze G.R., & Rogava A.D., Phys. Plasmas (accepted) (2000).
20. Northrop T.G. & Hill J.R., J. Geophys. Res., **83**, 1 (1983).
21. Elmegreen B.G., Ap. J., **312**, 626 (1987).
22. Chagelishvili G.D., Chanishvili R.G., Lominadze J.G., & Sokhadze Z.A., In *High Energy Astrophysics: American and Soviet Perspectives* (Washington, National Academy Press, 1991).
23. Gebhardt T. & Grossmann S., Phys. Rev. E, **50**, 3705 (1994).
24. Baggett J.S., Driscoll T.A. & Trefethen L.N., Phys. Fluids, **7**, 833 (1995).
25. Mahajan S.M. & Rogava A.D., Ap. J., **518**, 814 (1999).

Dust acoustic and gravity modes in barometric equilibrium

A.A. Shaikh[†], J.R. Bhatt[‡] and N.N. Rao[‡]

[†] C.U. Shah Science College, Ashram Road, Ahmedabad-380 014, Gujarat, India
[‡] Theoretical Physics Division, Physical Research Laboratory, Navrangpura, Gujarat,
Ahmedabad-380 009, India

Abstract. This paper deals with mode in non rotating horizontally stratified dusty atmosphere. We have studied dust gravity and acoustic modes in presence of a barometric equilibrium in an unmagnetized dusty plasma. All the dust grains are considered to have same and fixed negative (positive) charge on them. The electron and ions are considered to obey isothermal equations of state while the dust grain are following an adiabatic equation of state. This work can be considered as a generalization of single fluid treatment of gravity-acoustic modes. The analysis of dispersion relation suggests that the dust grains can introduce a new decay length of the perturbation. Therefore we find that the dispersion characteristics of single fluid treatment is strongly modified in the presence of charged dust grain. Finally we discussed importance of our results in the context of planetary and stellar environments.

INTRODUCTION

There has been a lot of activities and developments in theoretical and observational aspect of dusty plasma since last decade [5,10,15,23] due to its potential applications in technology and space science. The infusion of knowledge of pristine plasma has played an important role in formulating problems and methodology in dusty plasma. However, the presence of highly charged and massive dust grains can introduce a new time and spatial scales [e.g. 11-13 and ref.therein] which could give rise to several new phenomenon. One of the crucial difference between multi-components plasma and dusty plasma is that charge on dust grain is not constant. Magnitude of the charge on dust grain can vary with passage of waves in the dusty plasma as the current falling on the grain's surface varies with the wave motion. This scenario has been studied with the help of equation of charge dynamic which introduces a new time scale and instabilities in dusty plasma [4,22 and ref. therein].

In present work, we study the response of dusty plasma in a stratified barometric equilibrium in absence of magnetic field. It is well known that such equilibrium could support acoustic and gravity modes. It must be noted that such modes are extensively studied in various branches of physics such as planetary science

CP537, *Waves in Dusty, Solar, and Space Plasmas*, edited by F. Verheest, et al.
© 2000 American Institute of Physics 1-56396-962-9/00/$17.00

[3,7,9,24], solar physics, astro physics [1,2,14]. Hines [7] has studied these modes to understand the atmospheric oscillations in the presence of gravity. Robert [14] discussed acoustic gravity mode to outlined some aspect of Helioseismology. On the other hand Nye and Thomas [2] have studied this mode to explain running penumbra waves. Adam [1] also studied this mode pertinent to critical levels in solar MHD. Before we re-address problem in the presence of charged dust grain, let us recapitulate what Acoustic Gravity Wave (AGW) is. In lower planetary atmosphere pressure and density gradients are responsible for generation of acoustic and gravity mode respectively. When the planetary gravitational force and density gradient are in equilibrium, in a stratified neutral atmosphere, the resulting wave is acoustic gravity wave (AGW) [3]. In AGW temperature and density gradient of the neutral fluid are playing important role. This scenario has been analyzed in a dusty environment by Shukla and Shaikh [16]. It is found that a linear dust acoustic gravity waves can be generated and which resembles AGW of earlier studies. It was found [16] that dust gravity wave can become unstable and the instability depends on equilibrium values of dust density gradient and pressure of dust and ion species. It is demonstrated that the instability leads to formation of vortices in the dusty plasma. The acoustic gravity vortices in the literature of lower atmosphere without consideration of charged dust grain (for pristine plasma) has been studied by earlier researchers [17-20]. However there are some essential differences with the work carried out in ref.[16] and and the present study:
(1) electrons are not completely depleted.
(2) electrons and ions are obeying Maxwell-Boltzmann distribution law.
(3) We do not consider temperature gradients in either species of dusty plasma.

First difference bring forth the importance of electron dynamics in dusty plasma. By retaining the finite electron density, it is possible to have positively charged dust grain in the system [8]. The assumption of Maxwell-Boltzmann distribution law for electrons and ions is consistent with the time scales of dust dynamics. It must be noted that presence of finite streaming velocity gives rise to instability [16].

This paper addresses the analytical study of dust acoustic and gravity mode in the barometric equilibrium of an unmagnetized dusty plasma. The relevant set of equations are given in section 2. We derive dispersion relation in the linear limit in section 3. Finally we end up with discussion of results in section 4.

BASIC EQUATIONS

We consider unmagnetized tri-component dusty plasma comprising electrons, ions and massive charged dust grains. For simplicity, we assume the following
(i) all the dust grains are point charge and having uniform mass,
(ii) sizes of dust(in microns or sub microns) grain and inter dust grain spacing are very small compare to scale lengths involved in problem,
(iii) for collective phenomena density of charge dust grain and Debye's length obey the condition $n_d \lambda_D^3 \gg 1$, where λ_D is Debye length,

(iv) we assume that electrons and ions are isothermal i.e.

$$\frac{n_i}{n_{io}} = -\frac{e\nabla\Phi}{T_i} \tag{1}$$

$$\frac{n_e}{n_{eo}} = \frac{e\nabla\Phi}{T_e} \tag{2}$$

while dust grains are obeying the adiabatic equation of state.

Barometric dusty atmospheric oscillations are governed by following set of equations,

$$\partial_t n_d + \nabla \cdot (n_d V_d) = 0 \tag{3}$$

$$m_d n_d D_t V_d = m_d n_d g - \nabla P + n_d m_d e Z_d \nabla\Phi \tag{4}$$

$$D_t \left(P_d \rho_d^{-\gamma} \right) = 0 \tag{5}$$

where $D_t = \partial_t + V \cdot \nabla$, n_d, V_d, eZ_d, P, m_d, ρ_d, $\nabla\Phi$ and g are dust density, dust streaming velocity, fixed charge on dust grain, total pressure, mass of dust grain, dust mass density, electric potential (prescribed) in the dusty atmosphere and gravity respectively. Quasi-neutrality condition would read as,

$$n_i = n_e + n_d Z_d \tag{6}$$

Barometric equilibrium is established when upward pressure gradient is balanced by downward gravity i.e.

$$-\nabla P_{do} - \nabla P_{io} - \nabla P_{eo} - (\rho_{do} + \rho_{io} + \rho_{eo}) g = 0 \tag{7}$$

On solving above equation, we obtain

$$n_d = n_{d0} \exp(-z/H) \tag{8}$$

where H is scale height, define as under

$$H = \delta_d \frac{v_{ds}^2}{g} + \frac{V^2}{g} \tag{9}$$

Where $\quad \delta_d = \frac{z n_{d0}}{n_{io}}, \quad V^2 = \frac{T}{m_d} \left(\frac{1 - \frac{n_{eo}}{n_{io}}}{1 + \frac{T_i}{T_e} \frac{n_{eo}}{n_{io}}} \right), \quad v_{ds}^2 = \frac{\gamma P_{do}}{\rho_{do}}.$

We have used quasi neutrality condition, Boltzmann approximation for electrons and assuming that $\frac{m_i}{m_d} \ll 1$.

DISPERSION RELATION

We consider an appropriate following expression of perturbations on basic equations, we obtain,

$$\tilde{p} = p_0 \exp[i(\omega t - k \cdot r)] \tag{10}$$

$$\tilde{\rho} = \rho_0 \exp[i(\omega t - k \cdot r)] \tag{11}$$

$$\tilde{v} = v_0 \exp[i(\omega t - k \cdot r)] \tag{12}$$

where $k \cdot r = k_x x + k_z z$. Now carrying out Fourier transform for above perturbation on continuity, momentum and equation of state we obtain

$$i\omega\tilde{\rho}_d - ik_x\rho_{d0}\tilde{v}_{dx} + \left(-ik_z - \frac{1}{H}\right)\rho_{d0}\tilde{v}_{dz} = 0 \tag{13}$$

while momentum equation takes the form as

$$\rho_{d0}\partial_t\tilde{v}_d = -\nabla\tilde{P} - V^2\nabla\tilde{\rho}_d + \tilde{\rho}g \tag{14}$$

Now X and Z components of momentum equation are

$$i\omega\rho_{d0}\tilde{v}_{dx} - ik_x(\tilde{p}_d + V^2\tilde{\rho}_d) = 0 \tag{15}$$

$$\tilde{\rho}_d\left(g + V^2\left(-ik_z - \frac{1}{H}\right)\right) + \tilde{P}_d\left(-ik_z - \frac{1}{H}\right) + i\omega\rho_{d0}\tilde{v}_{dz} = 0 \tag{16}$$

Adiabatic equation of dust grain

$$\tilde{\rho}_d\left(-i\omega v_{ds}^2\right) + \tilde{P}_d(i\omega) + (\gamma - 1)g\rho_{d0}\tilde{v}_{dz} = 0 \tag{17}$$

From equations (13), (15), (16) and (17) we obtain matrix relation, leads to desire dispersion relation

$$\begin{pmatrix} i\omega & 0 & -ik_x & -ik_z - \frac{1}{H} \\ -iV^2k_x & -ik_x & i\omega & 0 \\ g + V^2\left(-ik_z - \frac{1}{H}\right) & -ik_z - \frac{1}{H} & 0 & i\omega \\ -i\omega v_{ds}^2 & i\omega & 0 & (\gamma - 1)g \end{pmatrix} \begin{pmatrix} \frac{\tilde{\rho}_d}{\rho_{d0}} \\ \frac{\tilde{P}_d}{\rho_{d0}} \\ \tilde{v}_{dx} \\ \tilde{v}_{dz} \end{pmatrix} = 0 \tag{18}$$

With some algebraic gimmick we arrived at following final dispersion relation

$$\omega^4 - \omega^2\left\{\left(V^2 + v_{ds}^2\right)\left(k_x^2 + k_z^2\right) - ig\gamma k_z - \frac{V^2}{H^2} - 2ik_z\frac{V^2}{H}\right\} + (\gamma - 1)k_x^2g^2 = 0 \tag{19}$$

Dispersion relation (19) is the dusty version of dispersion relation for barometric equilibrium. It is has to be noted that one can retrieve the dispersion relation, for pristine plasma, obtained in literatures [1,2,7,9,14,21,24] from our dispersion relations (19). However in literature it is discussed in various context in different branches of physics.

RESULT AND DISCUSSION

Dispersion relation (19) for normal mode is well accounted, in the absence of charged dust, with earlier results pertinent to running penumbra wave [2], oscillation on the sun [14] existence of critical levels in MHD and oscillation near earth [7,9]. It is clear from equation (19) that for an ideal medium (lossless), it is complex. Whether frequency (ω) and wave vectors k_x and k_z are real or complex depends on one's investigation. As there is no source of free energy we have to make dispersion relation (19) real. So we assume that k_x is real, ω is also real and therefore k_z has to be complex. To test this we assume, further, $k_z = k_{zr} + ik_{zi}$. On substituting it in relation (19) we obtain two equations corresponding to imaginary and real part respectively. Imaginary part gives

$$\omega^2 k_{zr} \left\{ 2V_1^2 k_{zi} + \gamma g + 2\frac{V^2}{H} \right\} = 0 \tag{20}$$

Now three possibilities are open up
(1) For time independent case, k_x is zero while k_z is finite. This gives the the steady horizontal dust wind with sheared in vertical direction for finite v_{xd} with no vertical component of velocity.
(2) For imaginary k_z one can find surface wave and
(3) final possibility gives damping scale i.e.

$$k_{zi} = - \left\{ \frac{\gamma g + 2\frac{V^2 v_{ds}^2}{\gamma g}}{2V_1^2} \right\} = H_d^{-1} \tag{21}$$

In the light of third possibility we modify the the fashion of perturbation by incorporating k_{zi} i.e. perturbed quantities vary as $\exp[i(\omega t - k_z z - k_x x) \pm k_{zi}]$, where plus sign for dust pressure and density while negative sign for components of velocity. Here choice of sign is justifiable by law of conservation of energy in a lossless medium. The real part of dispersion relation (19) can be written in a convenient forms with dust acoustic frequency (ω_{ad}) and the dust *Brunt-Väisälä* frequency (ω_{gd}),

$$\omega^4 - \omega^2 V_1^2 \left\{ k_x^2 + k_{zr}^2 + \frac{\omega_{ad}}{V_1^2} \right\} + \omega^2 V_1^2 k_x^2 = 0 \tag{22}$$

$$k_0^2 = \frac{k_{zr}^2}{D_{ad}} + \frac{k_x^2}{\frac{D_{ad}}{D_{bd}}} \tag{23}$$

$$\frac{n_z^2}{D_{ad}} + \frac{n_x^2}{\frac{D_{ad}}{D_{bd}}} = 1 \tag{24}$$

88

Where $V_1^2 = V^2 + v_{ds}^2$, $k_0^2 = \frac{\omega^2}{V_1^2}$, $D_{ad} = 1 - \frac{\omega_{ad}^2}{\omega^2} \frac{v_{ds}^2}{V_1^2}$, $D_{gd} = 1 - \frac{\omega_{gd}^2}{\omega^2}$.

$\omega_{ad} = \frac{v_1^2}{4H_d^2} - \frac{\gamma g}{2H_d}$, $\omega_{gd} = (\gamma - 1)^{\frac{1}{2}} \frac{g}{V_1}$, $n_x = \frac{k_x}{k_0}$, $n_z = \frac{k_z}{k_0}$.

It is clear from model that the modifications, compare to that of pristine plasma, are introduced by dust pressure and density. Here electron and ions are isothermal which could be justified by their high mobility. Also because of their override tendency to attain thermal equilibrium, the atmospheric temperature reached to a constant value for the height of interest. Moreover partial pressure and density are the exponentially decaying function in lower planetary dusty atmosphere. Because of charged dust grain scale height (damping scale) is strongly modified. Thus we point out that in barometric dusty environment all the perturbed quantities are varying as $\exp[i(\omega t - ik \cdot r) \pm \frac{z}{H_d}]$. The physical inference of this, is based on law of conservation of energy [24]. The kinetic energy/perturbational energy depend on ρ_0 and v^2. Here ρ_0 varies with $\exp\left(-\frac{z}{H_d}\right)$ i.e. ρ_0 decrease with height while v^2 varies with $\exp\left(\frac{z}{H_d}\right)$ i.e. increase with height. In this scenario kinetic energy remain constant and mode is called internal dust gravity mode.

Although our results are based on local theory but suggestive enough of what happens in realistic situation. Here we have demonstrated presence of charged dust grains gas significantly altered the scale height. Expression of new scale hight/damping scale is obtained (Eq.21) which clearly shows that it depends on, apart from temperature and gravity of planet, sign and magnitude of charged dust grain. we have cited references of several works on acoustic and gravity modes in barometric equilibria occurred in various branches of physics such as solar plasma, astrophysics, planetary science [1,2,6,7,9,14,21,24]. Our work indicates that introduction of highly charged dust grains can strongly modify the dispersion characteristics acoustic and gravity modes with modified scale length.

We would also like to clarify the virtual dissimilarity of our result with that of classical plasma. The dispersion relation obtained, of course for pristine plasma, in astro and solar physics literature [1,2,14,21] has imaginary term, like us, in dispersion relation. Reason, for this, is simple, we have started problem with imaginary k_z so it is manifested accordingly in final result. On the other hand if one is taking k_z as complex than the final dispersion relation has no imaginary term. If we put $k = (k \sin\theta, 0, k \cos\theta)$ in relation (22) we obtain vertical and horizontal propagations. Also it can be shown that, on plotting the graph of $k^2 V_1^2 \rightarrow \omega^2$ from relation (22), the poles of graph are modified. A problem can, further, be discuss with magnetic field and incorporation of neutral population in the model. Former would give the coupling of Alfvén and dust acoustic velocities [2] while later would give effect of neutral dynamics on existing modes. These could be our future work.

Acknowledgments

Two of the authors (A.A.S. and N.N.R.) deeply acknowledge financial support provided by the conveners/financial committee to present this work in the Workshop on Dusty, Solar and Space Plasmas at Leuven. A.A.S. also acknowledges discussions with Profs. P.K. Shukla and A.C. Das.

REFERENCES

1. Adam, J.A., *Solar Physics*, **52**, 293-307 (1977).
2. Nye, A.H. and Thomas, J.H., *Solar Physics*, **38**, 399-413 (1974).
3. Beer, T., *Atmospheric Waves*, Adam Hilger, London, 1974, p. 41.
4. Bhatt, J.R. and Pandey, B.P., *Phys. Rev.*, **E50**, 3980 (1994).
5. Goertz, C., *Rev. Geophys*, **27**, 271 (1989).
6. Hargreaves, J., *Solar-Terrestrial environment : An Introduction to Geospace*, Cambridge University Press, 1992.
7. Hines, C.O., *Can. J. Phys*, **38**, 1444 (1960).
8. Horanyi, M. and Travens, T.E., *Nature*, **381**, 293, (1996).
9. Kelley, M.C., *The Earth's Ionosphere*, Academic Press, New York, 1989.
10. Mendis, D.A. and Rosenberg, M., *Annu. Rev. Astron. Astrophysics*, **32**, 419 (1994).
11. Rao, N.N., Shukla, P.K. and Yu, M.Y., *Planet. Space Sci.*, **38**, 5431 (1990).
12. Rao, N.N., *Planet. Space Sci.*, **41**, 21, (1993).
13. Rao, N.N., *Planet. Space Sci.*, **53**, 317, (1995).
14. Robert, B. *Solar and Planetary Plasma Physics*, Ed. B. Buti, World Scientific, Singapore, 1990.
15. Shukla, P.K., Mendis, D.A. and Desai, T. (Eds.), *Advances in Dusty Plasmas*, World Scientific, Singapore, 1996.
16. Shukla P.K. and Shaikh A.A., *Physica Scripta*, **T75**, 247 (1998).
17. Stenflo, L., *Phys. Lett. A*, **186**, 133 (1994).
18. Stenflo, L., and Stepanyants, Yu.A, *Ann. Geophysics*, **13**, 973 (1995).
19. Stenflo, L., *Phys. Lett. A*, **222**, 378 (1996).
20. Stenflo, L., *Physica Scripta*, **T75**, 306 (1998).
21. Sturrock, P.A., *Plasma Physics: an introduction to the theory of astrophysical, geophysical and laboratory plasma*, Cambridge University Press, 1994.
22. Tsytovich, V.N. and Havnes, O., *Comm. Plasma Phys. Contr. Fusion*, **15**, 267, (1993).
23. Verheest, F., *Space Sci. Rev.*, **77**, 267 (1996).
24. Yeh, K.C., and Liu, *Theory of Ionospheric Waves*, Academic Press, New York, 1972, ch. 8.

Janus faces of Jeans instabilities

Frank Verheest*, Vladimir M. Čadež† and Gerald Jacobs*

* *Sterrenkundig Observatorium, Universiteit Gent, Krijgslaan 281, B-9000 Gent, Belgium*
† *Belgisch Instituut voor Ruimte-Aëronomie, Ringlaan 3, B-1180 Ukkel, Belgium*

Abstract. Self-gravitating clouds have been shown by Jeans to be unstable to harmonic perturbations whose wavelength exceeds some critical value involving the mass density and some thermal velocity or equivalent information. Based upon the assumption that the unperturbed cloud is initially uniform, the Jeans instability is non-oscillatory and purely growing. However, Newtonian gravitation precludes strictly homogeneous equilibria, but a way out is offered, in theory, by considering local perturbations, small compared to the inhomogeneity scale lengths. While in itself plausible, this procedure can in most cases not be tested for internal consistency, because real knowledge about the equilibrium is lacking, and is therefore called the Jeans swindle.

The severe limitations of such an approach lead to an unavoidable dichotomy, and an example of a plasma will be discussed where the computations can be done explicitly, both for the stationary as well as for the perturbed state, showing that the system is stable at all wavelengths compatible with the equilibrium inhomogeneity. Nevertheless, the present state of affairs does not allow self-consistent equilibria to be worked out in more complicated configurations, like in dusty plasmas with external magnetic fields. This typically leads to the Jeans swindle being used a little longer than desirable.

INTRODUCTION

Cosmic dust is a well-known and common constituent of many heliospheric and astrophysical media. Prime examples in the solar system are circumsolar dust rings, noctilucent clouds, cometary comae and tails, and rings of the Jovian planets. Among astrophysical applications interstellar dust clouds come to mind. Dust grains may be charged or neutral, depending on the nearby sources of radiation like stars and/or the presence of charged particles as in the solar wind. The combination of charged dust and plasma is referred to as a dusty plasma [1–3], and for certain of these, containing rather heavy charged grains, it was hypothesized that the intergrain gravitational force could become of the order of the intergrain electrostatic force, or at least large enough to figure in the description. This has revived the interest in Jeans instabilities and self-gravitational effects, reconsidered in a novel context. Interstellar dust clouds can be very large, with widely varying grain diameters, and it is these dust clouds that we keep in mind.

CP537, *Waves in Dusty, Solar, and Space Plasmas,* edited by F. Verheest, et al.
© 2000 American Institute of Physics 1-56396-962-9/00/$17.00

Before we can study how charged dust modifies the stability or instability, however, we have to recall the ambiguities connected with the proper description of gravitational phenomena. These remind us of the two faces of the Roman god Janus, looking in opposite directions.

To start with, the Jeans instability is a basic phenomenon in gravitating systems and has been known for nearly a century since Jeans in 1902 obtained the instability criterion for harmonic waves whose wavelength exceeds some critical value. His study assumes that the unperturbed gaseous cloud is initially uniform [4], and then the instability is non-oscillatory and purely growing.

However, Newtonian gravitation in extended mass systems precludes truly homogeneous equilibria [5], and disciples of Jeans have skirted around this difficulty by considering local perturbations, with wavelengths small compared to the inhomogeneity scale lengths. This could be acceptable, were it not that in most cases the internal consistency cannot be tested, because knowledge about the equilibrium is lacking, hence the name Jeans swindle.

We will in the next section start with an example of a plasma where the computations can be done explicitly, for the stationary as well as for the perturbed state, showing that the system is stable in the direction of the inhomogeneity [6]. Thereafter, we will briefly review recent work involving self-gravitational effects in dusty plasmas, based upon the Jeans swindle. Unfortunately, the present state of affairs does not allow anything more consistent, and this will typically remain so, until more equilibria are worked out in complex systems as dusty plasmas.

REVISITING JEANS

An attempt will be made to be fully consistent, to determine the basic stationary state before small perturbations are studied. Models that can be worked out are far from obvious, as recurrent discussions attest [7,8]. We will look at a self-gravitating, isothermal plasma cloud described by single-fluid magnetohydrodynamic (MHD) equations, because a multispecies description is much too complicated [6].

Basic equations and stationary state

The starting point is the basic set of MHD equations of continuity, induction, momentum and energy, augmented by the gravitational Poisson's equation,

$$
\frac{\partial \rho}{\partial t} + \boldsymbol{\nabla} \cdot (\rho \mathbf{v}) = 0, \qquad\qquad \frac{\partial \mathbf{B}}{\partial t} = \boldsymbol{\nabla} \times (\mathbf{v} \times \mathbf{B}),
$$

$$
\frac{\partial \mathbf{v}}{\partial t} + \mathbf{v} \cdot \boldsymbol{\nabla}\mathbf{v} + \frac{1}{\rho}\boldsymbol{\nabla}p + \boldsymbol{\nabla}\phi = \frac{1}{\mu_0 \rho}(\boldsymbol{\nabla} \times \mathbf{B}) \times \mathbf{B}, \qquad (1)
$$

$$
\frac{\partial p}{\partial t} + \mathbf{v} \cdot \boldsymbol{\nabla}p = c_s^2\left(\frac{\partial \rho}{\partial t} + \mathbf{v} \cdot \boldsymbol{\nabla}\rho\right), \qquad \nabla^2\phi = 4\pi G\rho.
$$

Here ρ, $\mathbf{v}$ and p represent the fluid mass density, velocity and pressure, respectively, $\mathbf{B}$ is the magnetic field and ϕ the gravitational potential. Pressure changes have been assumed adiabatic, and c_s is the thermal velocity.

For the stationary state we find first of all that the Poisson equation for the equilibrium gravitational potential,

$$\nabla^2 \phi_0 = 4\pi G \rho_0, \tag{2}$$

precludes any truly homogeneous equilibrium. We will therefore start with as simple as possible a configuration that is compatible with the basic equations, and introduce some plausible physical assumptions: (i) The plasma is treated as a perfect gas, so that c_s is a true constant. (ii) The equilibrium magnetic field lines are straight, so that $\mathbf{B}_0 = B_0 \mathbf{e}_x$. Gauss's law $\nabla \cdot \mathbf{B}_0 = \partial B_0 / \partial x = 0$ then indicates that the strength of the magnetic field B_0 cannot depend on x, and hence $\mathbf{B}_0 \cdot \nabla \mathbf{B}_0 = \mathbf{0}$. (iii) Furthermore, the magnetic field is assumed to vary in such a way that the ratio β of the plasma pressure to the magnetic pressure $B_0^2 / 2\mu_0$ remains constant. In this model the Alfvén speed V_A is constant, and the magnetic field and plasma pressure are stronger in regions with a higher plasma density ρ_0. This is rather realistic and may occur in highly conductive plasmas with frozen-in magnetic fields. The magnetohydrostatic equilibrium balance equation

$$\nabla p_0 + \rho_0 \nabla \phi_0 = \frac{1}{\mu_0} (\nabla \times \mathbf{B}_0) \times \mathbf{B}_0 \tag{3}$$

then reduces to

$$\left(1 + \frac{1}{\beta}\right) \nabla_\perp p_0 + \rho_0 \nabla_\perp \phi_0 = \mathbf{0}, \tag{4}$$

where the subscript $\perp$ refers to the directions across the equilibrium field. Together with (2), this yields a single equation for ρ_0,

$$\nabla_\perp^2 \ln \rho_0 + \frac{8\pi G}{V_A^2 + 2c_s^2} \rho_0 = 0. \tag{5}$$

This shows that the solutions depend upon y and z through $(y \sin \alpha + z \cos \alpha)/L$, and the typical scale length L is defined by $L^2 = (V_A^2 + 2c_s^2)/4\pi G \rho_{00}$. Here ρ_{00} is the density in the center of the cloud. For the discussion we take $\alpha = 0$ and proceed with density profiles $\rho_0(z)$ having an extremum at $z = 0$, because the gravitational force vanishes at the center of the cloud. Whether such a configuration is stable against perturbations has to be investigated afterwards, also in the lateral directions that are now assumed to be homogeneous.

Even though the stationary state is uniform along the equilibrium field, we would like to stress that this is not using the Jeans swindle, because the present choice is fully compatible with all equilibrium constraints. It might well turn out that such a state is Jeans unstable in the lateral directions, but that is quite different from starting with a completely uniform equilibrium. Proceeding then with equivalent one-dimensional equilibrium variations, the solutions are

$$\rho_0(z) = \rho_{00} \operatorname{sech}^2(z/L), \qquad B_0(z) = B_{00} \operatorname{sech}(z/L), \tag{6}$$

and the central field strength B_{00} is given through $B_{00}^2 = \mu_0 V_A^2 \rho_{00}$ as V_A is constant.

Linear perturbations

This basic state now undergoes linear perturbations whose amplitudes are z dependent only. Other perturbations will be discussed elsewhere [6]. The linearized equations for the perturbed quantities become

$$\frac{\partial \rho}{\partial t} + \nabla \cdot (\rho_0 \mathbf{v}) = 0,$$

$$\frac{\partial \mathbf{v}}{\partial t} + \frac{1}{\rho_0} \nabla p + \frac{\rho}{\rho_0} \nabla \phi_0 + \nabla \phi = \frac{1}{\mu_0 \rho_0} (\nabla \times \mathbf{B}_0) \times \mathbf{B} + \frac{1}{\mu_0 \rho_0} (\nabla \times \mathbf{B}) \times \mathbf{B}_0,$$

$$\frac{\partial \mathbf{B}}{\partial t} = \nabla \times (\mathbf{v} \times \mathbf{B}_0), \tag{7}$$

$$\frac{\partial p}{\partial t} + \mathbf{v} \cdot \nabla p_0 = c_s^2 \left(\frac{\partial \rho}{\partial t} + \mathbf{v} \cdot \nabla \rho_0 \right),$$

$$\nabla^2 \phi = 4\pi G \rho.$$

In what follows, $\nabla = \mathbf{e}_z \partial/\partial z$ and all variables in (7) are Fourier transformed in time. In order to write the resulting equations in an uncluttered form, we introduce a normalized magnetic field $\mathbf{b} = \mathbf{B}/B_0$, the total pressure $P = p + V_A^2 \rho_0 b_x$, a Lagrangian displacement ξ in the z direction defined through $v_z = -i\omega\xi$, and a new variable $\eta = \rho_0 \xi$. Using $\rho + d\eta/dz = 0$ and $d\phi/dz = -4\pi G\eta$ allows us to reduce the remainder of (7) to

$$\frac{d\phi_0}{dz} \frac{d\eta}{dz} + \omega^2 \eta = \frac{dP}{dz} - 4\pi G \rho_0 \eta, \qquad V_{ms}^2 \frac{d\eta}{dz} = \frac{V_A^2 \eta}{2} \frac{d\ln \rho_0}{dz} - P, \tag{8}$$

where V_{ms} is the magnetosonic velocity given through $V_{ms}^2 = V_A^2 + c_s^2$. By eliminating P from (8) and with the help of the explicit expressions (6) for the equilibrium quantities, we finally obtain a single differential equation

$$\frac{d^2}{dZ^2} (\eta \cosh Z) - \left(\mathcal{K} - 2 \operatorname{sech}^2 Z \right) (\eta \cosh Z) = 0, \tag{9}$$

where we have put $Z = z/L$ and $\mathcal{K} = 1 - \omega^2 L^2 / V_{ms}^2$. Equation (9) can easily be solved analytically under some special conditions. For example, simple analytical solutions are obtained in the quasi-uniform region $|Z| \gg 1$, where the coefficients in (9) become practically constant, and in the case of $\mathcal{K} = 0$ and $\mathcal{K} = 1$, when the solutions are given in terms of associated Legendre functions.

Standard assumption of a uniform basic state

Before going on, however, it is very instructive to point out the pitfalls associated with the Jeans swindle, used in the traditional treatment of Jeans instabilities, also in later attempts to generalize these to dusty plasmas. We repeat the standard derivation, when the initial basic state is assumed uniform everywhere and the gravitational Poisson equation is only considered for the perturbations. For a uniform initial state the set (8) reduces to a system with constant coefficients,

$$\frac{dP}{dz} = \omega^2\eta + 4\pi G\rho_{00}\eta, \qquad V_{ms}^2\frac{d\eta}{dz} = -P. \tag{10}$$

Because of the uniform equilibrium, the z dependence of the perturbations can be taken as $\exp(ikz)$, so that (10) immediately yields the dispersion law

$$\omega^2 = k^2V_{ms}^2 - \omega_J^2, \tag{11}$$

where the Jeans frequency has been used, defined here through $\omega_J^2 = 4\pi G\rho_{00}$. Now (11) indicates the possibility for gravitational instabilities to set in when k becomes small enough to render ω^2 negative. In this case the Jeans length λ_J is given through $\lambda_J^2 = \pi V_{ms}^2/G\rho_{00}$.

The conclusion regarding the instability is thus based on the assumption that the medium be uniform, at least locally, i.e. over a distance of several wavelengths. On the other hand, we have seen that the basic state cannot intrinsically be uniform on a larger scale, due to the ever-present gravitational forces. According to (6), the characteristic length of the nonuniformity is L, and unfortunately this is smaller than λ_J, as shown by the following inequality

$$\frac{\lambda_J^2}{L^2} = 2\pi^2\frac{2V_A^2 + 2c_s^2}{V_A^2 + 2c_s^2} > 1. \tag{12}$$

This indicates that the required condition for a medium to be locally uniform is not valid for unstable perturbations that satisfy the condition $\lambda > \lambda_J > L$. In the opposite case that $\lambda \ll L$, the local approximation is valid, of course, but k is now so large that the only correct way of writing the dispersion law (11) is

$$\omega^2 \simeq k^2V_{ms}^2, \tag{13}$$

i.e. without the gravitational term! Consequently, since the standard assumption of a uniform basic state is only valid locally, the perturbations propagate as stable, fast MHD modes, almost unaffected by gravitation, with a phase speed V_{ms}.

Local and global solutions

Going back to the full model, (9) can have local solutions provided the term containing $\mathrm{sech}^2 Z$ may be neglected. This occurs if $|\mathcal{K}| \gg 2\,\mathrm{sech}^2 Z$, i.e. when z/L is sufficiently large, depending on $\mathcal{K}$. The solution of (9) is then of the form

$$\eta \cosh(z/L) \simeq \exp(\pm ikz). \tag{14}$$

The wavenumber k itself is given by the expression $k^2 = -\mathcal{K}/L^2$, which dispersion law can be rewritten as

$$\omega^2 = k^2 V_{ms}^2 + \frac{V_{ms}^2}{L^2}. \tag{15}$$

Hence (15) refers to a more general local solution and, contrary to (13), it also includes gravitational effects through L. The validity of (15) requires that $|z| \gg \lambda$ and $|z| \gg L$, and (15) indicates that the system is stable to gravitational instabilities at $|z| \gg L$, so that perturbations propagate as modified fast magnetosonic waves. The frequency ω has a lower bound at the cut-off frequency V_{ms}/L below which the wave becomes evanescent and can not propagate anymore. Finally, the total pressure perturbation P follows from (8) and (14) as

$$P = C_1 \left(\frac{c_s^2}{L} \tanh \frac{z}{L} - ikV_{ms}^2 \right) \frac{e^{ikz}}{\cosh \frac{z}{L}} + C_2 \left(\frac{c_s^2}{L} \tanh \frac{z}{L} + ikV_{ms}^2 \right) \frac{e^{-ikz}}{\cosh \frac{z}{L}}, \tag{16}$$

where $C_{1,2}$ are integration constants. This shows that wave amplitudes are z dependent in the domain $|z| \gg L$, so that the total pressure perturbation decreases with $|z|$ as $\mathrm{sech}(z/H)$.

In the previous paragraph linear perturbations were treated locally through a restricted application of spatial Fourier analysis. The system under investigation could also oscillate as a whole at certain allowed eigenfrequencies, having its energy conserved. Thus, the perturbations have to be localized in the sense that they do not transport energy out of the system, i.e. the energy density of the perturbations has to tend to zero at sufficiently large $|z|/L$. As local solutions like (16) are asymptotic expressions of the corresponding global solutions at $|z| \gg L$, this is clearly achieved if the wavenumber k is taken imaginary, $k = i\kappa$. Consequently, the solutions we are looking for should have the following asymptotic behaviour at $|z|/L \gg 1$,

$$P_\pm = \pm C_\pm \left(\frac{c_s^2}{L} + \kappa V_{ms}^2 \right) \exp(\mp \kappa z) \, \mathrm{sech} \frac{z}{L}, \tag{17}$$

where the $\pm$ signs refer to whether $z > 0$ or $z < 0$. Here κ is given through

$$\kappa^2 = \frac{1}{L^2} - \frac{\omega^2}{V_{ms}^2} \tag{18}$$

and ω has to satisfy the inequality $\omega < V_{ms}/L$. These global solutions themselves may be either symmetric or antisymmetric in z, due to the symmetry of the basic state, and hence C_+ and C_- can differ only in sign. Moreover, one could argue that the global oscillations of the cloud cause displacements of plasma that are antisymmetric with respect to the center of the cloud $z = 0$. At two symmetric points with respect to the center, the fluid particles move simultaneously in opposite directions, either towards or away from the center, whereas the center ($z = 0$) remains at rest. As to the total pressure perturbation, it is a symmetric function of z, yielding either a local compression or rarefaction at each pair of symmetric points. The final conclusion is now that C_+ and C_- should have opposite signs.

The cloud can therefore oscillate stably as a whole with any of the frequencies from the continuous range given by $V_{ms}/L > \omega > 0$. For the two limiting values, $\omega = 0$ and $\omega = V_{ms}/L$, (9) has global solutions in a closed form for the whole range of z, given in terms of associated Legendre functions. For any other value of ω, the global solutions of (9) can only be obtained numerically or discussed asymptotically.

What remains to be studied here is the stability of the equilibrium (6) to oblique or parallel perturbations, with respect to the external magnetic field. Furthermore, bending of this magnetic field promises to include further complications.

SELF-GRAVITATION OF DUSTY PLASMAS

Various derivations of and modifications to Jeans instabilities in dusty plasmas are found in recent papers, to be quoted further on, and even a book [9]. All these share the common use of the Jeans swindle at some stage or another. The critical velocity occurring in the Jeans lengths is the dust-acoustic velocity c_{da} [10] for modes that propagate parallel [11–17] or obliquely [18] to the external magnetic field in dusty plasmas composed of very light electrons and ions, together with heavy and cool charged dust grains. On the other hand, when the propagation occurs strictly perpendicular to the external magnetic field, the characteristic velocity appearing in the expressions of the Jeans length is V_{ms} [19,20], as expected from the derivation in the preceding section.

SUMMING UP

The fundamental discussion remains, about the validity and real use of the basic model assumption of a homogeneous equilibrium, when considering wave perturbations that turn out to give rather large critical lengths before the gravitational instability could set in. If the wavelengths are small enough for the waves to be stable, the Jeans frequencies only serve to lower the effective frequencies and no great harm is done. Efforts should now go into trying to develop proper stationary states for complex plasmas, where self-gravitation modifies the equilibria, although it promises to be far from straightforward. Only then can the (in)stability of these configurations be studied with full confidence.

Mention should also be made of a few results which purport to deal with inhomogeneous dusty plasmas in the presence of self-gravitational forces [21,22]. These treatments are not really satisfactory, because the equilibrium is not determined in a self-consistent manner. Although there is thus an improvement over fully homogeneous models, the derivations of the perturbed states are still using the Jeans swindle, and in the end something better will be needed.

Acknowledgments

The Bijzonder Onderzoeksfonds (Universiteit Gent) is thanked for research (FV and GJ) and visitor (VČ) grants, and the Belgian Institute for Space Aeronomy for its kind hospitality (VČ). Stimulating discussions with M.A. Hellberg are gratefully acknowledged.

REFERENCES

1. Mendis, D.A. and Rosenberg, M., *Annu. Rev. Astron. Astrophys.* **32**, 419–463 (1994).
2. Horányi, M., *Annu. Rev. Astron. Astrophys.* **34**, 383–418 (1996).
3. Verheest, F., *Waves in dusty space plasmas*, Kluwer, Dordrecht, 2000.
4. Jeans, J.H., *Astronomy and cosmogony*, Cambridge University Press, Cambridge, 1929.
5. Fridman, A.M. and Polyachenko, V.L., *Physics of gravitating systems I & II*, Springer, New York, 1984.
6. Čadež, V.M., Verheest, F. and Jacobs, G., *Mon. Not. R. Astron. Soc.*, submitted (2000).
7. Nakano, T., *Publ. Astron. Soc. Japan* **40**, 593–604 (1988).
8. Gehman, C.S., Adams, F.C. and Watkins, R., *Astrophys. J.* **472**, 673–683 (1996).
9. Bliokh, P., Sinitsin, V. and Yaroshenko, V., *Dusty and self-gravitational plasmas in space*, Kluwer, Dordrecht, 1995.
10. Rao, N.N., Shukla, P.K. and Yu, M.Y., *Planet. Space Sci.*, **38**, 543–546 (1990).
11. Bliokh, P.V. and Yaroshenko, V.V., *Sov. Astron.* **29**, 330–336 (1985).
12. Avinash, K. and Shukla, P.K., *Phys. Lett. A* **189**, 470–472 (1994).
13. Pandey, B.P., Avinash, K. and Dwivedi, C.B., *Phys. Rev. E* **49**, 5599–5606 (1994).
14. Pandey, B.P. and Lakhina, G.S., *Pramana* **50**, 191–204 (1998).
15. Meuris, P., Verheest, F. and Lakhina, G.S., *Planet. Space Sci.* **45**, 449–454 (1997).
16. Verheest, F. and Shukla, P.K., *Physica Scripta* **55**, 83–85 (1997).
17. Rao, N.N. and Verheest, F., *Phys. Lett. A* **268**, 390–394 (2000).
18. Verheest, F., Jacobs, G. and Hellberg, M.A., *Physica Scripta* **T84**, 171–174 (2000).
19. Verheest, F., Meuris, P., Mace, R.L. and Hellberg, M.A., *Astrophys. Space Sci.* **254**, 253–267 (1997).
20. Verheest, F., Hellberg, M.A. and Mace, R.L., *Phys. Plasmas*, **6**, 279–284 (1999).
21. Shukla, P.K. and Rahman, H.U., *Planet. Space Sci.*, **44**, 469–472 (1996).
22. Mamun, A.A., Salahuddin, M. and Salimullah, M., *Planet. Space Sci.*, **47**, 79–83 (1999).

Electrostatic waves in dusty self-gravitational flows immersed in an electron-ion plasma

Victoria V. Yaroshenko[1]

Sterrenkundig Observatorium, Universiteit Gent, Krijgslaan 281, B-9000 Gent, Belgium

Abstract. In view of various occurrences in space of bounded dusty plasmas, we consider the wave processes in an individual cylindrical dusty self-gravitational flow and in the system of dusty self-gravitating plasma streams interacting through both electric and gravitational forces. Furthermore, we include the effect of an ambient electron-ion plasma, which can play a decisive role in the interaction between dusty flows.

INTRODUCTION

Dusty flows quite often occur in space, and typical examples include planetary rings, cometary comae and tails, asteroids zones, etc. Dust particles are electrically charged by various processes and interact through both electrical and gravitational fields. If the density of dust grains is sufficiently high, these grains are involved in collective processes and form a mixture that is referred to as a dusty self-gravitational plasma (DSGP). The analysis of collective phenomena in a DSGP often refers to an unbounded medium and demonstrates quite a number of unusual oscillating properties [1,2]. This oversimple treatment, however, cannot serve as an example of real dusty flows, which are usually separated in space. For this reason, we consider wave processes in separated cylindrical flows of a DSGP. Furthermore, we include an ambient electron-ion plasma, which can play a decisive role in the interaction between dusty flows. In this case the overall picture of wave processes in the dusty plasma flows changes substantially. It is important that the electric fields are greatly weakened owing to the Debye screening, while the forces of gravitation remain unchanged. As a result, self-gravitation can prove essential at long distances, even if electric forces prevail at short distances.

---

[1] Permanent address: Institute of Radio Astronomy of National Academy of Science of Ukraine, Chervonopraporna 4, Kharkov, Ukraine 61002, E-mail: yarosh@ira.kharkov.ua

CP537, *Waves in Dusty, Solar, and Space Plasmas,* edited by F. Verheest, et al.
© 2000 American Institute of Physics 1-56396-962-9/00/$17.00

To analyse the modification of the plasma processes in the DSGP flows we investigate electrostatic modes in an individual flow as well as in a multistream plasma system.

LOW-FREQUENCY ELECTRIC AND GRAVITATIONAL FIELDS INSIDE AND OUTSIDE THE DUSTY FLOW

We consider a cylindrical dusty particle flow with radius a. Let the density of the grains (each having mass M and charge Q) in the flow be N. We assume that the entire space (including the region occupied by the beam) is filled with an electron-ion plasma. But the presence of dust reduces the electron density inside the plasma beam to such an extent that we can consider the dusty plasma composed of warm ions and heavy charged grains. Outside of the dusty plasma, the medium is just a usual electron-ion plasma. Note that both plasma regions are electrically neutral in the equilibrium state. Let the densities of the external plasma ions and electrons (whose charges are $\pm e$ and masses are m_e, m_i) be n_e, n_i. We will be interested in the waves whose frequencies are low enough that the electron and ion densities obey Boltzmann distributions

$$n_{e,i} \cong n_0\left(1 \pm e\Psi_E / T\right),$$

where Ψ_E is the electric potential, and n_0, T are the unperturbed density and temperature of the plasma microparticles. We shall also introduce a gravitational potential Ψ_G.

Another simplifying assumption is that there exists a constant, infinitely strong magnetic field aligned with the flow axis (z-axis). This allows us to consider only longitudinal displacements of the charged particles, so that the space-charge waves in the dust plasma component can be described by the conventional set of equations, which consists of the equation of motion

$$\frac{\partial V}{\partial t} + V_0 \frac{\partial V}{\partial z} = -\frac{Q}{M}\frac{\partial \Psi_E}{\partial z} - \frac{\partial \Psi_G}{\partial z} + \frac{F_{ext}}{M}, \tag{1}$$

the continuity equation

$$\frac{\partial N}{\partial t} + \frac{\partial}{\partial z}\left(N_0 V + N V_0\right) = 0, \tag{2}$$

the Poisson equation for the electrostatic potential inside the flow

$$\Delta\Psi_E = -4\pi\left[QN + en_i\right] \cong -4\pi QN - \lambda_{Di}^{-2}\Psi_E, \tag{3}$$

and the Poisson equation for the gravitational potential

$$\Delta\Psi_G = 4\pi GMN. \tag{4}$$

Here G is the gravitational constant and $\lambda_{Di} = \sqrt{T/4\pi n_0 e^2}$ the ion Debye radius.

In the acceleration associated with the external force F_{ext}, we also take into account only the z-components of the electric and gravitational field, viz.

$$-(Q/M)\partial \Psi_{E,ext}/\partial z - \partial \Psi_{G,ext}/\partial z.$$

We consider the wave perturbations induced by a Fourier component of the external force $F_{ext}(\omega, k)\exp[i(kz - \omega t)]$. It is convenient to present the Poisson equation inside and outside the DSGP flow in the form

$$\frac{1}{r}\frac{d}{dr}\left(r\frac{d\Psi_E}{dr}\right) - p_i^2 \Psi_E = -4\pi NQ, \qquad (r < a)$$

$$\frac{1}{r}\frac{d}{dr}\left(r\frac{d\Psi_E}{dr}\right) - p^2 \Psi_E = 0, \qquad (r > a)$$

(5)

where $p_i^2 = k^2 + 1/\lambda_{Di}^2$, $p^2 = k^2 + 1/\lambda_D^2$ and $\lambda_D = \sqrt{T/8\pi n_0 e^2}$.

Neglecting variations in N over the beam cross section, we can readily obtain general solutions to equations (5) in terms of the modified Bessel functions I_ν and K_ν:

$$\Psi_E = C_1 I_0(p_i r) + C_2 K_0(p_i r) + 4\pi QN/p_i^2, \qquad r \le a;$$

$$\Psi_E = C_3 I_0(p\,r) + C_4 K_0(pr), \qquad r \ge a.$$

The integration constants C_i are found from the boundary conditions $\left(\Psi_E(0) \ne \infty, \; \Psi_E(\infty) = 0\right)$ and from the condition that $\Psi_E(r)$ and $\Psi_E'(r)$ be continuous at $r = a$. As a result, we obtain

$$\Psi_E = \frac{4\pi QN}{p_i^2}\left[1 - p\,\frac{K_1(pa)}{W}I_0(p_i r)\right], \qquad r \le a; \qquad (6)$$

$$\Psi_E = \frac{4\pi QN}{p_i}\frac{I_1(p_i a)}{W}K_0(pr), \qquad r \ge a \qquad (7)$$

with $W = p_i K_0(pa)I_1(p_i a) + p K_1(pa)I_0(p_i a)$.

From (6), we can readily evaluate Ψ_E at the beam axis $(r \to 0)$ $\Psi_E(0) = -2\pi QNa^2 \ln(pa)$ and thus formulate the condition for the flow to be "filamentary", $\left|\dfrac{\Psi_E(0) - \Psi_E(a)}{\Psi_E(0) + \Psi_E(a)}\right| \sim -\dfrac{1}{\ln(pa)} \ll 1$, when the variations in $\Psi_E(r)$ over the beam cross section should be small. As a result, we arrive at the condition $pa \ll 1$ and, for long-wavelength perturbations $k \ll \lambda_D^{-1}$ (i.e. $p \approx \lambda_D^{-1}$), we obtain $a \ll \lambda_D$, which indicates that the beam radius should be much smaller than the plasma Debye radius. Under these conditions, the potential Ψ_E can be approximated by

$$\Psi_E \cong -2\pi Q N a^2 \ln(pa), \quad r \leq a \tag{8}$$

$$\Psi_E \cong 2\pi Q N a^2 \sqrt{\frac{\pi}{2pr}} \exp(-pr), \quad r \gg a. \tag{9}$$

The Poisson equation for Ψ_G can be solved in an analogous way and we get

$$\Psi_G = 2\pi G M N a^2 \ln(ka), \quad r \leq a, \tag{10}$$

$$\Psi_G = -2\pi G M N a^2 \sqrt{\frac{\pi}{2kr}} \exp(-kr), \quad r \gg a. \tag{11}$$

As can be seen from (10), the condition for the beam to be filamentary is less restrictive than that for the electrostatic field: the potential Ψ_G can be assumed to be uniform over the entire beam cross section if $ka \ll 1$.

WAVELIKE DISTURBANCES IN A NARROW DSGP FLOW

Substituting the potentials (8) and (10) into equations (1) and (2) we arrive at

$$V(kV_0 - \omega) + 2ka^2 N\left(-\frac{Q^2}{M}\ln(pa) + M\ln(ka)\right) = -k\left(\frac{Q}{M}\Psi_{E,ext} + \Psi_{G,ext}\right),$$

$$VkN_0 + N(kV_0 - \omega) = 0. \tag{12}$$

Here, as before, the functions and the related Fourier amplitudes are denoted by the same symbols. These equations have the solution

$$N = \frac{k^2 N_0}{D}\left(\frac{Q}{M}\Psi_{E,ext} + \Psi_{G,ext}\right),$$

$$V = -\frac{k(kV_0 - \omega)}{D}\left(\frac{Q}{M}\Psi_{E,ext} + \Psi_{G,ext}\right), \tag{13}$$

where $D = (\omega - kV_0)^2 - \frac{k^2 a^2}{2}\left(\ln(ka)\omega_G^2 - \ln(pa)\omega_p^2\right)$ is the determinant of the system (12) and $\omega_p = (4\pi Q^2 N_0/M)^{1/2}$ and $\omega_G = (4\pi G M N_0)^{1/2}$ are the plasma and Jeans frequencies of the grains.

First, note that a purely gravitational perturbation ($\Psi_{E,ext} = 0$) gives rise to an alternating electric field in the beam. In fact, substituting N from (13) into the formula $E_z = \partial \Psi_E / \partial z = -2ik\pi a^2 QN \ln(pa)$, yields

$$E_z = \frac{-2i\ln(pa)k^3 \pi a^2 QN_0}{D} \Psi_{G,ext}.$$

Furthermore, self-gravitation manifests itself in free oscillations of the DSGP flow. The dispersion relation can be written from the condition $D = 0$, namely

$$(\omega - kV_0)^2 - \frac{k^2 a^2}{2}\left(\ln(ka)\omega_G^2 - \ln(pa)\omega_p^2\right) = 0. \tag{14}$$

It then follows that

$$\omega = kV_0 \pm \left[\frac{k^2 a^2}{2}\left(\ln(ka)\omega_G^2 - \ln(pa)\omega_p^2\right)\right]^{1/2} = kV_0 \pm \omega_{p,eff}.$$

As is to be expected for the narrow beam, the plasma parameters are controlled by the linear density of the grains $\sim \pi a^2 N [cm^{-1}]$, resulting in modification of the dispersion properties of the medium. In particularly, longitudinal waves can appear even in a cold dusty plasma without drift motion ($V_0 = 0$). For this simplest case the longitudinal phase velocity can be estimated as $V_{ph} \sim \left(a/\sqrt{2}\right)\left(|\ln(pa)|\omega_p^2 - |\ln(ka)|\omega_G^2\right)^{1/2}$. In an unbounded uniform self-gravitating plasma, longitudinal waves do not propagate under such conditions, and only vibrations of frequency $\omega = \left(\omega_p^2 - \omega_G^2\right)^{1/2}$ exist.

For $|\ln(pa)|\omega_p^2 > \ln(ka)\omega_G^2$ the dispersion equation has the form of a usual dispersion relation but with a reduced effective plasma frequency due to the effect of self-gravitation. An interesting feature of the SGDP flow is the existence of a critical wavelength λ_{cr} of perturbations,

$$\lambda_{cr} \cong \exp\left(-\frac{Q^2}{GM^2}\ln\left(\frac{a}{\lambda_D}\right)\right)$$

which corresponds to a zero effective plasma frequency of the beam $\omega_{p,eff} = 0$, and separates the domains of stable and unstable disturbances. Note that in the unbounded self-gravitational plasma with $\omega_p^2 < \omega_G^2$ all the disturbances will be unstable. Therefore, the spatial restriction of the DSGP leads to new criteria for the wave stability and increases the role of self-gravitation at least in the long wavelength limit. Indeed, if $|\ln(pa)|\omega_p^2 < \ln(ka)\omega_G^2$, then the effective plasma frequency will become imaginary, leading to a Jeans instability of the grains with the growth rate

$v = \left(ka/\sqrt{2} \right)\left[-\ln\left(pa \right)\omega_p^2 + \ln\left(ka \right)\omega_G^2 \right]^{1/2}$. The growing perturbations drift with the velocity V_0.

WAVE DISTURBANCES IN A SYSTEM OF SEVERAL COLLINEAR FLOWS OF THE SGDP

Now, we analyse a more complicated system that consists of several narrow cylindrical dusty flows of the same radius a. Let the flows be oriented along the z-axis and lie in the xz plane, and the distance between neighbouring beams be l. Since the entire space (including the regions occupied by the flows) is filled with a plasma, the flows interact with each other through both electric and gravitational fields. We turn to formulas (8) - (9) to represent the electric potential of the resulting field inside the j-th flow as

$$\Psi_{E,j} \cong -2\pi N_j Q a^2 \ln\left(pa \right) + \sum_{j \neq i} 2\pi Q N_i a^2 \sqrt{\frac{\pi}{2p\left| x_j - x_i \right|}} \exp\left(-p\left| x_j - x_i \right| \right), \quad (15)$$

where the first term on the right-hand side represents the potential in the j-th flow and the second term accounts for the contribution of the remaining flows, whose centres are at the points $x = x_i \left(i \neq j \right)$, to the electric field at the point $x = x_j$. The number of electrically interacting flows is of the order of λ_D / l.

Similarly, the gravitational potential can be written as

$$\Psi_{G,j} \cong 2\pi G\, M N_j a^2 \ln\left(ka \right) - \sum_{j \neq i} 2\pi\, G M N_i a^2 \sqrt{\frac{\pi}{2k\left| x_j - x_i \right|}} \exp\left(-k\left| x_j - x_i \right| \right). \quad (16)$$

Note, that the number of beams interacting gravitationally is larger and is proportional to $\sim \lambda / l \gg \lambda_D / l$ (needless to say that this is valid only for long-wavelength perturbations with $\lambda \gg \lambda_D$).

A complete set of equations describing the longitudinal disturbances consists of the linearized equations of motion and continuity equations for grains in the flows [equations (1) and (2) should be written for each of the flows] and also of the equations (15) and (16), which couple the dust densities N_i in different flows with the potentials $\Psi_{E,j}$ and $\Psi_{G,j}$ of the flow involved. The number of equations in the set is very large, and therefore a straightforward analysis is hardly possible.

As the simplest example, we consider the coupling between two flows of the SGDP. Let V_{01} and V_{02} be the unperturbed velocities of the grain flows, and N_{01} and N_{02} their unperturbed densities. We assume that the external force $-\left(Q / M \right)\partial \Psi_{E,ext} / \partial z - \partial \Psi_{G,ext} / \partial z$ operates only in the first flow. One can easily obtain that the perturbed density in the flows satisfies the set of algebraic equations:

$$N_1 D_1 - N_2 \frac{k^2 a^2}{2}\left[\omega_{p1}^2 \sqrt{\frac{\pi}{2pl}}\exp(-pl) - \omega_{G1}^2\sqrt{\frac{\pi}{2kl}}\exp(-kl)\right]$$

$$= k^2 N_{01}\left(\frac{Q}{M}\Psi_{E,ext} + \Psi_{G,ext}\right), \tag{17}$$

$$N_2 D_2 - N_1 \frac{k^2 a^2}{2}\left[\omega_{p2}^2 \sqrt{\frac{\pi}{2pl}}\exp(-pl) - \omega_{G2}^2\sqrt{\frac{\pi}{2kl}}\exp(-kl)\right] = 0, \tag{18}$$

where D_1 and D_2 are the determinants of the sets of equations describing the disturbances in each of the dusty flows, viz.

$$D_i = \left(\omega - kV_{0i}\right)^2 + \frac{k^2 a^2}{2}\left(\ln(ka)\omega_{Gi}^2 - \ln(pa)\omega_{pi}^2\right), \quad i = 1,2.$$

Although the external force acts only on the first beam, the density disturbances

$$N_1 = \frac{k^2 N_{01} D_2}{D_0}\left(\frac{Q}{M}\Psi_{E,ext} + \Psi_{G,ext}\right),$$

generate the electric and gravitational fields in the second beam, which, in turn, perturb its density

$$N_2 = \frac{k^4 a^2 N_{01}}{2D_0}\left(\frac{Q}{M}\Psi_{E,ext} + \Psi_{G,ext}\right) \times \left[\omega_{p2}^2 \sqrt{\frac{\pi}{2pl}}\exp(-pl) - \omega_{G2}^2\sqrt{\frac{\pi}{2kl}}\exp(-kl)\right],$$

$$\tag{19}$$

with

$$D_0 = D_1 D_2 - \frac{\pi k^4 a^4}{8}\left[\omega_{p1}^2 \sqrt{\frac{1}{pl}}\exp(-pl) - \omega_{G1}^2\sqrt{\frac{1}{kl}}\exp(-kl)\right]$$

$$\times \left[\omega_{p2}^2 \sqrt{\frac{1}{pl}}\exp(-pl) - \omega_{G2}^2\sqrt{\frac{1}{kl}}\exp(-kl)\right]$$

the determinant of the set (17)- (18).

According to (19), the density perturbations in the first beam induce the density perturbations in the second beam via electric and gravitational fields, so that we can write $N_2 = N_2^E + N_2^G$. Moreover, the ratio

$$\frac{N_2^E}{N_2^G} \sim \frac{\omega_{p2}^2}{\omega_{G2}^2}\sqrt{k\lambda_D}\exp\left[-l(p-k)\right] \sim \frac{Q^2}{GM^2}\sqrt{\frac{\lambda_D}{\lambda}}\exp\left[-\frac{l}{\lambda_D}\left(1 - \frac{\lambda_D}{\lambda}\right)\right]$$

decreases as the distance l between the flows increases. One can easily see that, as l increases, the electromagnetic interaction between the beams drops sharply, so that the wave perturbations of the density of the first beam perturb the density of the second beam mainly via gravitational fields.

Hence, Debye screening can substantially weaken the electric coupling between beams of dust grains of the SGDP, without affecting the gravitational forces, which thus might become dominant at long distances $l \gg \lambda_D$, even though the electric fields play a governing role in the interaction between closely spaced flows.

CONCLUSIONS

A study of the low-frequency plasma modes in an individual plasma flow demonstrates significant modification of the dispersion properties of the dusty media, like the growth of the dispersion in the long wavelength limit, and substantially new conditions for the wave stability. In this work, we have shown that eigenwaves are characterised by decreasing effective plasma or Jeans frequencies. Moreover, there is always a critical wavelength which corresponds to "zero" effective plasma frequency. A question of considerable physical interest is that of the excitation of electromagnetic waves in a DSGP flow by an external force. In such a medium, electric disturbances can be excited even by external gravitational forces (due to for example, a massive body flying nearby), which is impossible for a conventional plasma.

Self-gravitational effects are of considerable importance for a multistream model, too. It is important that the electric fields are greatly weakened due to the Debye screening, while the forces of gravitation remain unchanged. As a result, self-gravitation dominates at long distances, even if electric forces prevail at short distances. As a result, an external (e.g. gravitational) perturbation in one stream can excite electrostatic waves in the whole multistream system.

ACKNOWLEDGEMENT

This work was partially supported by the Bijzonder Onderzoeksfonds of the RUG through a foreign visitor grant. Useful discussions with F. Verheest and M.A. Hellberg are gratefully acknowledged.

REFERENCES

[1] Gisler G.R., Rushdy Ahmand Q. and Wollman E.R., *IEEE Trans. On Plasma Sci.* **20**, 922-927 (1992).
[2] Bliokh P.V., Sinitsin V.G., and Yaroshenko V.V. *Dusty and Self-Gravitational Plasmas in Space* Kluwer Acad. Publ., Dordrecht, 1995, pp. 220-232.
[3] Verheest F. *Waves in Dusty Space Plasmas* Kluwer Acad. Publ., Dordrecht, 2000, pp. 196-214.

PART 2: SOLAR PLASMAS

Magnetohydrodynamic Waves
in Laboratory and Astrophysical Plasmas

J.P. Goedbloed

FOM-Institute for Plasma Physics, Association Euratom-FOM,
P.O. Box 1207, 3430 BE Nieuwegein, the Netherlands
(also @ Free University Amsterdam & Utrecht University)

Abstract.
The study of magnetohydrodynamic waves and instabilities of both laboratory and astrophysical plasmas has been conducted for many years starting from the assumption of static equilibrium. Recently, there is an outburst of interest for plasma states where this assumption is violated. In fusion research, this interest is due to the importance of neutral beam heating and pumped divertor action for the extraction of heat and exhaust needed in future tokamak reactors. Both result in rotation of the plasma with speeds that do not permit the assumption of static equilibrium anymore. In astrophysics, observations in the full range of electromagnetic radiation has revealed the primary importance of plasma flows in such diverse situations as coronal flux tubes, stellar winds, rotating accretion disks, and jets emitted from radio galaxies. These flows have speeds which substantially influence the background stationary equilibrium state, if such a state exists at all. Consequently, it is important to study both the stationary states of magnetized plasmas with flow and the waves and instabilities they exhibit. We will present new results along these lines, extending from the discovery of gaps in the continuous spectrum and low-frequency Alfvén waves driven by rotation to the nonlinear flow patterns that occur when the background speed traverses the full range from sub-slow to super-fast. The solutions obtained may bridge the gap between insights from linear and nonlinear analyses.

I INTRODUCTION

Magnetohydrodynamic (MHD) waves control the dynamics of plasma, the main constituent of the universe. They occur as the natural response of plasmas to global excitation. Their frequency and wave forms are determined by the overall magnetic confinement geometry and the detailed distribution of the density, the pressure, the magnetic field, and the background velocity of the plasma. Hence, measurement of the spectrum of MHD waves gives direct information on the internal state of the plasma, provided a theoretical model is available to solve the forward as well as the inverse spectral problems. This activity has been called *MHD Spectroscopy* [1]. The suggested analogy with quantum mechanical spectroscopy results from the

CP537, *Waves in Dusty, Solar, and Space Plasmas,* edited by F. Verheest, et al.
© 2000 American Institute of Physics 1-56396-962-9/00/$17.00

	Internal state	Spectrum	Wave equation
MHD:	$\rho, p, \mathbf{B}\ (, \mathbf{v})$ $\xrightarrow{\text{forward}}$ $\xleftarrow{\text{inverse}}$	$\{\omega_i; \boldsymbol{\xi}_i\}$	$\mathbf{F}(\boldsymbol{\xi}) = \rho\dfrac{\partial^2 \boldsymbol{\xi}}{\partial t^2} = -\rho\omega^2\boldsymbol{\xi}$
QM:	V	$\{E_i; \psi_i\}$	$H\psi = i\hbar\dfrac{\partial\psi}{\partial t} = E\psi$

FIGURE 1. MHD spectroscopy compared with QM spectroscopy: Force operator $\mathbf{F}$ for plasmas and Hamilton operator H for atoms depend on *the internal state variables*.

mathematical similarity of the respective wave equations, which involve eigenvalue problems of linear operators (Fig. 1). The terminology also entails a program, viz. to improve the accuracy of our knowledge of plasmas, both in the laboratory and in astrophysics. Here, helioseismology (which could be considered as one of the forms of MHD spectroscopy) may serve as a luminous example.

The well-known phase and group diagrams of the three MHD waves (Fig. 2) already show the anisotropy that is present in homogeneous plasmas. Therefore, it is not too surprising that toroidal plasmas exhibit an *extremely rich spectral structure* of the waves and instabilities (global Alfvén waves, continuous spectra, TAE modes, ballooning modes, etc.). The standard approach for the study of these phenomena in tokamaks is to split the problem in that of *the static equilibrium + the linear waves and instabilities* [2], as described by the equations

$$\mathbf{j} \times \mathbf{B} = \nabla p, \qquad \mathbf{F}(\boldsymbol{\xi}) = -\rho\omega^2\boldsymbol{\xi}. \tag{1}$$

This approach has been successfully followed during 40 years of intensive research.

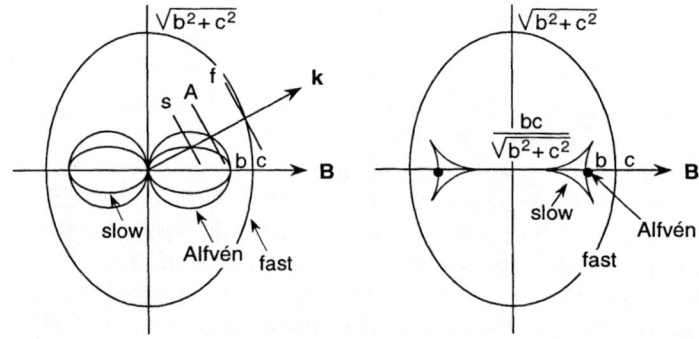

FIGURE 2. Phase and group diagram of the three MHD waves.

However, astrophysical plasmas are *never* in static equilibrium, they are all dominated by flow. This background flow is usually *transsonic*: it crosses all critical MHD speeds (slow/Alfvén/fast) and exhibits shocks. In recent tokamak research, heating by neutral beams has resulted in toroidal flow of the plasma, whereas the introduction of a divertor for exhaust removal has produced poloidal flows that frequently are supersonic. Obviously, these flows influence tokamak stability significantly. Hence, *plasmas with background flow* becomes a relevant and exciting common research theme of laboratory and astrophysical plasma physics.

For the study of MHD waves in plasmas with background flow, again, the standard approach is followed. Now, the equations describing the stationary background equilibrium ($\mathbf{v} \neq 0$) are much more complicated:

$$\nabla \cdot (\rho \mathbf{v}) = 0, \tag{2}$$

$$\rho \mathbf{v} \cdot \nabla \mathbf{v} + \nabla p = \mathbf{j} \times \mathbf{B}, \qquad \mathbf{j} = \nabla \times \mathbf{B}, \tag{3}$$

$$\mathbf{v} \cdot \nabla p + \gamma p \nabla \cdot \mathbf{v} = 0, \tag{4}$$

$$\nabla \times (\mathbf{v} \times \mathbf{B}) = 0, \qquad \nabla \cdot \mathbf{B} = 0. \tag{5}$$

The study of the linear waves and instabilities also requires a substantially more involved spectral problem [3]:

$$\mathbf{F}(\boldsymbol{\xi}) + 2i\rho\omega \mathbf{v} \cdot \nabla \boldsymbol{\xi} + \rho\omega^2 \boldsymbol{\xi} = 0, \tag{6}$$

$$\mathbf{F} \equiv \mathbf{F}_{\text{static}} + \nabla \cdot \left[\rho(\mathbf{v} \cdot \nabla \mathbf{v})\boldsymbol{\xi} - \rho\mathbf{v}\mathbf{v} \cdot \nabla \boldsymbol{\xi} \right]. \tag{7}$$

At this point, we already wish to express concern about this split. We will see in the following section why.

II WAVES IN PLASMA WITH BACKGROUND FLOW

We have seen that spectral theory for static equilibria is a very powerful instrument, analogous to quantum mechanics for atomic systems, and that MHD spectroscopy could offer accurate internal information on plasmas, similar to that obtained in helioseismology. The short-wavelength limit determines the spectral structure resulting in *three singular continuous spectra* [4],

$$\text{slow} : \{\omega_S^2(x)\}, \quad \text{Alfvén} : \{\omega_A^2(x)\}, \quad \text{fast} : \omega_F^2 \ (= \infty). \tag{8}$$

For equilibria with flow the local Doppler shifted frequencies, $\tilde{\omega} \equiv \omega - \mathbf{k} \cdot \mathbf{v}$, give rise to *new continuous spectra*:

$$\Omega_S^{\pm} = \pm\omega_S + \mathbf{k} \cdot \mathbf{v}, \tag{9}$$

$$\Omega_A^{\pm} = \pm\omega_A + \mathbf{k} \cdot \mathbf{v}, \tag{10}$$

$$\Omega_F^{\pm} = \pm\infty. \tag{11}$$

Moreover, the problem is no longer self-adjoint: *overstable modes* are possible.

In general, the spectral structure with flow becomes extremely involved. However, for weak inhomogeneity, the three types of waves remain distinguishable (Fig. 3): there are now forward and backward slow, Alfvén, and fast waves, with each subspectrum clustering at its own continuum given by Eqs. (9)–(11). The continua are organised around the central frequency range $\{\Omega_0(x)\}$, where $\Omega_0 \equiv \mathbf{k} \cdot \mathbf{v}$ is the local Doppler shift. In contrast to spectra for static equilibria, which are symmetric about the marginal frequency $\omega = 0$, the spectrum is now completely asymmetric (for simplicity, this is not shown in the figure).

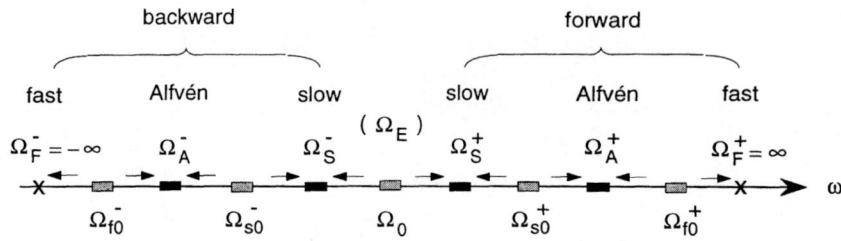

FIGURE 3. Schematic MHD spectrum of plasma with flow.

For strong inhomogeneity, the continua may become unstable, new gaps appear, etc. A significant example are the new global modes induced by toroidal flow found by van der Holst et al. [5]. These modes occur at low-frequency. Due to toroidal flow, a tiny gap in the continuum at marginal stability in the static case is Doppler shifted and much widened. In this new gap, a global Alfvén wave appears. This mode is still to be detected experimentally. It should provide a useful diagnostic for MHD spectroscopy in the presence of background plasma flow.

Let us now turn to the question posed in the introduction. Perturbations of the flow propagate along space-time manifolds called *characteristics* (Fig. 4). The MHD group diagram shown in Fig. 2(b) represents a snapshot of the spatial part of such a characteristic. Note that the Lagrangian time derivative

$$\frac{D}{Dt} \equiv \frac{\partial}{\partial t} + \mathbf{v} \cdot \nabla \tag{12}$$

introduces temporal phenomena (waves and instabilities) through $\partial/\partial t$, whereas the spatial derivative ∇ becomes relevant in plasmas with a background flow. The

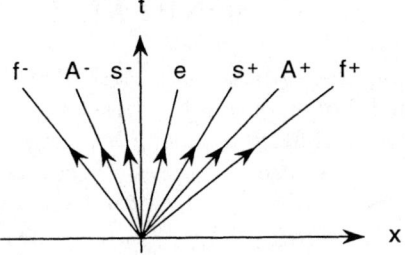

FIGURE 4. Characteristics (stressing the temporal part) in MHD with flow.

spatial part of the characteristics then betrays their wave-like origin so that *linear waves and non-linear stationary equilibria are no longer a separate issue.*

To illustrate this, consider the example of sound in a stationary flow about a point source (Fig. 5). Note that, in the case of subsonic flow, eventually all space is perturbed. The relevant PDE is *elliptic*: there are *no* spatial characteristics. On the other hand, after transition to supersonic flow, only a sector is perturbed. The relevant PDE is now *hyperbolic*: there are *two* spatial characteristics. These characteristics are just the shock fronts, where the wave fronts constructively interfere, separating the perturbed from the unperturbed regions.

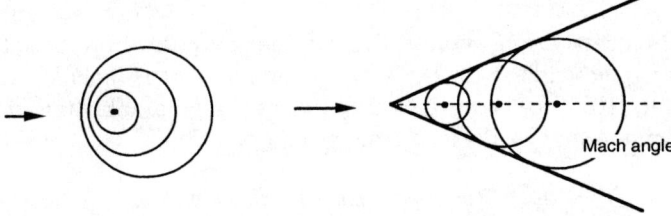

FIGURE 5. Transition from *subsonic* ($v < c$) to *supersonic* ($v > c$) flow in gas dynamics: stressing the spatial part of the characteristics.

To get the spatial characteristics (or caustics) in MHD one should construct the tangents to the Friedrichs group diagram of Fig. 2(b). Of course, due to the magnetic field anisotropy, this creates much more intricate patterns than in gas-dynamics where the elementary waves are just spheres. Beautiful pictures of the resulting fronts for homogeneous plasmas can be found in Refs. [6]. However, for our subject, this is not enough. For waves and instabilities of magnetically confined plasmas, it is essential to account for the inhomogeneity of the background equilibrium state. As we will see, additional complexities of plasmas compared to ordinary gases are the occurrence of three characteristic speeds, of forbidden flow regimes and singularities. Hence, rather than studying the waves, we should *reconsider the background equilibrium state itself*.

113

III TRANSSONIC MHD FLOW

To appreciate the intricacies of transsonic MHD flow, it is necessary to consider stationary states that are inhomogeneous in at least two directions, giving rise to 2D nonlinear PDEs. Apart from that, we will simplify the problem as much as possible without loosing the basic structure. The present section is based on Refs. [7] and [8], where the full analysis is to be found.

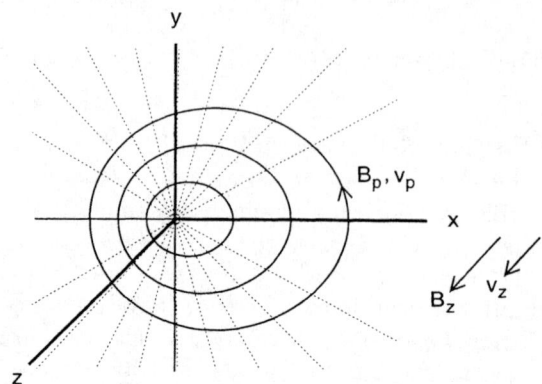

FIGURE 6. Geometry of translation symmetric stationary state.

Consider the translation symmetric stationary equilibrium depicted in Fig. 6. Toroidicity is neglected here for the sake of simplicity. It is included in forthcoming publications [9], [10]. The essential variables are the poloidal magnetic field $\mathbf{B}_p$ and velocity $\mathbf{v}_p$, which may be derived from streamfunctions:

$$\mathbf{B}_p = \mathbf{e}_z \times \nabla\psi \quad \Leftarrow \quad \text{poloidal flux} \quad \psi(x,y) \,, \tag{13}$$

$$\rho\mathbf{v}_p = \mathbf{e}_z \times \nabla\chi \quad \Leftarrow \quad \text{stream function} \quad \chi(\psi) \,. \tag{14}$$

As compared to static equilibria, which are fully described by the poloidal magnetic flux dependence alone, an enormous increase of possibilities is obtained from the appearance of the square of *the poloidal Alfvén Mach number*:

$$\mu(x,y) \equiv \frac{\rho v_p^2}{B_p^2} = \frac{\chi'^2}{\rho} \,. \tag{15}$$

Note that it is not a flux function because of the density ρ in the denominator.

A complicating detail is the appearance of five arbitrary flux functions: χ, H (Bernoulli function), S (entropy), I (momentum flux), Ω (electric field). Fortunately, they combine to three generic functions: $\Pi_{1,2,3}(\psi)$. The basic problem then is, for a given choice of the Π's, to determine the poloidal distribution of $\psi(x,y)$ & $\mu(x,y)$.

This problem may be cast in a variational principle stating that the stationary states are obtained by minimizing the following Lagrangian:

$$\delta \int \mathcal{L}\, dV = 0\,, \quad \mathcal{L} \equiv \tfrac{1}{2}(1-\mu)|\nabla\psi|^2 - W(\psi,\mu)\,, \tag{16}$$

$$W \equiv \frac{\Pi_1(\psi)}{\mu} - \frac{\Pi_2(\psi)}{\gamma\mu^\gamma} + \frac{\Pi_3(\psi)}{1-\mu}\,. \tag{17}$$

This yields a nonlinear PDE for the determination of the magnetic flux $\psi(x,y)$ and an algebraic equation, the Bernoulli equation, for the Mach number $\mu(x,y)$:

$$\nabla\cdot[\,(1-\mu)\,\nabla\psi\,] + \frac{\partial W}{\partial\psi} = 0\,, \tag{18}$$

$$\tfrac{1}{2}|\nabla\psi|^2 + \frac{\partial W}{\partial\mu} = 0\,. \tag{19}$$

Note that the factor $1-\mu$ changes sign at the poloidal Alfvén speed. However, this is just one of the contributing features. More important is the intricate way in which the variable $\mu(x,y)$ is obtained from Eq. (19) and then to be inserted in Eq. (18) to appear in front of the highest derivative of the flux equation.

Reduction to tractable size is achieved by choosing *a master profile* $\pi \equiv \psi^{2-2/\lambda}$,

$$\Pi_1 = \pi(\psi)\,, \quad \Pi_2 = A\,\pi(\psi)\,, \quad \Pi_3 = B\,\pi(\psi)\,, \tag{20}$$

and assuming *self-similarity* in the polar coordinates r, θ,

$$\mu^{-1} = X(\theta)\,, \qquad \psi = r^\lambda Y(\theta)\,. \tag{21}$$

A system of ODEs for X and Y is then obtained:

$$\frac{dX}{d\theta} = \pm\frac{H}{J}\sqrt{2F}\,, \tag{22}$$

$$\longrightarrow \text{Trajectory:} \quad \frac{dY}{dX} = \frac{J}{H}\,. \tag{23}$$

$$\frac{dY}{d\theta} = \pm\sqrt{2F} \tag{24}$$

The different *flow regimes* follow from algebraic conditions in X and Y:

$F = 0$ – Bernoulli Boundary (fast & slow),

$\Delta = 0$ – Characteristics (none: $\mathcal{E}$ / two: $\mathcal{H}$),

$J = 0$ – Limiting Line characteristic,

where Δ is a factor that determines the direction of the characteristics.

These conditions produce a phase diagram with seven flow regimes (Fig. 7):

$$\mathcal{H}_{ff}\,(1^+) \quad [\![\,Fast\ LL\,]\!] \quad \mathcal{H}_f\,(1^-)\,, \ \mathcal{E}_f\,(2) \tag{25}$$

$$[\![\,Alfvén\ gap\,]\!] \tag{26}$$

$$\mathcal{E}_s\,(3)\,, \ \mathcal{H}_s\,(4^-) \quad [\![\,Slow\ LL\,]\!] \quad \mathcal{H}_{ss}\,(4^+)\,, \ \mathcal{E}_{ss}\,(5)\,, \tag{27}$$

where $\mathcal{E}$ and $\mathcal{H}$ indicate ellipticity and hyperbolicity of the Eqs. (18) and (19). The Bernoulli condition $F \geq 0$ gives rise to fast and slow flow domains, the $J = 0$ singularity splits these domains in superfast (ff) and fast (f) subdomains, and in slow (s) and subslow (ss) subdomains, whereas the $\Delta = 0$ condition gives rise to the vertical lines separating elliptic and hyperbolic regions.

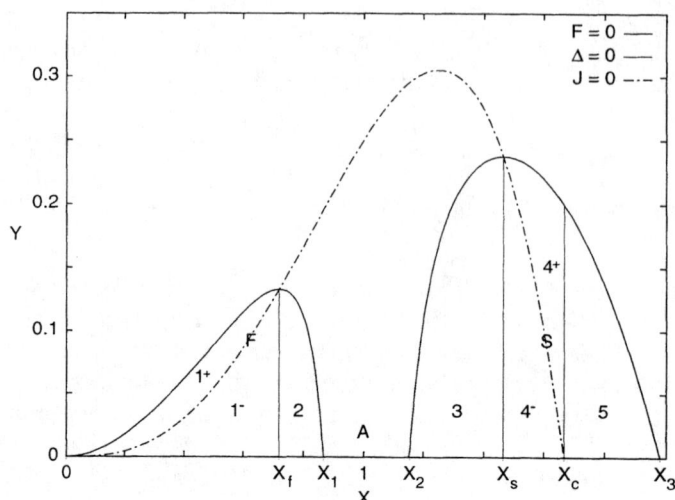

FIGURE 7. Four main flow domains due to Alfvén gap (A) & fast (F) / slow (S) limiting line.

Let us now consider the flow patterns that correspond to the trajectories in the X–Y phase plane (not shown in Fig. 7). These trajectories may either intersect the limiting line in the X–Y plane or not. In the first case, poloidal flows are obtained that are periodic, i.e. the poloidal velocity (expressed by the variable μ) may increase when θ increases and then decreases again to return to its initial value, or vice versa. For the poloidal magnetic field variable, this would to be the ordinary behavior for tokamaks. However, this is not the generic behavior in the presence of rotation. Instead, trajectories usually intersect the limiting line in the X–Y plane and then show a very peculiar behavior in the corresponding physical x–y plane (note the difference between upper case and lower case variables!). An example of this more frequent behavior is shown in Fig. 8. A slow flow $(\mathcal{E}_s, \mathcal{H}_s)$, with increasing speed in the bottom part of the figure and decreasing speed in the upper part, hits a limiting line characteristic and is then 'reflected' to manifest subslow flow $(\mathcal{E}_s, \mathcal{H}_s)$ in the same sector of the poloidal plane. Streamlines and characteristics appear to be blocked: Two solutions in the same place!? Obviously, one cannot accept such simultaneous presence of two different solutions in the same place. Apparently, the smooth crossing of the limiting line in the X–Y phase plane does not correspond to acceptable solutions in the physical x–y plane.

At this point, our conclusion should be that the occurrence of limiting line char-

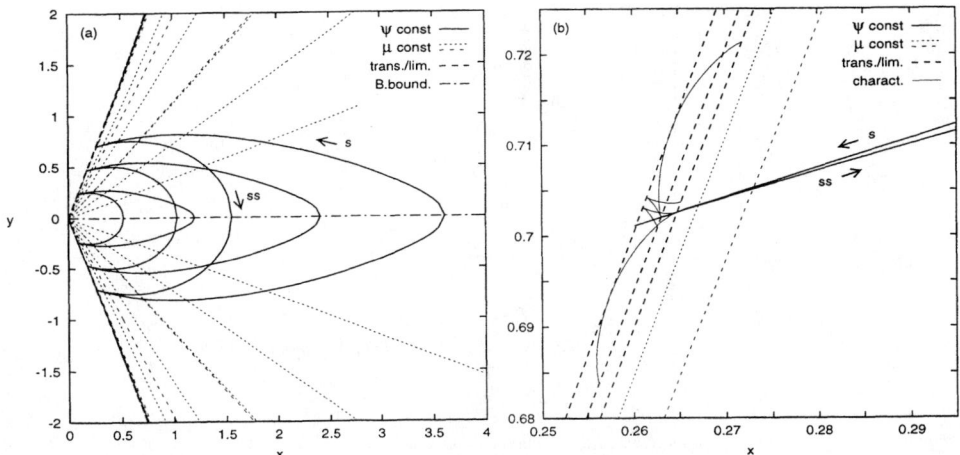

FIGURE 8. Flow 'reflected' by the limiting line: (a) Slow and subslow flow patterns in the full sector, (b) Blowup of the flows and characteristics close to the limiting line.

acteristics indicate that singular discontinuous flows are to be considered, i.e. we should permit *shocks*. By means of an extensive analysis of the permitted jumps of the physical variables, we arrive at the following shock conditions:

– The flux variable Y and the toroidal field parameter B should be continuous.

– The Mach variable X and the entropy parameter A may jump, but A increases. Elimination of A_2 results in the *distilled jump & entropy conditions*:

$$f(\hat{X}_1, \hat{X}_2, \hat{Y}, A_1) = 0, \tag{28}$$

$$g(\hat{X}_1, \hat{X}_2, \hat{Y}, A_1) \geq 0; \tag{29}$$

– The variables in front of the shock should stay within the *Bernoulli boundaries*:

$$F(\hat{X}_1, \hat{Y}, A_1, B) \geq 0. \tag{30}$$

The procedure to find the permitted kinds of shocks is then, for given $\hat{Y}$ and A_1, to plot f in the $\hat{X}_1$-$\hat{X}_2$ plane and to cut out the forbidden entropy and Bernoulli parts. An illustrative result of this procedure is shown in Fig. 9: just three pieces of the jump curve remain. These pieces precisely correspond to singular behavior with the features of slow, Alfvén, and fast wave polarizations: The three shocks permit the solutions to jump across the limiting lines. They may be considered to be the *non-linear counterparts of the three linear MHD waves*.

IV   CONCLUSIONS

• The structure of the spectrum of MHD waves hinges on the three continua where the perturbations become singularly localised.

117

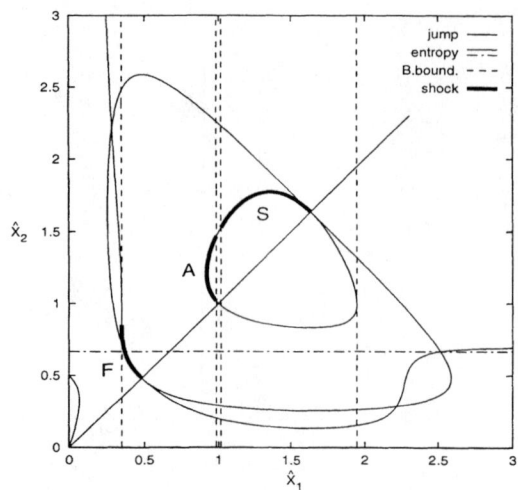

FIGURE 9. Fast, Alfvén, slow shocks

- With flow, this structure becomes much more intricate through the space-dependent Doppler shifts.
- In tokamaks, wide gaps in the continua with *new global modes induced by toroidal flow* have been found.
- With rotation, the singular structure transfers to the equilibrium state: it exhibits precisely three types of 'singularities' where the variables are discontinuous.
- Consequently, *linear waves and non-linear stationary equilibria are no longer a separate issue in magnetohydrodynamics*.

REFERENCES

1. Goedbloed, J.P., Huysmans, G.T.A., Holties, H.A., Kerner, W., and Poedts, S., *Phys. Controlled Fusion* **35**, B277 (1993).
2. Bernstein, I.B., Frieman, E.A., Kruskal, M.D., and Kulsrud, R.M., *Proc. Roy. Soc. A* **244**, 17 (1958).
3. Frieman, E. and Rotenberg, M., *Rev. Mod. Phys.* **32**, 898 (1960).
4. Goedbloed, J.P., *Phys. Fluids* **18**, 1258 (1975).
5. van der Holst, B., Beliën, A.J.C., and Goedbloed, J.P., *Phys. Rev. Lett.* **84**, 2865 (2000); *Phys. Plasmas*, submitted (2000).
6. Kulikovskiy, A.G., Lyubimov, G.A., *Magnetohydrodynamics* (Addison Wesley, Reading, 1965); Jeffrey, A., *Magnetohydrodynamics* (Oliver & Boyd, Edinburgh, 1966).
7. Goedbloed, J.P., Lifschitz, A., *Phys. Plasmas* **4**, 3544 (1997).
8. Lifschitz, A., Goedbloed, J.P., *J. Plasma Phys.* **58**, 61(1997).
9. Goedbloed, J.P., Lifschitz, A., to be published (2000).
10. Beliën, A.J.C., Goedbloed, J.P., van der Holst, B., EPS Budapest, 12–16 June 2000.

Three-fluid 2.5D MHD models of waves in solar coronal holes and the relation to SOHO/UVCS observations

L. Ofman

Raytheon ITSS/NASA Goddard Space Flight Center, Code 682,
Greenbelt, MD 20771, USA

Abstract. The physical properties of the minor ions in the corona provide clues on the coronal heating and solar wind acceleration mechanism. Recent observations show that protons and minor ions are hot ($T_p > 10^6$ K, $T_i > 10^8$ K) and anisotropic in coronal holes. A possible cause of the large perpendicular motions is unresolved Alfvénic fluctuations in the solar wind. Using the three-fluid 2.5D MHD model I have shown that that the unresolved Alfvénic fluctuations lead to apparent proton temperature and anisotropy consistent with UVCS observations. I show the calculated dependence of the apparent kinetic temperatures of protons and O^{5+} ions with heliocentric distance and compare the results to the UVCS observations.

INTRODUCTION

Recent SOHO Ultraviolet Coronagraph Spectrometer (UVCS) observations of O^{5+} and show that Ly-α extreme ultraviolet (EUV) emission show that the protons reach 3.5×10^6 K, and O^{5+} ions exceed temperatures of 10^8 K in coronal holes [1]. The physical properties of the minor ions in the corona provide clues on the coronal heating and solar wind acceleration mechanism. Using 2.5D 2-fluid MHD model it has been shown that unresolved wave motion may account for the enhanced perpendicular proton temperatures observed by UVCS [2–4].

Most of the previous multi-fluid studies of the solar wind considered the effects of α-particles in a 1D model. Thus, the self consistent effect of waves, and of coronal hole structure were not included. In the present study I use 3-fluid electron-proton-ion plasma with a broad band Alfvén waves source in the millihertz frequency range and include the effects of Coulomb friction, important for minor ions. I consider the effects of waves on the observed line width temperatures of protons and O^{5+} ions in a structured coronal hole.

CP537, *Waves in Dusty, Solar, and Space Plasmas*, edited by F. Verheest, et al.

THREE–FLUID 2.5D MHD MODEL

Neglecting electron inertia ($m_e \ll m_p$), relativistic effects ($V \ll c$), assuming quasi-neutrality ($n_e = n_p + Zn_i$), and neglecting the ion-cyclotron terms ($\omega \ll \omega_{ci}$) the normalized 3-fluid MHD equations are

$$\frac{\partial n_k}{\partial t} + \nabla \cdot (n_k \mathbf{V}_k) = 0, \tag{1}$$

$$n_k \left[\frac{\partial \mathbf{V}_k}{\partial t} + (\mathbf{V}_k \cdot \nabla)\mathbf{V}_k \right] = -E_k \nabla p_k - E_e \frac{Z_k n_k}{A_k n_e} \nabla p_e - \frac{n_k}{F_r r^2}$$
$$+ Z_k a_k \frac{n_k}{n_e} (\nabla \times \mathbf{B}) \times \mathbf{B} + \mathbf{F}_{k,coul}, \tag{2}$$

$$\frac{\partial \mathbf{B}}{\partial t} = -\nabla \times \mathbf{E}, \quad \mathbf{E} = -\mathbf{V}_e \times \mathbf{B} + \frac{1}{S}\nabla \times \mathbf{B} \tag{3}$$

$$\mathbf{V}_e = \frac{1}{n_e}(n_p \mathbf{V}_p + Zn_i \mathbf{V}_i - b\nabla \times \mathbf{B}), \tag{4}$$

$$\frac{\partial T_k}{\partial t} = -(\gamma_k - 1)T_k \nabla \cdot \mathbf{V}_k - \mathbf{V}_k \cdot \nabla T_k + (\gamma_k - 1)S_k, \tag{5}$$

where the index $k = p, i$ (in Equation (5) $k = e, p, i$), and S_k is the heating term.

The 2.5D model is obtained by assuming $\partial/\partial\phi = 0$ in the above equations. The thermal forces, viscosity, and the heating terms are neglected in this study. Ohmic heating of electrons due the dissipation of the Alfvén waves is calculated from the current density j, but not included in Equation (5). The Coulomb friction terms $\mathbf{F}_{k,coul}$ are [5–7]

$$\mathbf{F}_{p,coul} = m_e \nu_{ep}(\mathbf{V}_p - \mathbf{V}_e)\Phi(X_{ep}) + m_e \nu_{ei}(\mathbf{V}_i - \mathbf{V}_e)\Phi(X_{ei})$$
$$+ \nu_{pi}(\mathbf{V}_i - \mathbf{V}_p)\Phi(X_{pi}) + \nu_{pe}(\mathbf{V}_e - \mathbf{V}_p)\Phi(X_{pe}) \tag{6}$$

$$\mathbf{F}_{i,coul} = \frac{Z}{A}m_e \nu_{ep}(\mathbf{V}_p - \mathbf{V}_e)\Phi(X_{ep}) + \frac{Z}{A}m_e \nu_{ei}(\mathbf{V}_i - \mathbf{V}_e)\Phi(X_{ei})$$
$$+ \nu_{ip}(\mathbf{V}_p - \mathbf{V}_i)\Phi(X_{ip}) + \nu_{ie}(\mathbf{V}_e - \mathbf{V}_i)\Phi(X_{ie}), \tag{7}$$

where the collision frequencies between the species are [5,7]

$$\nu_{kl} = \tau_A \frac{16\sqrt{\pi}}{3} \frac{n_l m_l}{m_k + m_l} \left(\frac{2k_b T_{kl}}{\mu_{kl}} \right)^{-3/2} \frac{Z_k^2 Z_l^2 e^4}{\mu_{kl}^2} ln\Lambda \tag{8}$$

$$\mu_{kl} = \frac{m_k m_l}{m_k + m_l}; \quad T_{kl} = \frac{m_k T_l + m_l T_k}{m_k + m_l} \tag{9}$$

$$\Phi(X_{kl}) = (1 + 0.74 X_{kl}^3)^{-1}; \quad X_{kl} = |V_l - V_k| \left(\frac{2k_b T_{kl}}{\mu_{kl}} \right)^{-1/2}, \tag{10}$$

where Z_k is the charge number, $ln\Lambda$ is the Coulomb logarithm, and k_b is the Boltzmann constant.

In the above equations I use the following normalization: $r \rightarrow r/R_\odot$, where $R_\odot$ is the solar radius; $t \rightarrow t/\tau_A$; $V \rightarrow V/V_A$; $B \rightarrow B/B_0$; $n_k \rightarrow n_k/n_{e0}$; $S = \tau_r/\tau_A$ the Lundquist number; $\tau_r = 4\pi R_\odot^2/\nu c^2$ the resistive time scale, where ν is the resistivity, and c is the speed of light; $\tau_A = R_\odot/V_A$ the Alfvén time scale; $E_{e,p} = (k_b T_{e,p,0}/m_p)/V_A^2$ the electron or proton Euler number; $E_i = (k_b T_{i,0}/m_i)/V_A^2$ the ion Euler number; $F_r = V_A^2 R_\odot/(GM_\odot)$ the Froude number, where G is the universal gravitational constant and $M_\odot$ is the solar mass; $a_k = n_{p0}/n_{e0}/A_k + m_e/m_k + A_k n_{i0}/n_{e0}$; $b = cB_0/(4\pi e n_{e0} R_\odot V_A)$, where A_k is the atomic mass of the species, and B_0 is the magnetic field at the coronal boundary.

The typical parameters of the model are $\gamma_p = 1.1$, $\gamma_e = 1.05$, $\gamma_{O5+} = 1.0$, $T_{p0} = 1.6\times10^6$ K, $T_{O5+0} = 12\times10^6$ K, $T_{e0} = 10^6$ K, $n_{e0} = 10^8$ cm^{-3}, $n_{O5+0} = 6\times10^4$ cm^{-3}, $S = 10^4$, $B_0 = 7$ G which gives $V_A = 1521$ km s^{-1}. The initial densities of the protons, ions, and electrons are

$$n_{p0}(\theta, r) = \left[1 - (1 - n_r)\, e^{-[(\theta-\pi/2)/\theta_c]^4}\right] n(r)/n_r, \tag{11}$$

$$n_{i0} = \epsilon n_{p0}, \tag{12}$$

$$n_{e0} = n_{p0} + Z n_{i0}. \tag{13}$$

where $\epsilon = 6 \times 10^{-4}$, $n_r = 0.3$, and $n(r)$ and $V_r(r, t = 0)$ are given by Parker's [8] solution.

The boundary conditions at $r = 1R_\odot$ are incoming characteristics approximated by zero order extrapolation.

$$B_r(1, \theta) = B_{r,0}, \; V_{k,\theta} = B_\theta = 0, \; T_p = T_{p0}, \; T_e = T_{e0}, \; T_i = T_{i0}. \tag{14}$$

At $r = r_{max}$, and at $\theta = \theta_{max}$ I use open boundary conditions, with symmetry boundary conditions imposed at $\theta = \pi/2$.

The broad band Alfvén wave driver is modeled by

$$B_\phi(t, \theta, r = 1) = -V_d/V_{A,r} F(t, \theta) \tag{15}$$

$$F(t, \theta) = \sum_{i=1}^{100} a_i sin(\omega_i t + \Gamma_i(\theta)) \tag{16}$$

where $a_i = i^{-1/2}$, $\omega_i = \omega_1 + (i - 1)\Delta\omega$, $\Delta\omega = (\omega_N - \omega_1)/(N - 1)$, and $\Gamma_i(th)$ is the random phase. The frequencies in the broad band driver are in the millihertz range $[\omega_0, \omega_N]$. In the present study I use $V_d = 0.02 V_A$, $\omega_0 = 0.013$ Rad s^{-1}, and $\omega_N = 10\omega_0$. The coronal boundary is divided into eight regions that produce the Alfvén waves with a different $\Gamma_i(\theta)$.

Simulated line profile

The calibrated observed line profile is usually given in terms of radiative intensity as a function of wavelength. In order to convert the wavelength to velocity the Doppler's shift relation is used

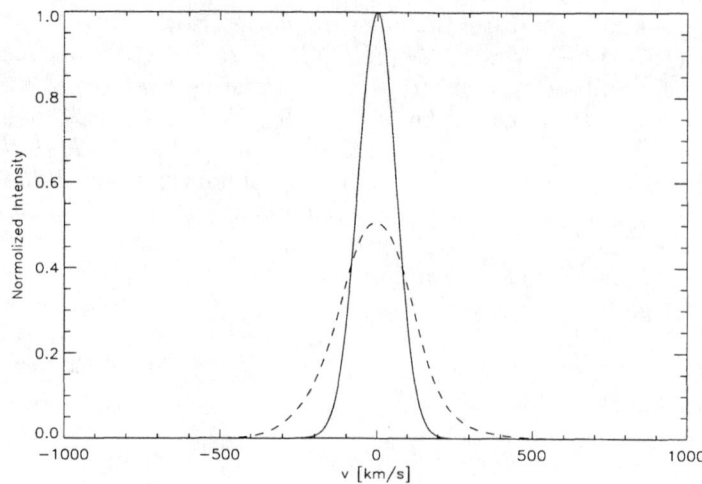

FIGURE 1. The thermal (solid line) and simulated (dashed line) line profile at $3R_\odot$

$$\frac{\Delta\lambda}{\lambda_0} = \frac{v}{C}, \tag{17}$$

where λ_0 is the wavelength of the peak intensity of the emission line, and $\Delta\lambda$ is the wavelength shift due to the Doppler effect.

The simulated line intensity is obtained by integrating the thermal emission shifted by the nonthermal wave velocity $w(t)$ for the duration τ of the simulated observation:

$$I(v) = \int_0^\tau \rho e^{-(v-w(t))^2/v_{th}^2}\, dt, \tag{18}$$

where v_{th} is the kinetic thermal speed of the particles, and the emission is weighted by the density of the species for the resonantly scattered line. The effective temperature is obtained by determining the $v_{1/e}$ of the simulated line width broadened by the nonthermal motion. An example of a simulated emission line profile broadened by nonthermal wave motions at $3R_\odot$ is shown in Figure 1.

NUMERICAL RESULTS AND COMPARISON TO UVCS OBSERVATIONS

In Figure 2 I show the spatial dependence of the solutions at $t = 39.4\tau_A = 5.0$ hours. The ϕ-component of the proton and ion velocity, the Ohmic heating associated with the broad band Alfvén waves, and the radial velocity of the protons are shown in the $\theta - r$ plane. It is evident that the Alfvénic fluctuations propagate into the coronal hole in protons and ions. The amplitude of the Alfvénic fluctuations in ions is smaller than the proton amplitude by a factor of $Z/A = 5/16$. It is also evident that the regions of largest Ohmic heating are associated with the highest amplitude Alfvén waves, with phase mixing effects at the boundary of the low density region of the coronal hole.

The radial component of the proton velocity exhibits compressional fluctuations due to the nonlinear effects of the Alfvén wave pressure, in agreement with single fluid models [9,10]. On average, largest radial velocity occurs in regions of highest Alfvén wave amplitude. However, the radial velocity amplitude predicted by the model at $3R_\odot$ is about half the velocity deduced by UVCS at this height [11]. The reason for this discrepancy may be due to the particular choices of the Alfvén wave spectrum (that affects the acceleration profile) and the other model parameters. Single fluid studies show that low frequency waves (<1 mHz) are needed to obtain the fast solar wind in coronal holes [10]. This part of the spectrum is not included in the present study, and the modeling of fast solar wind streams is left for a future work.

In Figure 3 I show the radial dependence of the effective proton and O^{5+} temperature in the central (low density) region of the coronal hole. The effective temperatures were obtained by integrating Equation (18) for protons and ions with the corresponding $< v_\phi(t) >$. The velocities were averaged in time and in space (θ) to simulate the effects of several hour exposure of the UVCS observations. The dashed lines show the time-averaged kinetic temperature for the protons and ions that was obtained from the solution of the polytropic energy equation. The dotted line shows the electron temperature.

It is evident that the radial dependence of the effective proton temperature is in good qualitative agreement with UVCS observations [1,11]. Due to the anisotropic velocity of the Alfvén waves the apparent peak proton anisotropy is about 2. However, the Alfvén waves can not explain the large perpendicular temperature of O^{5+} ions reported by UVCS [1,11]. The effective O^{5+} temperature in the model is only about 50% larger than the kinetic temperature. The amplitude of the heavy ion response to the Alfvénic fluctuation is reduced by a factor of $Z/A = 5/16$ compared to the response of the protons. Coulomb friction does not affect significantly the amplitude of the perpendicular ion fluctuations. The main effect of the Coulomb friction terms is to increase the average ion outflow speed and abundance due to collisions with outflowing protons [12].

SUMMARY AND CONCLUSIONS

Nonthermal broadening of EUV emission lines in solar coronal holes provide clues for the coronal heating and acceleration mechanism of the fast solar wind. Using the 3-fluid 2.5D MHD model I study the role of MHD waves and coronal structures on the accelerations of the solar wind. The 3-fluid model allows to investigate the self-consistent acceleration and heating of multi-temperature electron-proton-ion plasma by broad band nonlinear waves.

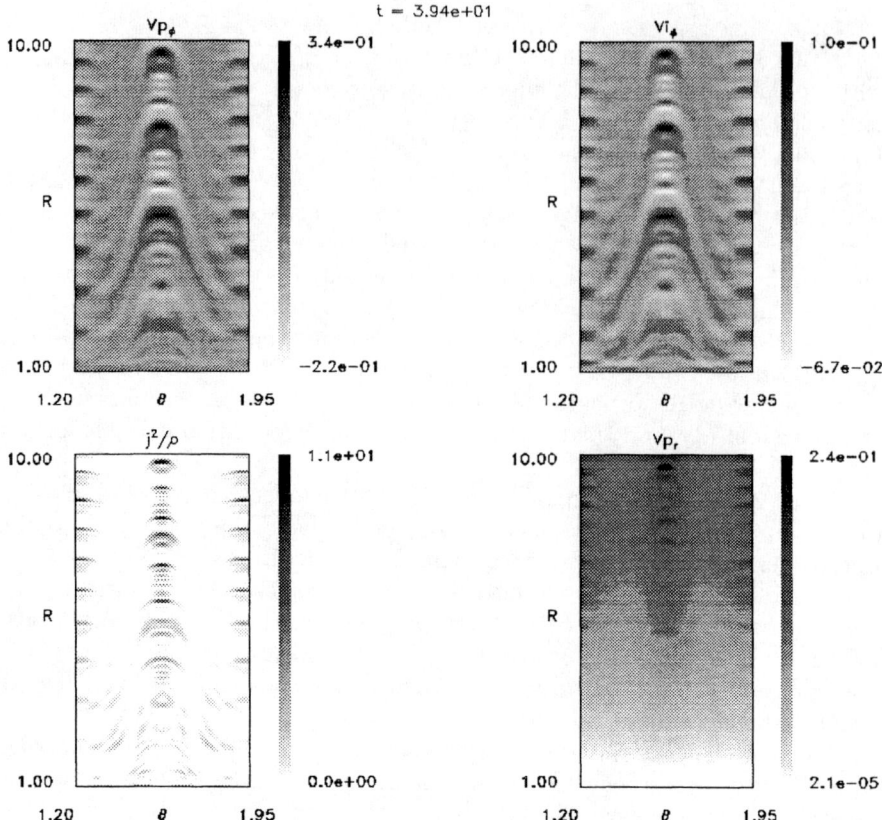

FIGURE 2. The spatial dependence of the ϕ-component of the proton and ion velocity, the Ohmic heating associated with the broad band Alfvén waves, and the radial velocity of the protons are shown. The velocities are scaled by $V_A = 1521$ km s^{-1}

I calculate the apparent proton and ion temperature and find that for reasonable Alfvén wave spectrum and coronal parameters (constrained by observations near the limb) the wave motions can account for large fraction of the observed proton line width and anisotropy. Thus, the acceleration and heating of the protons in coronal holes can occur due to low frequency MHD waves in agreement with single fluid studies [9,10].

The broad O^{5+} lines can not be explained by unresolved MHD wave motions. Due to the low abundance and collision rate of the O^{5+} ions it is possible that kinetic effects such as ion-cyclotron heating, may be responsible for the high minor ion temperature and anisotropy deduced from UVCS observations.

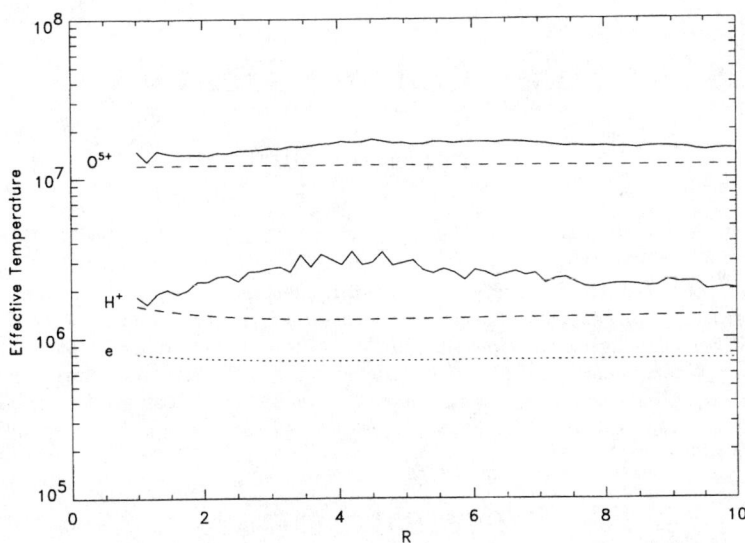

FIGURE 3. The thermal (dashed line) and effective (solid line) temperatures due to unresolved wave motions for protons and O^{5+} ions obtained from the 3-fluid model. The dotted line is the electron temperature.

ACKNOWLEDGMENTS

This work was supported by NASA SR&T (NASW-98004) Program. The author would like to thank J.M Davila for insightful discussions.

REFERENCES

1. Kohl, J.L., et al., *Astrophys. J.*, **501**, L127 (1998).
2. Ofman, L., Davila, J.M., *Astrophys. J.*, **476**, L51 (1997).
3. Davila, J.M., Ofman, L., *Sp. Sci. Rev.*, **87**, 287 (1999).
4. Ofman, L., Davila, J.M., Solar Wind Nine, S. Habbal et al. (eds), *AIP Conference Proceedings*, **471**, 405-408, AIP, New York (1999).
5. Braginskii, S. I., *Rev. Plasma Phys.*, **1**, 205 (1965).
6. Nakada, M.P., *Solar Phys.*, **14**, 457 (1970).
7. Li X., Esser, R., Habbal, S.R., Hu, Y.-Q., *J. Geophys. Res.*, **102**, 17419 (1997).
8. Parker, E.N., Interplanetary Dynamical Processes, New York, Interscience (1963).
9. Ofman, L. & Davila, J.M., *Astrophys. J.*, **476**, 357 (1997).
10. Ofman, L., Davila, J.M., *J. Geophys. Res.*, **103**, 23677 (1998).
11. Antonucci, E., *Proceedings 8th SOHO Workshop*, ESA SP-446, 53 (1999).
12. Ofman, L., *Geophysical Research Letters*, in press (2000).

Solar Corona Heating

Francesco Califano

Istituto Nazionale Fisica della Materia, Sez. A, Dip. Fisica, Università di Pisa, Italy

Abstract. Despite of the large number of models and mechanisms proposed in the literature, the problem of the heating of the solar corona is still unsolved and represents one of the challenge of solar physics. In this context, a basic question to be addressed by any viable theoretical model concerns understanding the mechanism capable of transferring "efficiently" the energy from the large injection scales to the much smaller dissipative scales in an almost ideal plasma where these scales differ by many order of magnitude.

INTRODUCTION

The solar corona is a very hot plasma at a typical temperature of $\sim 2 \div 3 \ 10^6 K$, much higher than the underlying layers, the chromosphere and the photosphere, so that nonthermal energy must be transported into the corona and dissipated in situ [1]. The high temperature coronal plasma mainly radiates in the soft x-rays range corresponding to typical wavelengths of the order of $10 - 100 \ A°$. This radiation is not homogeneous neither in space nor in time, with a wide range of spatial and time scales. Large scale structures of typical length of $10^8 - 10^{10} \ cm$ persist during a characteristic time much longer than the Alfvén time, while small bright points of dimension of the order of $10^6 \ cm$ evolve on a few Alfvén times and in some cases even more rapidly. High resolution observations have now given an image of the solar corona as a rapidly evolving dynamic plasma where energetic phenomena occur mainly on very "small" scales. If the large scale motions observed, for example, at a photospheric level can be described by the MHD equations, dissipative effects (heating) are most probably of kinetic nature. This is the direct consequence of the high conductivity of such a hot plasma where the dissipative coefficients, as for example ohmic resistivity and/or viscosity, are so small that it is hard to conceive how the energy injected on the large scales and low frequencies by the photospheric and sub-photospheric random motions can be efficiently converted into heat in the solar corona.

As a first indication, the role of dissipative effects in a fully ionized plasma can be rationalized by the value of the Lundquist (or magnetic Reynolds) number, i.e. by the ratio of the characteristic time scale of the resistive processes, $\tau_r = l^2/\eta$ and the characteristic time scale of the ideal processes (the Alfvèn time), $\tau_A = l/c_A$. Here l, η and c_A (the Alfvèn speed) are a characteristic length, velocity and ohmic resistivity of the system. Assuming $n \simeq 10^9 cm^{-3}$ as a typical plasma

CP537, *Waves in Dusty, Solar, and Space Plasmas*, edited by F. Verheest, et al.
© 2000 American Institute of Physics 1-56396-962-9/00/$17.00

density in the solar corona, the ohmic resistivity can be roughly estimated as $\eta \simeq 10^{13} T^{-3/2} cm^2 \ s^{-1}$; therefore, for a plasma temperature $T \simeq 10^6 \ K$, the Lundquist number of the large scale structures in the solar corona turns out to be very large, $S \sim 10^{15}$. In these conditions, the dissipative terms which are proportional to S^{-1} in the MHD equations are too small with respect to the inertial terms and the system can be considered as non-dissipative. However, the dissipative terms are characterized by the highest derivatives in the equations, and the only chance to transform part of the energy into heat (i.e. to dissipate) is to efficiently transfer the energy towards the small scales where dissipation becomes efficient. Here efficiently means at a rate independent of the numerical value of the resistive and/or viscous coefficients.

This is the crucial problem of the solar corona heating problem: *how is it possible to transfer the energy on the very small diffusive length scales of the order of* $1 \div 10^3 cm$ *where the local Lundquist number is of order of unity ?* Furthermore, such dissipative scales are comparable to the characteristic kinetic length scales, as for example the electron skin depth or the Debye length, adding further difficulties in the study of the dissipative mechanism.

In a MHD system, the possibility of building up a strong nonlinear energy cascade with a corresponding extended inertial spectrum, has been proved numerically but only in the limit of a homogeneous system with large amplitude initial perturbations [2], or by assuming the initial condition of a 3D magnetic sheared layer where magnetic reconnection develops [3] (we recall that in 2D the occurrence of the tearing mode do not generate turbulence and eventually leads to the coalescence of the magnetic islands [4] instead of producing small scale dissipative structures). However, the initial conditions used in these cases seems not to adapt to the solar corona where the external driver is given by the slow motions of the magnetic field lines rooted in the dense photospheric plasma or by low-frequencies, long wavelengths MHD waves of relatively moderate amplitude.

The main difficulty related to the understanding of the coronal heating mechanism is given by the impossibility of performing "sufficiently" high resolution observations (i.e. observations at typical wavelengths of the order of the dissipative length scales), capable to shed light on the physical dissipative process at work in the solar corona. As a consequence, no realistic models, even very simplified, have been developed so far, while a number of "conceptual" models, starting from the available data, try to roughly describe the small scale dynamics and to derive all the possible consequences on the mean dynamics in order to fit the observational large scale constraints.

Nevertheless, some basic ideas have now reached a large general consensus. First of all, photospheric and sub-photospheric random motions are considered as the energy source of the heat deposed into the corona since the corresponding energy flux (of about $10^7 \ ergs \ cm^{-2} \ s^{-1}$ [5]) flowing outwards from the photosphere towards the outer layers of the sun is large enough to compensate for the radiative and conductive energy losses in the corona. The second important point now accepted in the solar physics community is the key role of the magnetic field as (1) the link between the energy source, the photosphere, and the region where the energy is converted into heat, the corona, as well as (2) the main agent in the energy transfer process from the large injection scales to the small dissipative scales. The strict connection between the magnetic field and coronal heating events has been experimentally affirmed by satellite observations combined with magnetogram images, which show in the solar corona a large number of x-ray emitting plasma regions strongly correlated to the magnetic field coronal structures. Roughly speaking, in the x-ray range, the corona appears brighter everywhere the magnetic field is concentrated [6]. However, even if there is a large consensus on the fundamental role of the magnetic field in the energy dissipation process, a strong debate is still open on the precise mechanism, as outlined by the impressive quantity of literature on the subject. Magnetic reconnection [7], resonant absorption [8] and phase mixing [9] of MHD waves, non-linear (turbulent) MHD interactions [10], tangential discontinuities (or ideal MHD singularities) [11], development of MHD turbulence in coronal loops [12], Alfvèn waves [13], high frequency Alfvèn waves [14], and so on are only the mostly known processes invoked to understand the nature of the mechanism acting to dissipate the energy in the solar corona. A different approach to the problem of the heating of the solar corona is the so called "nanoflare heating picture" where one consider the formation of a current sheet (or any small scale dissipative process) just as a single, energetically insignificant case of a very large number of events which statistically contribute to a continuous heating of the global system. Then, at least as a first approach, one could even ignore the complicate details of the small scale dissipative processes, and look at their statistical distribution by using a simplified phenomenological theory [15].

By considering the specific case of the solar corona, an important parameter of the problem is the characteristic time τ_{pht} of the photospheric random perturbations with respect to the Alfvèn time τ_A. In the limit of "slow" perturbations, $\tau_{pht} \gg \tau_A$, the large scale coronal structures can evolve through a series of magnetostatic force-free equilibria. The typical case is that of a coronal loop continuously stressed at its footpoints. Eventually strong current sheets are generated near the separatrix and magnetic energy is released, for example, via magnetic reconnection. However, further dynamical investigations of the evolution of such magnetostatic configurations are necessary in order to estimate the characteristic time needed to generate the current sheets and to dissipate the magnetic energy.

In the limit, instead, of "fast" photospheric perturbations, $\tau_{pht} \ll \tau_A$, most of the energy is converted into MHD waves which propagate outwards. However, only Alfvèn waves due to their highly anisotropic dispersion relation are able to reach the solar corona, the other MHD wave modes being dissipated or reflected at lower altitudes. Then, the problem is how to dissipate the energy carried by the Alfvén waves due to their strong dissipative inefficiency in a perfectly homogeneous plasma, even when considering nonlinear interactions. For this reason, wave heating theories faced with the problem of how to speed up the dissipative effects, have mostly considered the interaction of Alfvén waves with an inhomogeneous supporting medium. In this context, the most promising mechanism are, as

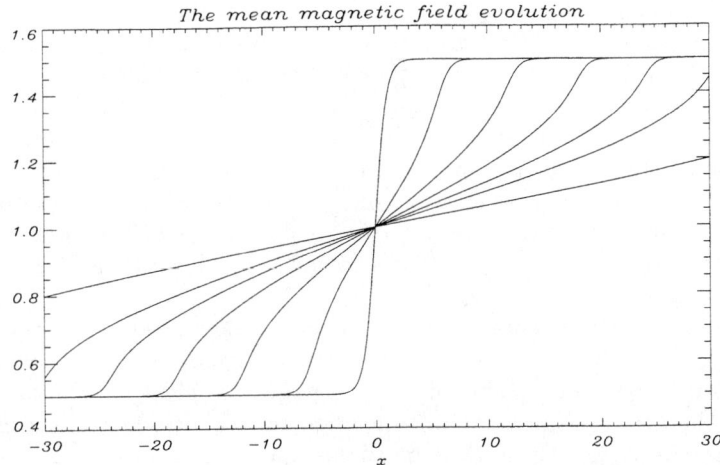

FIGURE 1. The progressive smoothing of the inhomogeneity of the magnetic field in a resistive decay case (energetically isolated system). The initial profile of the magnetic field is $< B_z(x, t = 0) >= 1 + 0.5 \tanh(x)$.

already cited, the phase mixing and the resonant absorption which occur whenever the wave mode propagates along a magnetic field inhomogeneous in the transverse direction. In particular, the phase mixing mechanism is driven by the spatial variation of the phase speed which transversely corrugates more and more the wave front, so producing smaller and smaller spatial scales as time evolves eventually reaching the dissipative length scale in a characteristic time which scales with the Lundquist number as $\sim S^{1/3}$. The resonant absorption, instead, is a normal mode solution driven by the resonance between the frequency of the eigenmode and the local (in space) value of the Alfvén speed; as a result, the energy is concentrated in a thin boundary layer of typical width $\sim S^{1/3}$ at the point where $\omega/k = c_A$. The interesting feature of the resonant absorption mechanism is that the dissipation rate of the corresponding normal mode solution is independent of S (see [16] and references therein). However, ab initio numerical simulations have shown that the transient time necessary to set up this dissipative mechanism is exceedingly long if only linear interactions are considered [17].

A first attempt to investigate the influence of nonlinear interactions has been done in Ref. [18] where the decay of moderate amplitude, long wavelengths MHD perturbations in a nonuniform medium has been addressed. In particular, even if limited to relatively small Lundquist numbers ($S \sim 10^3$), it has been suggested that the time necessary to develop a current sheet (the resistive analog of the ideal singularity) is independent of the actual value of the Lundquist number, as conjectured twenty years ago [19]. Another important new result of that investigation is the resonant interaction of long wavelengths Alfvèn waves with the inhomogeneous field which opens the possibility of tapping and dissipating the energy contained in the equilibrium, the waves acting mostly as a catalyst; it is worth to stress that in this work the initial amplitude of the waves is not infinitesimal, but still much smaller (two order of magnitudes) than the equilibrium field. However, in that *resistive decaying* study, no energy was introduced into the system after the initial time thus preventing the system to reach a time-asymptotic regime. The background field evolves towards an increasingly uniform state thus reducing and eventually stopping the resonant driving mechanism. The typical smoothing of the magnetic shear occurring in a resistive decaying simulation is shown in Fig. 1 where we plot the equilibrium field, initially modeled by the function $< B_z >= 1 + 0.5$ $\tanh(x)$, at different times.

In order to relax the assumption of an energetically isolated system, and to inject the energy on the large scale of the system, as suggested by the observations, we have decoupled the mean inhomogeneous magnetic field from the fluctuations and imposed that the initial magnetic shear do not change in time, even in the full nonlinear regime [20]. Even if the mechanism which sustains the initial equilibrium is not specified, we consider that our simplified model is an interesting and sufficiently general prove of the possibility (1) of tapping the energy contained in the large scale fields and (2) of reaching an asymptotic regime in a characteristic time independent of the dissipative coefficients with (3) a dissipative rate also independent of the dissipative coefficients. It is worth to stress that no neutral lines are present in the initial configuration which is therefore stable with respect to tearing modes. In the following, this simplified, but sufficiently general, model is briefly discussed.

THE MODEL

We address the important point of principle that, under typical conditions for the solar corona, it is possible to transfer and to dissipate the energy stored in the large scale field at a rate independent of the dissipative coefficients by a rather general mechanism of resonant interaction between moderate amplitude, long wavelength Alfvèn waves and the background inhomogeneity. Indeed, the possibility of building up a strong nonlinear energy cascade with a consequent extended inertial spectrum has been already demonstrated numerically [2], but only in the unrealistic limit (at least for the solar corona) of a homogeneous mhd system with large amplitude initial perturbations.

As suggested by the observations where the large scale fields seem to maintain their inhomogeneous shape for times much longer than the Alfvèn time, we fix the background field in time by assuming that an unspecified "external mechanism" is at work to compensate the modifications resulting from the dynamical nonlinear evolution of the system. The energy injection by this "external mechanism" is however small (a few percent) with respect to the initial free magnetic energy. Since the system is no longer energetically isolated, we refer to this case as a driven dissipative evolution; on the other hand, as discussed in the previous Section, in the absence of such external energy injection, the available free energy is limited

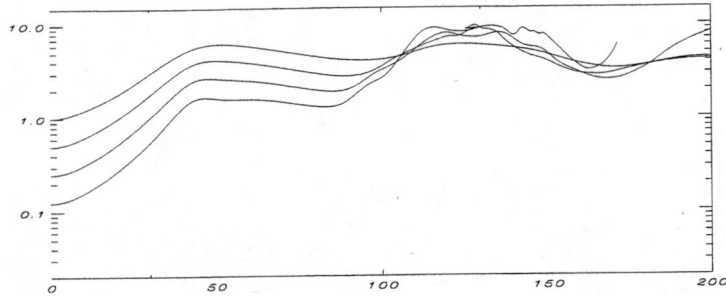

FIGURE 2. The time evolution of the dissipation rate for $S = 100,\ 200,\ 400,\ 800$.

and the magnetic shear would be progressively smoothed out (see Fig. 1) as a consequence of the transfer of the free energy towards the small scales (dissipative decay) [18,21].

We consider the incompressible magnetohydrodynamic equations in the Elsasser formulation and we assume that the Reynolds number is equal to the Lundquist number. We use the spatial scale of variation of the initial magnetic field, $\bar{l}$, the average Alfvén speed, c_A, and a typical density, $\bar{\varrho}$, as the characteristic length, velocity and density of the problem. Then, the dimensionless equations read:

$$\frac{\partial \mathbf{z}^\pm}{\partial t} = (\pm \mathbf{B}_0(x) - \mathbf{z}^\mp) \cdot \nabla(\pm \mathbf{B}_0(x) + \mathbf{z}^\pm) - \nabla\Pi + \frac{1}{S}\nabla^2 \mathbf{z}^\pm, \tag{1}$$

$$\nabla \cdot \mathbf{z}^\pm = 0, \quad z^\pm = \mathbf{v} \pm \mathbf{b}, \quad \mathbf{B}_0(x) = [1 + 0.5\tanh(x)]\mathbf{e}_z, \tag{2}$$

where $\mathbf{z}^\pm$ are the Elsasser variables and $\mathbf{v}$ and $\mathbf{b}$ are the velocity and the magnetic field fluctuations, respectively. In these equations $\mathbf{B}_0$ is the time independent equilibrium field, Π is the total kinetic + magnetic pressure and S is the Lundquist number. The Eqs. (1)-(2) are integrated in the 2-D domain (x, z) of dimension $[-L_x, L_x] \times [0, 2\pi/k_{0,z}]$, where $L_x \geq 2\pi/k_{0,z}$, and $k_{0,z}$ is the wavenumber of the initial perturbation. The time independent equilibrium magnetic field $\mathbf{B}_0(x)$ (see last term in Eq. 2), is perturbed by a "small" amplitude, long wavelength disturbance of the form $z_x^+ = 0.015\sin(k_{0,z}z)$, $k_{0,z} = 0.02$, and $z_x^- = z_z^\pm = 0$. We use periodic boundary conditions in the z direction and $\partial z_z/\partial x = 0$, $\partial z_x/\partial x = -\partial z_z/\partial z$ in the x inhomogeneous direction.

The numerical algorithm is a modified version of the explicit *Adams-Bashford III* and we use forth order finite differences in the x direction on a nonuniform grid and fast Fourier transform in the z direction (see [22] and [18] for further details).

NUMERICAL RESULTS

In Fig. 2 we show the time evolution of the dissipation rate *vs.* time for four runs with $S = 100,\ 200,\ 400,\ 800$. The decreasing values of the curves at $t = 0$

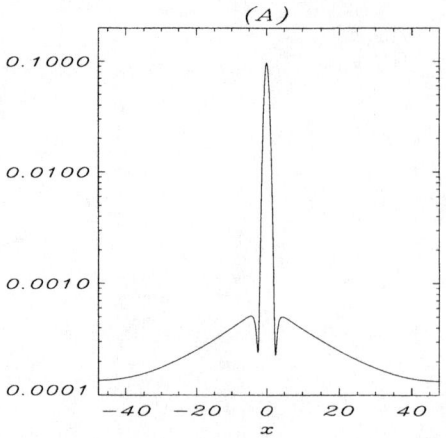

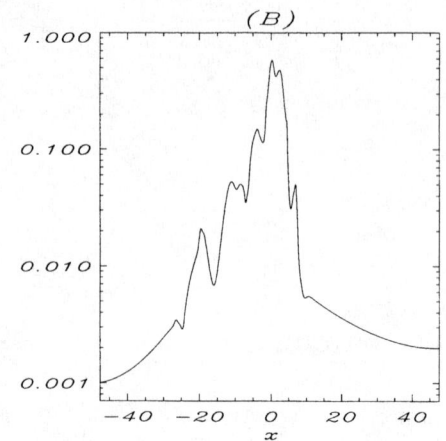

FIGURE 3. The averaged squared fluctuations $< (z_z^-)^2 >$ vs. x (here $<>$ mean averaged along the z-axis) during (A) the linear and (B) the nonlinear phase.

correspond to increasing values of the Lundquist number. This figure shows the existence of three distinct regimes, the linear one, $t < 50$, the intermediate regime, $50 < t < 110$, and the nonlinear regime, $t > 110$. In the linear phase, the evolution of a long wavelength perturbation, $k_{0,z} \ll 1$, in a nonuniform medium is driven by the resonant absorption, while the phase mixing mechanism is much less efficient. As a consequence, the energy is more and more concentrated in the center of the inhomogeneous region thus producing very strong peaks in the fluctuations (i.e. small spatial scale in the x direction), despite the initial perturbation is completely flat along the x-axis. A typical shape of the resulting squared fluctuations (averaged along the z-axis) is shown in Fig. 3, frame (A). On the other hand, no energy injection on smaller spatial scales along the magnetic field (i.e. along the z-axis) occurs, $k \sim k_{0,z}$. At the end of this phase, the dissipation rate which scales as S^{-1} at $t = 0$ is much increased and scales now as $S^{-7/2}$. More important, the local amplitude of the perturbation becomes now comparable with the equilibrium field, and the system enters in the intermediate regime where the nonlinear interactions start to dominate the plasma dynamics, producing a strong energy cascade along the magnetic field, so exciting larger horizontal wavevectors, $k \gg k_{0,z}$. The most important result of this intermediate phase is the build up of current sheets together with magnetic sheared regions (characterized by the presence of neutral lines) in a characteristic time independent of the Lundquist number. The system finally reaches the strongly nonlinear regime where the dissipation rate becomes independent of the Lundquist number, as shown in Fig. 2 for $t > 110$. In this phase strong current sheets (not shown here) are formed and dissipated close to the regions where neutral lines are dynamically generated. A typical magnetic sheared configuration observed in the simulations is shown in Fig. 4 (in a small part of the integration domain) where the arrows represent the total magnetic field (i.e.

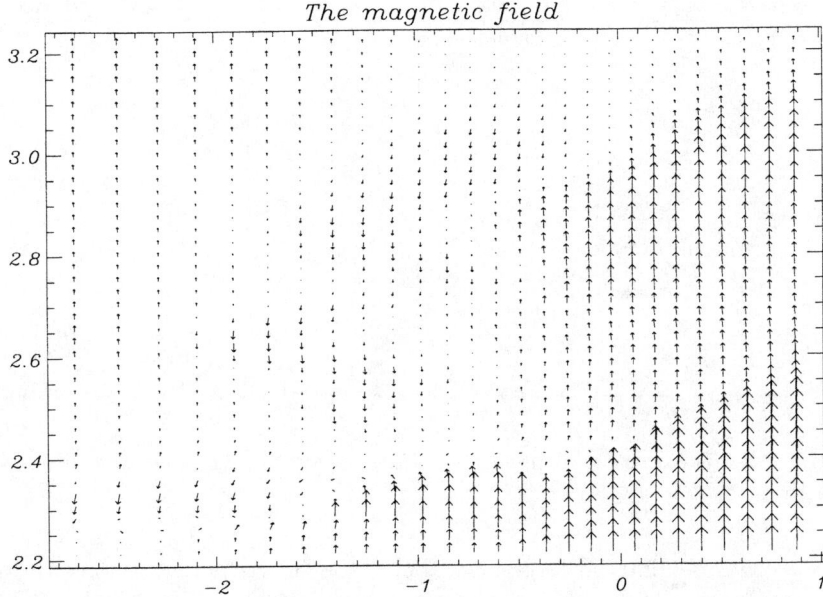

FIGURE 4. The total magnetic field represented by the arrows in a small fraction of the (x, y) plane during the strongly nonlinear phase. At $t = 0$ all the arrows are directed upward.

including the time independent field $\mathbf{B}_0(x)$). The corresponding typical shapes of the magnetic fluctuations along the inhomogeneous direction are shown in Fig. 5. These structures are reminiscent of the linear resonant eigenmodes (and also quite similar to the tearing mode).

CONCLUSIONS

The heating of the solar corona is an old, still unsolved problem where a realistic model including the large injection scales and the small dissipative scales is far to be realized. Here, in the framework of a wave theory approach, we have discussed one of the basic related problems of how to transfer and dissipate the energy stored in the large magnetic field independently of the numerical value of the dissipative coefficients. In our model we have assumed that the inhomogeneous large scale magnetic field is sustained by some unspecified external mechanism for times longer with respect to the few Alfvèn times needed to reach the asymptotic regime. As suggested by the observation, this configuration is perturbed by long wavelength Alfvèn waves of relatively moderate amplitude (a few percent with respect to the equilibrium). The main results of this forced anisotropic 2D MHD system are the following. First, due to the resonant interaction between the waves and the inhomogeneous background, it is possible to tap and to dissipate the free

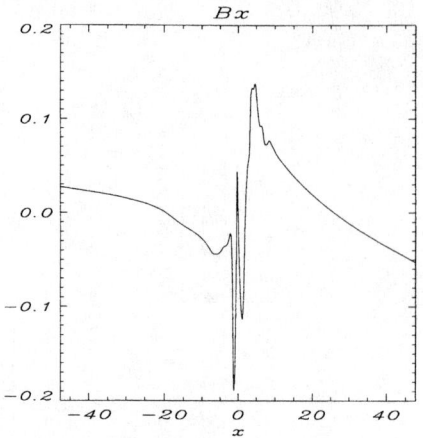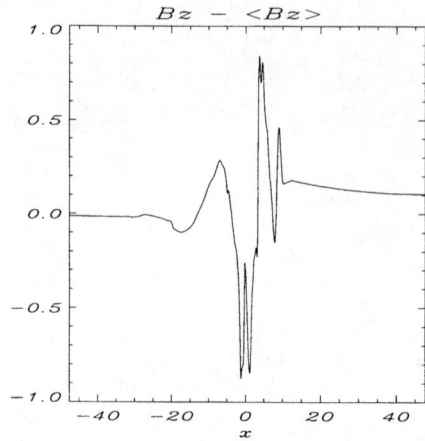

FIGURE 5. The x and z components of the magnetic fluctuations, first and second frame, respectively, *vs.* x at $z = 3.1$.

energy contained in the large scale field, the initial perturbations acting mostly as a catalyser. Second, the system naturally forms regions of sheared magnetic field (in a finite time) where strong current sheets dissipate the energy at a rate indipendent of the Lundquist number. Finally, the characteristic time necessary to reach this regime is also independent of the Lundquist number. As these properties were already suggested in the case of a self-consistent evolving large scale magnetic field (resistive decaying), we conclude that the mechanism is independent of the particular choice of the external forcing, at least for the values of the Lundquist number used in our simulations.

REFERENCES

1. Wheatland M.S, Sturrok P.A., *Astrophys. J.*, **482**, 510 (1997)
2. Fyfe D., Joyce G., Montgomery D., *J. Plasma Phys.*, **17**, 317 (1977); Biskamp D., Welter H., *Phys. Fluids B*, **1**, 1964 (1989); Politano H., Pouquet A., Sulem P., *Phys. Fluids B*, **1**, 2330 (1989)
3. Strauss H., *Astrophys. J.*, **326**, 412 (1988)
4. Malara F., Veltri P., Carbone V., *Phys. Fluids B*, **3**, 1801 (1991)
5. Withbroe G. L., and Noyes R. W., *Ann. Rev. Astr. Astrophys.*, **15**, 363 (1977)
6. Falconer D.A., Moore R.L., Porter J.G., Gary G.A., Shimizu T., *Astrophys. J.*, **482**, 519 (1997); Falconer D.A., Gary G.A., Moore R.L., Porter J.G., *Astrophys. J.*, **528**, 1004 (2000)
7. Giovanelli R. G., *Nature*, **158**, 81 (1946); Coppi B., *Phys. Lett.*, **11**, 226 (1964); Furth H. P., Killeen J., Rosenbluth M. N., *Phys. of Fluids*, **6**, 459 (1963); Parker E. N., *J. Geophys. Res.*, **62**, 509 (1957); Sweet P. A., *IAU Symp. 6 Electr. Phenom. in Cosmical Plasma*, ed. Lehnert B., (New York, Cambridge University Presse), p. 123

(1958); Biskamp D., *Phys. of Fluids*, **29**, 1520 (1986); Priest E. R., Forbes T.G., J. Geophys. Res., **97**, 16,757 (1992); Priest E. R., *Phys. of Plasmas*, **4**, 1945 (1997)

8. Hasegawa A., Uberoi C., *The Alfvén Wave*, (Washington DC: US Department of Energy) (1982); Lee, M., Roberts, B., *Astrophys. J.*, **301**, 430 (1986); Hollweg, J., *Astrophys. J.*, **312**, *880* (1987); Califano F., Chiuderi C. and Einaudi G., *Phys. Plasmas*, **1**, *43* (1994); Steinolfson, R.S., Davila, J.M., *Astrophys. J.*, **415**, *354* (1993); Ofman, L., Davila, J.M. and Steinolfson, R.S., *Geophys. Res. Letts.*, **21**, *2259* (1994)

9. Heyvaerts, J. and Priest, E.R., *Astr. Astrophys.*, **117**, 220 (1983)

10. Kraichnan R., *Phys. Fluids*, **8**, 1385 (1965); Frish U., Pouquet A., Leorat J., Mazure A., *J. Fluid Mech.*, **68**, 769 (1975); Pouquet A., Frish U., Leorat J., *J. Fluid Mech.*, **77**, 321 (1976); Dobrowolny M., Mangeney A., Veltri P., *Phys. Rev. Lett.*, **45**, *144* (1980); Matthaeus W., Zhou Y., *Phys. Fluids B*, **1**, 1929 (1989); Grappin R., Frish U., Leorat J., Pouquet A., *Astron. Astrophys.*, **102**, 6 (1982); Zhou Y., Matthaeus W., *J. Geophys. Res.*, **95**, 10291 (1990); Matthaeus W., Oughton S., Pontius D., Zhou Y., *J. Geophys. Res.*, **99**, 19267 (1994)

11. Parker E.N., *Astrophys. J.*, **174**, 499 (1972); Van Ballegooijen, A.A., 1985: *Astrophys. J.*, **298**, 421 (1985); Poedts S., Kerner W., *Phys. Rev. Lett.*, **66**, 2871 (1991); Bhattacharjee, A., Wang X., *Astrophys. J.*, **372**, 321 (1991); Longcope D.W., Sudan R.N., *Astrophys. J.*, **384**, 305 (1992); Parker E.N., *Astrophys. J.*, **407**, 342 (1993); Wang X., Bhattacharjee A., *Astrophys. J.*, **420**, 415 (1994); Amari T., Luciani J. F., *Phys. Rev. Lett.*, **84**, 1196 (2000)

12. Gomez D.O., Ferro Fontan C., *Astrophys. J.*, **394**, 662 (1992); Heyvaerts, J. and Priest, E.R., *Astrophys. J.*, **390**, 297 (1992); Dmitruk P., Gomez D.O., *Astrophys. J.*, **527**, L63 (1999)

13. Halberstadt G. , Goedbloed J.P., *Astr. Astrophys.*, **301**, 577 (1995); Belien A.J.C., Poedts S., Goedbloed J.P., *Phys. Rev. Lett.*, **76**, 567 (1996)

14. Axford W.I., McKenzie J., *Solar Wind Seven*, ed. Marsch E., Schwenn R., (Pergamon Press), p. 1 (1992); Tu C.-Y. Marsch E., *Solar Phys.*, **171**, 363 (1997); Ruzmaikin A., Berger M. A., *Astr. Astrophys.*, **337**, L9, (1998); Shukla P.K., Bingham R., McKenzie J.F., Axford W.I., *Solar Phys.*, **186**, 61 (1999)

15. Parker E.N., *Astrophys. J.*, **330**, 474 (1988); Lu E.T., Hamilton R.J., *Astrophys. J.*, **380**, L89 (1991); Einaudi G., Velli M., Politano H., Pouquet A., *Astrophys. J. Lett.*, **457**, L113 (1996); Vlahos L., Georgoulis M, Kluiving R., Paschos P., *Astron. Astrophys.*, **299**, 897 (1995)

16. F. Califano, Chiuderi C., Einaudi G., *Phys. Plasmas*, **1**, 43 (1994)

17. Malara, F., Veltri, P., Chiuderi, C., Einaudi, G., *Astrophys. J.*, **396**, 297 (1992); Einaudi, G., Chiuderi, C., Califano, F., *Adv. Space Res.*, **13**, 85 (1993)

18. Einaudi G., Califano F., Chiuderi C., *Astrophys. J.*, **472**, 853 (1996)

19. Matthaeus, W. H., Montgomery, D., *Ann. N. Y. Acad. Sci.*, **357**, 203 (1980)

20. Califano F., Chiuderi C., *Phys. Rev. E*, **60**, 4701 (1999)

21. Califano F., Chiuderi C., *Physica Scripta* T**75**, 197 (1998)

22. F. Califano, *Comp. Phys. Comm.*, **99**, 29 (1996)

Kelvin-Helmholtz instability and resonant flow instability in a 1-dimensional coronal plume model

J. Andries and M. Goossens

Centre for Plasma Astrophysics, Celestijnenlaan 200B, B-3001 Leuven, Belgium

Abstract. In a previous paper we have illustrated the concept of resonant flow instability of the trapped modes both in a 1-D slab model and a 1-D cylindrical model for a coronal plume in a cold plasma. We found that much larger values of the velocity shear are needed for Kelvin-Helmholtz than for resonant instability to occur. The aim of this paper is to study the effect of a non-zero plasma pressure on the eigenmodes of the plume structure. We show that the instability most probably to occur in coronal plumes is due to the resonant coupling of slow body modes to local resonant Alfvén waves. These instabilities could lead to disruption of the coronal plumes and to the mixing with interplume plasma.

INTRODUCTION

Plumes are bright quasi-radial rays between one and several $R_{\odot}$ in coronal holes. Inside 10 $R_{\odot}$ plumes flow much more slowly than interplume plasma. Being bright in white light, plumes are denser than interplume plasma. All this suggests that plumes should be observable in the interplanetary medium. However, the high speed solar wind coming from coronal holes is remarkably smooth (for references concerning these plume facts see [1]). Therefore, plume and interplume plasma must be mixed somewhere close to the Sun.

There are probably numerous processes which could lead to plume/interplume mixing. Nevertheless, Suess [2] states that it is easy to show that plumes will be subject to MHD Kelvin-Helmholtz shear instabilities beginning at around 10 $R_{\odot}$ and that these instabilities will otherwise lead to disruption of the plumes and to the mixing with interplume plasma. These Kelvin-Helmholtz instabilities could be a potential source for some of the Alfvénic fluctuations observed in the solar wind. However, no published simulations or detailed evaluations explicitly address parameters appropriate for coronal plumes, so considerable analysis remains to be done with respect to this hypothesis [2].

The focus of this manuscript is on another possible process, also triggered by the velocity shear in the mass flow and which could be responsible for the

CP537, *Waves in Dusty, Solar, and Space Plasmas,* edited by F. Verheest, et al.

plume/interplume mixing : the resonant flow instability [3]. The aim is to show how a global MHD wave trapped in the plume structure can become overstable due to resonant coupling to Alfvén waves in the presence of a background velocity shear. This resonant flow instability occurs for velocity shears below the onset value for the Kelvin-Helmholtz instability.

In [1] we recently illustrated the concept of the resonant flow instability in coronal plumes by putting $\beta = 0$ (references to earlier papers can be found in that paper). This is a good approximation in the low corona. However we will show that taking into account a small but non-zero β will create new modes and instabilities that set in at even lower velocity shears. We consider both a 1-D slab model and a 1-D cylindrical model for a coronal plume in which we ignore the geometric spreading of the coronal plume. Although the geometric spreading of the plume structure and the longitudinal gradients will certainly have an influence on the resonant flow instability mechanism, these features will not destroy the mechanism and should be taken into account in a next step. Moreover the stability analysis of the 1-D models can be seen as a local stability analysis of the plume structure around a certain height.

By deriving analytically the solution around the resonance in resistive MHD, the effect of the velocity shear on the damping rate can easily be investigated and it clearly shows how and when the resonant instability occurs. In the presence of a background flow it can be anticipated that the damping of the MHD mode due to resonant wave transformation is altered, since the flow does not only Doppler shift the continuum frequencies but it also affects the energy of the eigenmodes. At the resonance the flow could drain energy away from the mode, which additionally increases the wave damping, but the flow could also be an energy source so that the mode gains energy and becomes overstable (e.g [3]).

The paper is organized as follows. In Section 1 we describe the 1-D cylindrical model for the coronal plume. Section 2 is a summary of the conclusions drawn in [1]. The analysis is extended to non-zero β in section 3. The outcome of the eigenmode calculations is split up into two cases. First we take no non-uniform intermediate region between the plume and interplume regions into account, and show how Kelvin-Helmholtz instability occurs by aid of the analytical dispersion relation. Then we take into account the non-uniform region which will cause resonant instability. Section 4 draws the conclusions regarding coronal plumes.

I 1-D CYLINDRICAL MODEL

All equilibrium quantities depend on the r-coordinate only. Since in coronal holes the plasma pressure can be neglected in comparison to the magnetic pressure, we have assumed the plasma β parameter to be zero in [1]. Our present goal is to investigate the influence of non-zero β. The field lines are orientated in the direction of the z-axis and the magnitude of the magnetic field is taken to be uniform over the whole space, although a different magnitude of the magnetic field in the plume and interplume region would not change the equations. The higher density inside the plume causes a dip in the Alfvén and sound speed profile across the configuration. The characteristic length scale of the plume is given by $2d$ whereas the thickness of the transition layer between the plume and interplume regions is given by L. In both the interior region of the plume and the exterior interplume region the Alfvén and sound speed is assumed to be uniform. In the transition layer the Alfvén and sound speed change linearly. The background mass flow inside the plume is smaller than that outside the plume and changes discontinously at $x = d$. In the non-uniform layer the flow speed is equal to the external flow speed. All calculations are performed in a reference frame moving with the mass flow outside the plume. Hence in this reference frame the interplume region is static whereas the plume has a background mass flow V in the direction opposite to the orientation of the z-axis. We can Fourrier analyse with respect to the ϕ and z coordinates and time, k_z and m are the longitudinal and azimuthal wave number, ω is the frequency.

Because of the very high values of the Reynolds numbers for the solar coronal conditions, dissipation due to the finite electrical resistivity and viscosity can be ignored except in narrow layers of steep gradients (e.g. around resonances). Outside these dissipative layers the MHD equations can be reduced to the following two coupled first order differential equations for the perturbations of the normal component of the Lagrangian displacement ξ_x and of the total pressure perturbation P' [4]:

$$D\frac{\mathrm{d}r\xi_r}{\mathrm{d}r} = C_1 r\xi_r - C_2 r P' \tag{1}$$

$$D\frac{\mathrm{d}P'}{\mathrm{d}r} = C_3\xi_r - C_1 P' \tag{2}$$

where, for the present gravitationless cylindrical configuration:

$$D(r;\omega) = \rho(c^2 + v_A^2)(\Omega^2 - \omega_C^2)(\Omega^2 - \omega_A^2)$$
$$C_1(r;\omega) = 0$$
$$C_2(r;\omega) = \Omega^4 - \omega^2(c^2 + v_A^2)(\frac{m^2}{r^2} + k_z^2) + (\frac{m^2}{r^2} + k_z^2)c^2\omega_A^2$$
$$C_3(r;\omega) = D\rho(\Omega^2 - \omega_A^2)$$

The other perturbed quantities $(\rho_1, p_1,...)$ can be computed once ξ_x and P' are known. The sound speed and the Alfvén speed are defined as $c^2 = (\gamma p)/\rho$ and $v_A^2 = B^2/(\mu_0\rho)$ where the ratio of specific heats $\gamma = 5/3$, as usual. $\omega_A = k_z v_A$ and $\omega_C = \sqrt{c^2/(c^2 + v_A^2)}\omega_A$ denote the Alfvén and cusp frequency respectively. $\Omega = \omega - k_z U(r)$ denotes the frequency Doppler shifted by the background flow.

The set of ordinary differential equations (1-2) has mobile regular singularities at the positions r_A and/or r_C where $D(r)$ vanishes:

$$\omega = k_z U(r) \pm \omega_A(r_A) \quad \text{and/or} \quad \omega = k_z U(r) \pm \omega_C(r_C) \tag{3}$$

As both $\omega_A(r)$ and $\omega_C(r)$ are functions of r, they define two continuous ranges of frequencies referred to as the Alfvén continuum and the slow or cusp continuum respectively. These continua are now Doppler shifted due to the presence of the flow as indicated by equations (3). In the reference frame moving with the background flow in the interplume region and in the transition layer the Alfvén continuum is not Doppler shifted. A plasma β different from zero introduces the slow wave continuous spectrum besides the Alfvén spectrum and modifies the spectrum of trapped waves but the mechanism of the resonant flow instability remains the same as discussed in [1].

From a physical point of view, the conditions (3) mean that the eigenmodes resonantly interact with one of the two continua and also with the flow. A discussion about the mechanisms by which energy is exchanged between MHD waves and a streaming background plasma is for instance given by Walker [5]. This will be discussed in a subsequent paper. The interaction with the continua in the absence of the flow causes damping of the eigenmodes due to resonant wave transformation and we talk about quasi-modes with complex eigenfrequency $\omega = \omega_r + i\omega_i$ with $\omega_i < 0$ in this case ([6] and references therein). A flow not only Doppler shifts the continuum frequencies but it can also affect the energy of the eigenmodes. At the resonance, the flow can namely drain energy away from the modes, which additionally increases the wave damping, but the flow can also be an energy source in which case the modes gain energy and become overstable, i.e. $\omega_i > 0$.

In ideal MHD, the solutions diverge at the resonance points as defined by (3). These singular solutions are nothing spectacular but just tell us that a basic ingredient is missing in the mathematical formulation of the problem. In reality, dissipation is important in a narrow layer embracing the resonant magnetic surface [7,8]. To remove the singularity from the ideal MHD equations (1-2) we include, for regions close to the Alfvén resonances, the resistive terms into the equations. For very high Reynolds numbers, these resistive equations represent a singular perturbation problem which can be solved analytically in the dissipative layer and in two overlap regions (to the left and the right of the dissipative layer), where ideal MHD is valid too [6,9]. The dissipative solution around the Alfvén and cusp resonance can be found in [3] (see also [6]).

For the interior of the plume and the interplume region where all the equilibrium quantities are constant, the solutions can be found in terms of Bessel and Hankel functions.

The procedure for solving the eigenvalue problem for the plume oscillations in our model is explained in [1]. In the calculations, length, speed, density and magnetic field strength are non-dimensional and scaled with respect to d, v_{Ae}, ρ_e and B respectively. When L is different from zero, it is taken to be 0.1d.

II RESULTS FOR $\beta = 0$

First of all in absence of the nonuniform region we found in [1] that the threshold for KH-instability in the slab equivalent of the model is much larger than the Alfvén speed. We did not explicitly address any value to the threshold velocity because of the uncertainty about it in the cylindrical model. In that case we found no KH-instability within the range of the non-leaky modes. It was not clear if KH-instability would occur in the range of leaky modes. In the mean while this is clarified: for the cylindrical configuration instability will set in at the velocity shear for which the frequency will shift into the leaky region (see further discussion for $\beta \neq 0$). The boundary of this region is the external Alfvén frequency. On the other hand the frequency of the first mode as seen from a reference frame fixed to the plume-plasma is independent of the flow velocity and approximately the internal Alfvén frequency. The frequencies in the different frames are related by the Dopplershift formula. Therefore this mode will shift into the leaky region for velocity shears larger than the sum of the internal and external Alfvén velocity. A closer look to the results of the slab configuration reveals the same conclusion. It is clear that the previously not investigated effect of the external magnetic field is to raise the threshold from the internal Alfvén velocity to the sum of the internal and external Alfvén velocity.

In the presence of the non-uniform region we illustrated the resonant flow instability process. Instability sets in when the first forward mode shifts into the backward Alfvén continuum. The boundary of the continuum is the internal Alfvén frequency. And thus resonant instability will set in before KH-instability more precisely for $|V| > 2V_{Ai}$. In [1] the resonant flow instability process was explained in terms of negative energy waves. The flux going into the resonant layer was shown to be negative for high enough velocity shear. It was also noted that the threshold velocity for KH-instability is lowered by the presence of the non-uniform region.

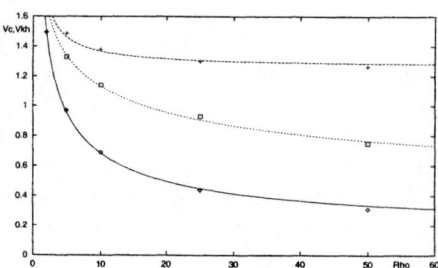

FIGURE 1. The dependence of the velocity thresholds on the density contrast (slab model): full line=RFI, dashed line=KHI with L=0, dotted line=KHI with L=0.1

We investigated the effect of the density contrast. The result is summarized in Figure 1 and can easily be explained by the expressions for the threshold velocities just derived, because of the relation between internal and external Alfvén speed determined by the density contrast. The KH-instability threshold depends on the density contrast as follows $1 + \frac{1}{\sqrt{\rho}}$ whereas for resonant flow instability this is $\frac{2}{\sqrt{\rho}}$.

III   RESULTS FOR $\beta \neq 0$

A   The eigenmodes when $L = 0$

In absence of the non-uniform region the dispersion relation can be written down analytically:

$$F_i(\Omega_i) \equiv \frac{\xi_i}{P_i'} = \frac{\xi_e}{P_e'} \equiv F_e(\Omega_e)$$

In which $\xi_{i,e}$ and $=P_{i,e}'$ are the internal and external solutions. For real ω F_i is always real whereas F_e is complex in the cut-off regions $[\omega_{Ce}, k_z c_e]$ and $[\omega_{Ae}, +\infty[$. We can therefore conclude that there can be no real frequency modes with frequencies in this cut-off regions. Normally modes in this region will be damped because they are carrying energy away (leaky modes). We will show that in the presence of a background flow it will be these cut-off regions that produce KH-instability when the real part of the frequency is forced into this regions by the Dopplershift.
F_i has a series of vertical asymptotes in the internal equivalent of the external cut-off regions, $[\omega_{Ci}, k_z c_i]$ and $[\omega_{Ai}, +\infty]$ making F_i range from $-\infty$ to $+\infty$ between two asymptotes and therefore making crossing with (the real part of) F_e inevitable. These intersections correspond with the slow and fast body modes trapped in the plume structure being reflected between the two boundaries and are not present in the single boundary layer problem. We have to notice that higher order slow modes are found at lower frequencies with ω_{Ci} as an accumulation point. This anti-Sturmian behaviour confirms that this are slow modes.
In Figure 2 we have plotted the oscillation frequency of the first slow and fast modes together with the imaginary part of the eigenfrequencies as function of the velocity shear V for the following values of the parameters: $m = 1$, $k_z = 5$, $\beta = 0.6$ and $\rho = 2$. The external cut-off regions are marked by the horizontal lines. The modes with positive frequency for $V = 0$ we call forward propagating waves (they may go backward with respect to the external reference frame they are moving forward with respect to a frame fixed to the internal background flow), while the modes with negative frequency for $V = 0$ are called backward propagating waves. Being shifted by the flow the modes disappear and become unstable as they have to cross the external cut-off regions. The effect of the flow can be introduced into the analytical dispersion relation very easily, by noting that the only influence is in a Dopplershift of the frequency inside the plume. We thus can follow these modes in their way to instability by a translation of F_i with respect to F_e as done by Bodo et al. [10,11]. In this way it can be seen that the instabilities occur as the two curves become tangent. In general we can state that instability occurs when forward modes meet backward external cut-off regions and that solutions vanish (become damped solutions) when they enter cut-off regions of the same direction. We conclude that there are two types of KH-instability, one associated with the slow cut-off region and one associated with the fast cut-off region. We will refer to them as slow and fast KH-instability.

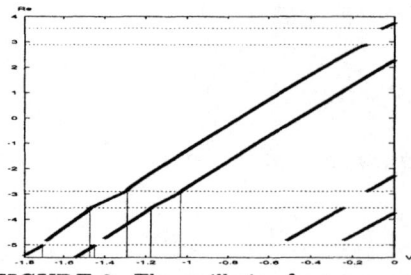
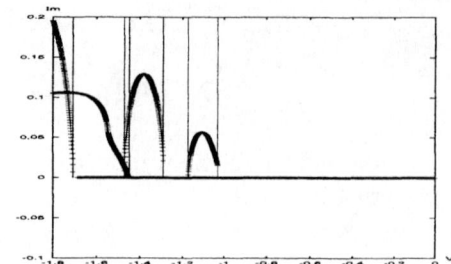

FIGURE 2. The oscillation frequency and corresponding imaginary part of the frequency of the first slow and fast wave as a function of the velocity shear for $L = 0$.

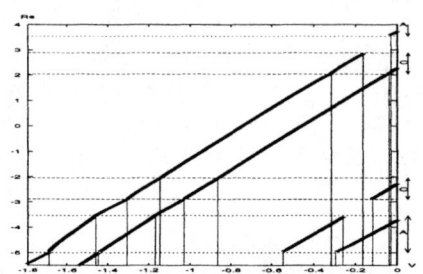
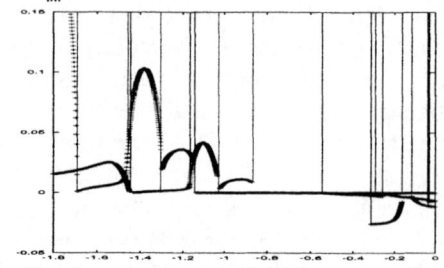

FIGURE 3. The oscillation frequency and corresponding imaginary part of the frequency of the first slow and fast wave as a function of the velocity shear for $L = 0.1$.

B Resonant instability of the eigenmodes when $L \neq 0$

When we introduce a nonuniform layer between the plume and interplume region, the modes with oscillation frequency within the range of the Alfvén or slow continuum resonantly couple to localized Alfvén or slow continuum modes, and, in absence of a velocity shear, they are damped due to the resonant wave excitation. In figure 3a we plot the oscillation frequencies of the ordinary and first fast and first slow modes as function of the velocity shear for the same parameters as before. In this figure we also indicate the upper and lower bounds of the slow continua $[-\omega_{Ce}, -\omega_{Ci}]$ and $[\omega_{Ci}, \omega_{Ce}]$ and of the Alfvén continua $[-\omega_{Ae}, -\omega_{Ai}]$ and $[\omega_{Ai}, \omega_{Ae}]$ by the horizontal lines. These are not Doppler shifted since there is no mass flow present in the nonuniform layer in the reference frame used. The fast cut-off region lies above the Alfvén continuum and the slow cut-off region is found between the cusp and Alfvén continuum. It is a pure coincidence that for the parameters chosen the upper bound of the slow cut-off region is equal to the internal Alfvén frequency. In figure 3b the corresponding imaginary parts of the eigenfrequencies are plotted as function of the velocity shear V.

For $V = 0$ the first fast modes lie in the Alfvén continuum and are damped. So are the slow waves due to coupling to slow resonant waves. As the velocity shear is increased the forward fast mode gets out of the Alfvén continuum and enters the cut-off region where it vanishes. Subsequently it enters the slow continuum and gets damped again. In the same way the backward slow mode vanishes when it comes out of the slow continuum and afterwards becomes damped again by Alfvén resonance this time. The interesting part appears where the forward modes get into the backward continua. In this case they will be amplified as we have already shown for Alfvén resonance of fast waves. The picture is completely the same.

We can see that the velocity thresholds for instability can be understood again in terms of the characteristic velocities. We must emphasize here that plasma-β used her is too high for the low corona, and that the characteristic velocities are highly dependent on β. In a low-β configuration the ranges for which slow KH-instability and slow resonant instability will occur will be very narrow, and therefore we can make no meaningful prediction concerning its occurrence in coronal plumes. Fast KH-instability will set in for $|V| > V_{Ai} + V_{Ae}$ and $|V| > V_{Ci} + V_{Ae}$, for the fast and slow body modes respectively. Alfvén resonant instability is operative for velocity shears in the ranges $2V_{Ai} < |V| < V_{Ai} + V_{Ae}$ and $V_{Ci} + V_{Ai} < |V| < k_z c_i + V_{Ae}$ for the first fast mode and the slow modes respectively.

IV CONCLUSIONS

We have shown that the range of frequencies for which slow and fast body modes occur can be determined by aid of the characteristic frequencies only. Furthermore we showed how KH-instability occurred by crossing the cut-off regions. Resonant instability will be present when crossing the continua. These considerations make it unnecessary to go through the numerical calculations for all different parameters, since the occurrence off instabilities is clearly understood by looking at the different frequency ranges. Our computations suggest that instability off coronal plumes is more likely to occur due to Alfvén resonant flow instability of slow modes than due to KH-instability. However important analysis remains to be done with respect to the dependence of this process on the velocity profile.

REFERENCES

1. Andries, J., Tirry, W.J., Goossens, M, ApJ **531**, 561-570 (2000)
2. Suess, S.T., in *Solar Jets and Coronal Plumes*, ESTEC, The Netherlands, ESA SP-421, 223 (1998)
3. Tirry, W.J., Čadež, V.M., Erdélyi, R., & Goossens, M., A&A **332**, 786 (1998)
4. Appert, K., Gruber, R., Vaclavik, J., *Physics of Fluids* **17**, 1471 (1974)
5. Walker, A.D.M., *J. Plasma Phys.*, submitted (1999)
6. Tirry, W.J., & Goossens, M., ApJ **471**, 501 (1996)
7. Sakurai, T., Goossens, M., & Hollweg, J.V., SPh **133**, 227 (1991)
8. Goossens, M., Ruderman, M.S., & Hollweg, J.V., SPh **157**, 75 (1995)
9. Erdélyi, R., Goossens, M., & Ruderman, M.S., SPh **161**, 123 (1995)
10. Bodo, G.,Rosner, R., Ferrari, A., & Knobloch, E., ApJ **341**, 631-649 (1989)
11. Bodo, G.,Rosner, R., Ferrari, A., & Knobloch, E., ApJ **470**, 797-805 (1996)

Absorption of fast magnetosonic waves in the solar atmosphere in the limit of weak nonlinearity

István Ballai[1], Róbert Erdélyi[2] and Marcel Goossens[1]

[1] Centre for Plasma-Astrophysics, K.U.Leuven, Celestijnenlaan 200B, B-3001 Heverlee, Belgium
[2] Space and Atmosphere Research Center, Dept. of Applied Mathematics, University of Sheffield, S3 7RH, Sheffield, England (UK)

Abstract. We study the resonant absorption of fast magnetoacoustic (FMA) waves in inhomogeneous weakly dissipative, isotropic and anisotropic plasmas. The equilibrium states on which the waves are superimposed is static or stationary and are assumed to be in a 1-D planar geometry. The equilibrium model consists of three layers with an inhomogeneous magnetised plasma surrounded by two homogeneous magnetised semi-infinite plasmas. The propagating FMA waves are partly absorbed and partly reflected by coupling to local nonlinear slow magnetohydrodynamic (MHD) waves in the inhomogeneous layer.

The dissipation acts only in a narrow layer called the *dissipative layer* which embraces the resonant magnetic surface. In linear theory it has been shown that in the vicinity of the resonant surface the energy density, the amplitudes of waves and the spatial gradients become large, suggesting that in this region nonlinearity might be important. The wave motions far from the dissipative layer are described by the linear, ideal MHD equations, while inside this layer the wave motions are governed by the full system of the dissipative, nonlinear MHD equations.

The coefficient of wave energy resonant absorption is derived assuming weak nonlinearity and long-wavelength approximation.

INTRODUCTION

A fundamental process in the solar atmosphere is the complicated interaction of the motions of the plasma with the magnetic fields. The solar atmosphere is a highly non-uniform and dynamic system which is a natural medium for magnetohydrodynamic waves. Waves, are able to transport momentum and energy which can be dissipated. One possibility for efficient dissipation is the resonant absorption which owes its existence to the coupling of global oscillations to local waves in non-ideal plasmas. [1] proposed resonant absorption of MHD waves as a mechanism for heating magnetic loops in the solar corona. Since then, resonant absorption became a very popular heating mechanism to explain the high temperature of the corona.

CP537, *Waves in Dusty, Solar, and Space Plasmas*, edited by F. Verheest, et al.
© 2000 American Institute of Physics 1-56396-962-9/00/$17.00

Linear theory shows that in the vicinity of resonant surface the amplitudes of the variables can be very large even when they are small far away from these surfaces. This observation implies that the linear theory can break down in this region.

Many studies in the field of resonant absorption considered only the sound and Alfvén waves as excellent candidates for plasma heating because of their wide continuum. However magnetoacoustic waves might also have an important contribution in explaining the coronal temperatures.

The MHD slow modes can be ignored in this context since they propagate to slowly to carry the required energy flux into corona, subject to the constrains imposed by the observed amplitudes of non-thermal motions in the corona and chromosphere which are of the order of $1 - 3 \times 10^6 \ cms^{-1}$ ([2], [3]). The FMA wave is the only one magnetic wave which is able to propagate across the magnetic field lines carrying energy due to its large group velocity. It is compressive and therefore subject to dissipation by viscosity, heat conduction, Landau and transit-time damping in the high-frequency limit where the Coulomb collisions are ineffective. However, in order to have an acceptable heating by fast MHD waves we have to impose a very restrictive condition, namely that these waves are not refracted by the steep rise of the Alfvén and/or slow wave speed with height and become evanescent. In addition only the high frequency waves can reach the corona (with period of a few tens of seconds).

The energy flux density of fast waves at the bottom of the corona that is required for significant heating is of the order of $10^5 \ erg \ cm^{-2} \ s^{-1}$. This value is not inconsistent with the upper limit on acoustic waves of $10^4 \ erg \ cm^{-2} \ s^{-1}$ ([4]), provided the coronal base magnetic field is sufficiently large ($\geq 10G$).

There is, however, another possibility. It is not necessary that the waves have to travel up from photospheric regions and transport energy from the huge kinetic energy reservoir (convection motion) which are then dissipated in the corona. FMA waves (and slow or Alfvén waves) can be generated locally in the corona by e.g., magnetic reconnection. It has been estimated by e.g., [5] that the shuffling the magnetic field lines in the solar atmosphere (by convective motions down in lower regions) builds up magnetic stresses which can be released through, e.g., reconnection providing magnetic energy to maintain the high temperatures in the solar corona. This mechanism was called nanoflarering.

The aim of the present paper is to study the nonlinear resonant interaction of FMA waves tunneled into slow dissipative layers in isotropic and anisotropic, static and steady state plasmas.

THE EQUILIBRIUM

As an application we study the nonlinear interaction of incident FMA waves with one-dimensional plasmas, i.e., the nonlinear coupling of fast waves and slow MHD waves. Monochromatic fast waves are impinging from the $x < 0$ homogeneously magnetised half-space (Region I, with magnetic induction B_e) and penetrate into the inhomogeneous region, $0 < x < x_0$ (Region II, B_0). This inhomogeneous region is bounded on its right by a semi-infinite subspace containing another homogeneously magnetised plasma (Region III, B_i). The magnetic field is unidirectional and parallel to the z axis. In what follows we use the subscripts 'e', '0' and 'i' to indicate equilibrium quantities in the three regions (regions I, II and III respectively). An equilibrium flow, v_0 is considered in the inhomogeneous and internal region. For sake of simplicity we take this velocity proportional to the cusp velocity in the external region. From the very beginning we restrict our analysis to the *long wavelength approximation*, i.e., the thickness of the inhomogeneous region is much less than the wavelength of the incoming fast wave.

Dissipation is one of the basic feature of the solar plasma. This deviation from the ideal medium is reflected by the dimensionless Reynolds number which under solar condition takes large values. In the process of resonant absorption this fact appears as a simplifying process in the sense that due to the large Reynolds numbers, the dissipation is relevant only in a very narrow region called it dissipative layer. One of the most used dissipative effects is the viscosity which is done in the most general form by the Braginskii's tensor([6]). The key quantity in our discussion is the product $\omega_i \tau_i$, where ω_i is the ion cyclotron frequency and τ_i is the mean collisional time of the ions. If $\omega_i \tau_i \ll 1$ then the Braginskii's tensor can be considered isotropic and the coefficient of viscosity is a scalar (photosphere and low chromosphere). If $\omega_i \tau_i \gg 1$ the full tensor has to be taken into account. Since $\omega_i \tau_i \gtrsim 10^5$ we can demonstrate that the leading term in the mentioned tensor is the first one. Hence, it seems a good approximation to describe viscosity by only the first term of the Braginskii's tensor.

The second dissipative process we use is thermal conduction which is connected to the heat conduction. Since in strongly magnetic plasmas (solar corona) the heat transport coefficient parallel to the magnetic field is much larger than in the perpendicular direction we limit our study to the parallel component of the thermal conductivity.

The magnetic diffusivity is related to the quantity $\omega_e \tau_e$ where ω_e and τ_e are the electron cyclotron frequency and mean collisional time for electrons. The coefficient of magnetic diffusivity is isotropic provided by the condition $\omega_e \tau_e \ll 1$ which is satisfied only in the lower photosphere.

To remove the Alfvén resonance we restrict our analysis to waves that are independent of y and we assume that the incoming fast waves are polarized in the xz-plane, i.e., $k_y = 0$. The dispersion relation for the impinging propagating fast waves is

$$\omega^2 = \frac{1}{2}|k|^2 \left\{ (v_A^2 + c_S^2) + \left[(v_A^2 + c_S^2)^2 - 4v_A^2 c_S^2 \cos^2 \theta \right]^{1/2} \right\} \qquad (1)$$

where θ is the angle between the direction of propagation and the background magnetic field.

Let us consider that the wave vector of the incoming wave is of the form of $\mathbf{k} = (\kappa_e k, 0, k)$. Having in mind that the equilibrium magnetic field is along the

z axis, we can note that $\cos^2\theta = \kappa_e^2/(\kappa_e^2 + 1)$. The dispersion relation (1) then becomes

$$\omega^2 = \frac{1}{2}|k|^2(1 + \kappa_e^2)\left\{(v_A^2 + c_S^2) + \left[(v_A^2 + c_S^2)^2 - 4v_A^2 c_S^2 \frac{\kappa_e^2}{\kappa_e^2 + 1}\right]^{1/2}\right\}$$ (2)

I GOVERNING EQUATIONS

As mentioned earlier the dissipation acts only inside a narrow layer which contains the ideal resonant position. Outside this dissipative layer, the plasma is described by the ideal MHD which means that here the plasma motion is described by the same equations for isotropic and anisotropic plasmas respectively. The governing equations can be written in form of a system of two coupled first order PDE for the total pressure perturbation, P, and the normal component of the velocity, u (see e.g., [7])

$$\frac{\partial u}{\partial x} = \frac{\mathcal{V}}{D}\frac{\partial P}{\partial \theta}, \qquad \frac{\partial P}{\partial x} = \frac{\rho_0 D_A}{\mathcal{V}}\frac{\partial u}{\partial \theta},$$ (3)

where

$$D = \frac{\rho_0 D_A D_T}{(\mathcal{V}^2 - c_S^2)(\mathcal{V}^2 - v_A^2)}, \quad D_A = \mathcal{V}^2 - v_A^2, \quad D_T = (v_A^2 + c_S^2)(\mathcal{V}^2 - c_T^2).$$ (4)

and the ideal slow resonant position is determined by the condition

$$\mathcal{V}^2 = c_T^2(x)$$ (5)

From a mathematical point of view the dissipative layer can be considered as a surface of discontinuity where the position of this surface coincides with the resonant position. In order to solve the system of governing equations for the whole domain of the outer region we need boundary conditions for u and P at the surface of discontinuity. These boundary connect the solutions at both sides of the resonant surface.

One connection formula is

$$[P] = 0,$$ (6)

where the square brackets indicate the jump in a quantity across the dissipative layer. This formula coincides with that in linear theory. However, in contrast to linear theory, the second connection formula can only be written in an implicit form. A second boundary condition is calculated for the normal component of the velocity u as

$$[u] = -\frac{\mathcal{V}}{k}\mathcal{P}\int_{-\infty}^{\infty}\frac{\partial q}{\partial \theta}d\sigma.$$ (7)

where we used the Cauchy principal part because the integral is divergent at infinity. In the above relation q can be considered as the dimensionless component of the perturbed velocity parallel to the equilibrium magnetic field lines and *sigma* is a variables which is of the order of unity in the dissipative layer.

The equation that governs the wave dynamics in the dissipative layer can be written in dimensionless form as

$$\sigma \frac{\partial q}{\partial \theta} - \Lambda_1^{(i)} q \frac{\partial q}{\partial \theta} + k \frac{\partial^2 q}{\partial \sigma^2} = \Lambda_2^{(i)} \frac{dP}{d\theta},$$ (8)

for isotropic plasma (see, e.g. [8]) and

$$\sigma \frac{\partial q}{\partial \theta} - \Lambda_1^{(a)} q \frac{\partial q}{\partial \theta} + k^{-1} \frac{\partial^2 q}{\partial \theta^2} = \Lambda_2^{(a)} \frac{dP}{d\theta},$$ (9)

for anisotropic plasmas ([9]). The coefficients Λ are given in terms of the equilibrium quantities. In the Eqs. (8) and (9) the second terms are the nonlinear terms, the third terms are the dissipative terms, while the terms in the RHS of the equations can be considered as driving terms and the form of the total pressure can be prescribed. The system of equations (3) and the two non-linear jump conditions (6) and (7), respectively, constitute a complete system of equations and boundary conditions. Note though, that we have to solve these *simultaneously* with Eqs. (8)–(9).

The ratios of the second (nonlinear term) to the third (dissipative term) terms in Eqs. (8)–(9) give the nonlinearity parameters $\tau^{(i)}$ and $\tau^{(a)}$. Our second assumption is to consider weak nonlinearity, i.e., $\tau^{(i)} \ll 1$ and $\tau^{(a)} \ll 1$. This assumption means that the nonlinearity can be considered as a perturbation to the linear regime and it gives only small corrections to the linear results. This approach allows us to use a regular perturbation method for solving the governing equations inside and outside the dissipative layer.

Applying successive orders of approximations we can show that the nonlinearity generates higher overtones in the outgoing wave in addition to the fundamental one. Collecting all monochromatic components we find the coefficient of wave energy absorption can be written in the form

$$\mathcal{A} = \mathcal{A}_L + \tau^{2(i),(a)} \mathcal{A}_{NL}$$ (10)

where $\mathcal{A}_L$ and $\mathcal{A}_{NL}$ are the linear and nonlinear coefficient of energy absorption.

We should note here two very important results. In contrast to the linear theory where the coefficient of energy absorption is independent of the type of the dissipation, in *nonlinear theory the coefficient of wave energy absorption depends on the dissipative effects*. In contrast to the static equilibrium where the absorption rates decrease due to nonlinearity, in steady state plasmas we will see that the nonlinearity can *increases* the absolute value of the coefficient of absorption.

II NUMERICAL RESULTS

In spite of the many detailed observations of the solar atmosphere the spatial structure of the bulk plasma motion is not yet resolved even in the simplest cases. Here we suppose that the motion is inhomogeneous and simply proportional to the local cusp speed. We are aware of the simplification and have considered different equilibrium. Though the obtained quantitative result might not be directly applicable to observations, the qualitative results still tell us a lot about the resonant dissipation mechanism.

Propagating FMA waves are impinging from the homogeneous magnetic (e.g. lower parts of the canopy) side (Region I) into the inhomogeneous magnetic part of the canopy (Region II and III) where they are resonantly coupled to the slow resonant continuum. Physical parameters are in dimensionless form. Length is measured in the thickness of the inhomogeneous layer (L) while all speeds are expressed in the unit of sound speed in Region I.

Figs. 1-2 show the relative and full absorption coefficients as a function of the flow strength parameter and the angle of incidence for isotropic and anisotropic plasmas. The relative absorption rate is defined as the ratio of the nonlinear correction of absorption for a given value of flow strength parameter and angle of incidence divided by the nonlinear correction in the static case. The relative absorption is an appropriate measure to estimate the influence of flow.

We can generally conclude that flow has a strong influence on the absorption rate which is in agreement with previous studies. When the plasma is anisotropic this effect is even more pronounced. Nonlinearity decreases the absorption rate resulting in over-reflection in the anisotropic case (Figs. 1d and 2d). When the absorption rate becomes less than zero for a given flow this means the outgoing wave has larger amplitude than its incoming counterpart. There is still absorption occurring but energy is transferred from the equilibrium flow to the waves. Figs. 2 also show that in the case of anti-parallel propagation only a fairly small amount of flow may have an effect on the resonant heating because around above $f < -0.2$ there is no tunneling of the FMA waves into the slow continuum because the slow continuum is Doppler-shifted to much.

III CONCLUSION

The present study applies the nonlinear theory of resonant tunneling of FMA waves into slow MHD continuum in isotropic and anisotropic plasmas using the theories originally derived by [8] and [9].

The effect of an equilibrium flow is derived both for isotropic and anisotropic plasmas. By means of numerical investigations we found how a steady flow can affect the efficiency of the absorption leading to a complex picture of coupling between the incoming FMA and the local inhomogeneous plasma.

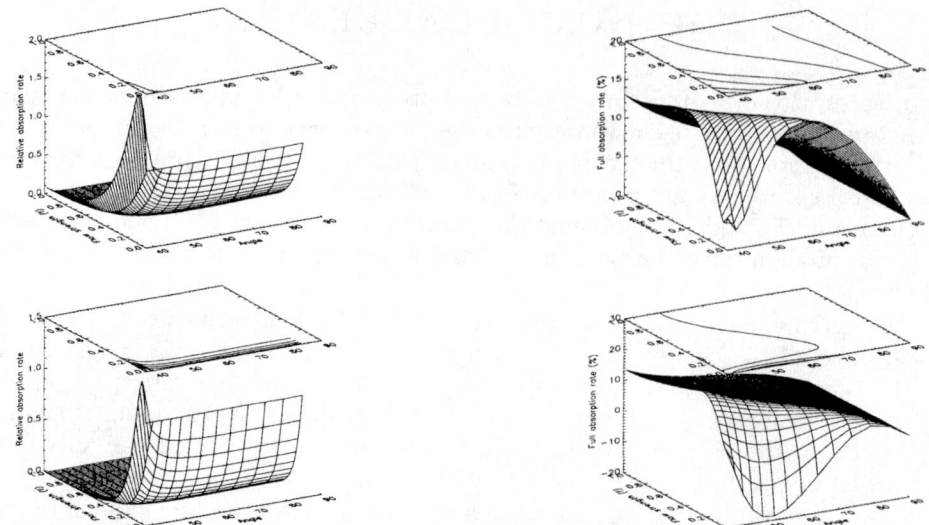

FIGURE 1. The relative, nonlinear and full absorption coefficient in an isotropic and anisotropic plasma as a function of the field-aligned inhomogeneous plasma flow strength parameter, f, and the angle of incidence, θ.

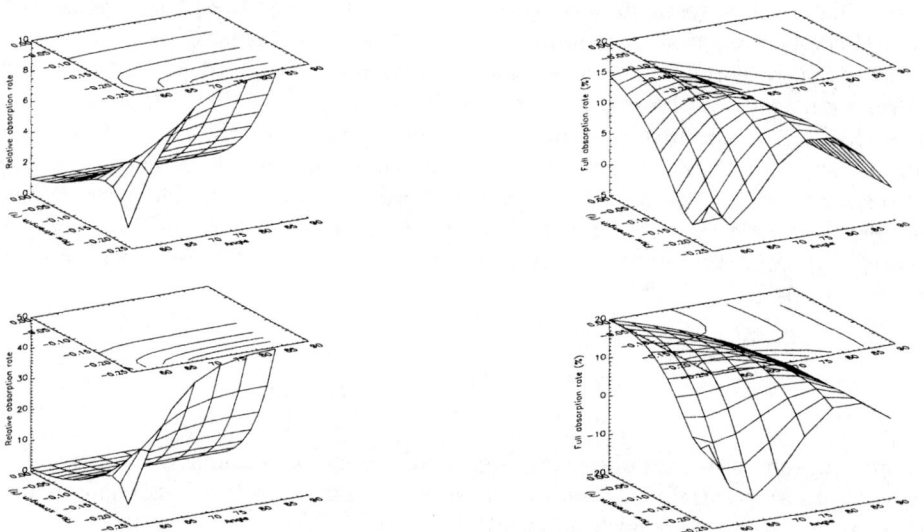

FIGURE 2. The same as Figure 1a-d but for an anti-parallel flow.

We applied the obtained results to cases appropriate to solar physics. The model studied in the present paper can be considered as an approximate model of the interaction of FMA waves in the solar chromosphere-magnetic canopy and/or low corona.

We have assumed that (i) the thickness of the slab containing the inhomogeneous plasma is small in comparison with the wavelength of the incoming fast wave; and (ii) the nonlinearity in the dissipative layer is weak.

We found, that:

(i) nonlinearity in slow dissipative layers generates higher harmonic contributions in the outgoing fast wave in addition to the fundamental harmonic. This result is similar to the one found by [10], [11] and [12] for sound waves.

(ii) An equilibrium flow in the slow dissipative layer can either increase or decreases the coefficient of the wave energy absorption. Thus, a field-aligned flow has an important effect on the resonant interaction of fast waves and nonlinear slow resonant layers.

(iii) Resonant flow instability (or over-reflection) has been found for a wide range of parameters.

IV   ACKNOWLEDGMENTS

I. Ballai acknowledges the financial support by 'Onderzoeksfonds K.U. Leuven'. R. Erdélyi acknowledges M. Kéray for patient encouragement. This research was also supported by the Hungarian National Science Fund, Nr. T032462.

REFERENCES

1. Ionson, J. A.: 1978, *Astrophys. J.*, 226, 650
2. Bruner, E.C. 1978, *Astrophys. J.*, 226, 1140
3. Bruner, E.C. and Poletto, G., 1981, Bull. AAS, 13, 835
4. Athay, R.G. and White, O.R. 1978, *Astrophys. J.*, 226, 1135
5. Parker, E.N.: 1988, *Astrophys. J.*, 330, 474
6. Braginskii, S.I.: 1965, *Rev. Plasma Phys.*, 1, 205
7. Ballai, I. and Erdélyi, R.: 1998, *Solar Phys.*, 180, 65
8. Ruderman, M. S., Goossens, M. and Hollweg, J. V.: 1997a, *Phys. Plasmas*, 4, 91
9. Ballai, I., Ruderman, M. S. and Erdélyi, R.: 1998, *Phys. Plasmas*, 5, 252
10. Ruderman, M. S., Hollweg, J. V. and Goossens, M.: 1997b, *Phys. Plasmas*, 4, 75
11. Ballai, I., Erdélyi, R. and Ruderman, M. S.: 1998, *Phys. Plasmas*, 5, 2264
12. Erdélyi, R. and Ballai, I.: 1999, *Solar Phys.*., 186, 67

Linear and nonlinear waves in dilute plasmas

István Ballai[1], Róbert Erdélyi[2] and Marcel Goossens[1]

[1] *Centre for Plasma-Astrophysics, K.U.Leuven, Celestijnenlaan 200B, B-3001 Heverlee, Belgium*
[2] *Space and Atmosphere Research Center, Dept. of Applied Mathematics, University of Sheffield, S3 7RH, Sheffield, England (UK)*

Abstract. Small-amplitude magnetohydrodynamic (MHD) waves are studied in a dilute collisionless plasma with an anisotropic pressure distribution. The parallel and perpendicular pressure are defined with the aid of two polytropic pressure laws. For specific values of the polytropic indices, previous results obtained with the usual Chew-Goldberger-Low (CGL) double-adiabatic (i) and double-isothermal (ii) models are recovered. The double-polytropic model can be considered as the counterpart of the single-polytropic model.

Dispersion relations for the linear waves are derived and analyzed in the presence of pressure anisotropy. The weakly nonlinear dynamics is shown to be governed by the Benjamin-Ono equation. The results are discussed in the CGL and double-isothermal limits.

I   INTRODUCTION

The study of waves in inhomogeneous plasmas is fundamental for solar and astrophysical plasmas. Waves are important in their own right as they reflect the stable dynamic behaviour of the plasma objects they occur in. They are also important since they transport momentum and energy. When part of their momentum and energy is dissipated, they can heat and accelerate plasma providing channels for heating the solar corona and accelerating the solar wind. Finally they can be used as probes for investigating the structure and composition of the plasma objects in which they are observed. Since many starts have a similar structure as our Sun, MHD waves are important for astrophysics.

So far the MHD waves were studied mostly in collision dominated plasmas where the waves are described by the ideal MHD equations with isotropic plasma pressure.

However the scalar gas pressure approximation is not valid in dilute plasmas such as coronal streamers, solar wind or magnetosphere where the mean free path for the particles is long compared to any other dimensions in the plasma. When the cyclotron frequency is much larger than the collisional frequency, the particles gyrate many times around a line of magnetic force in between two collisions, so

CP537, *Waves in Dusty, Solar, and Space Plasmas*, edited by F. Verheest, et al.
© 2000 American Institute of Physics 1-56396-962-9/00/$17.00

that there is an equipartition between the particles kinetic energies in the two independent directions, parallel and perpendicular to the equilibrium magnetic field line direction. The presence of the magnetic field induces a splitting in the pressure, introducing a parallel and a perpendicular component of the pressure.

A new approach in the problem of wave propagation was the theory developed by [1]. The usual MHD equations are found from the Boltzmann equation using an expansion in powers of the collisional mean free path. In this case the plasma is collision-dominated therefore the collisional term in the Boltzmann equation is the leading term, all the other terms being treated as perturbations. When the density is so small that the plasma can be considered as a collisionless medium, a new form of the Boltzmann equation can be found using an expansion in powers of the Larmor radius as here the role of the collisional term is played by the Lorentz force. This approximation can be considered as an adiabatic approximation since it depends on the Larmor frequency being large compared to other frequencies of the problem. In many cases (in magnetosphere or magnetosheat) the plasma expands and compress closer to an isothermal behaviour than the adiabatic (see, e.g., [2], [3]).

The wave properties in such structures were intensively studied in the recent past (see, e.g. [4], [5], [6] and [7]). The magnetoacoustic waves in this limit show a strange behaviour, namely for a certain range of plasma beta, $\beta_\parallel$ and $\beta_\perp$ the phase speed of the slow waves exceeds the phase speed of the Alfvén wave. These authors found that when this condition is fulfilled standing slow waves in a field aligned flow are then downstream facing instead of upstream facing, as they always are in the usual MHD model.

II   BASIC EQUATIONS

Let us consider an isolated magnetic slab of infinite extention in the z direction. The equilibrium magnetic field is parallel to the z axis and the nonuniformity is in the x direction.

The plasma motion is described by the ideal single-fluid MHD equation where a double-polytropic pressure law is used

$$\frac{D\rho}{Dt} + \rho\nabla \cdot \mathbf{v} = 0, \quad \nabla \cdot \mathbf{B} = 0, \quad \frac{\partial \mathbf{B}}{\partial t} = \nabla \times (\mathbf{v} \times \mathbf{B}) \tag{1}$$

$$\rho\frac{D\mathbf{v}}{Dt} = -\nabla\mathbf{P} + \frac{1}{\mu}\left[(\nabla \times \mathbf{B}) \times \mathbf{B}\right], \tag{2}$$

$$\frac{D}{Dt}\left(\frac{p_\perp}{\rho B^{\gamma_\perp - 1}}\right) = 0, \quad \frac{D}{Dt}\left(\frac{p_\parallel B^{\gamma_\parallel - 1}}{\rho^{\gamma_\parallel}}\right) = 0 \tag{3}$$

where $D/Dt = \partial/\partial t + \mathbf{v} \cdot \nabla$ is the convective operator and $\gamma_\parallel$, $\gamma_\perp$ are the parallel and perpendicular polytropic indices. These indices express the increase in the temperature upon the plasma compression. For $\gamma_\perp = 2$ and $\gamma_\parallel = 3$ the usual double adiabatic CGL expressions are recovered whereas for $\gamma_\parallel = \gamma_\perp = 1$ we obtain the isothermal behaviour. Eqs. (3) are a simpler case of the most general polytropic laws introduced by [8] which contain four polytropic exponents.

In the equation of momentum conservation $\mathbf{P}$ is the pressure tensor defined by

$$\mathbf{P} = p_\perp \hat{\mathbf{I}} + (p_\parallel - p_\perp)\mathbf{bb} \tag{4}$$

where $\hat{\mathbf{I}}$ is the unit dyadic and $\mathbf{b} = \mathbf{B}/B$ is the unit vector parallel to the magnetic field direction.

III   LINEAR WAVES

In this section we study the properties of linear waves and the influence of the pressure anisotropy on the wave propagation. Let us perturb the system and assume a harmonic dependence of the perturbation on time and coordinates of the form

$$f = f(x)exp[i(k_y y + k_z z - \omega t)]$$

The linearization of the system (1)–(3) yields a system of two coupled ODE for the perpendicular component of the total pressure and the x component of the velocity perturbation

$$\frac{d\delta P_\perp}{dx} = \frac{i\rho}{\omega} D_A \delta v_x \tag{5}$$

$$D_A(\omega^2 - k_z^2 c_{S\parallel}^2)\delta P_\perp - k_y^2 D_C \delta P_\perp = \frac{i\rho}{\omega} D_A D_C \frac{d\delta v_x}{dx} \tag{6}$$

where

$$D_A = \omega^2 - k_z^2 v_A^2(1 - \Gamma), \quad D_C = (c_{S\perp}^2 + v_A^2)(\omega^2 - k_z^2 c_T^2) \tag{7}$$

where $\Gamma = (p_\parallel - p_\perp)/v_A^2 \rho$ is the pressure anisotropy factor and $c_A^2 = B^2/\mu\rho$, $c_{S\perp}^2 = \gamma_\perp p_\perp/\rho$, $c_{S\parallel}^2 = \gamma_\parallel p_\parallel/\rho$ are the Alfvén, the perpendicular and parallel sound speeds and c_T^2 is the modified cusp speed defined as

$$c_T^2 = \frac{c_{S\perp}^2(c_{S\parallel}^2 - c_{S\perp}^2/\gamma_\perp^2) + c_{S\parallel}^2 v_A^2}{c_{S\perp}^2 + v_A^2} \tag{8}$$

For the isotropic case (i.e., when $c_{S\perp}^2 = c_{S\parallel}^2 = c_S^2$ and $\gamma_\perp = 1$ in 8) we obtain the classic cusp speed which is always positive. However, in anisotropic plasmas the cusp velocity can be negative for $c_{S\parallel}^2 > c_{S\perp}^4/[\gamma_\perp^2(v_A^2 + c_{S\perp}^2)]$ and this condition gives rise the mirror instability threshold. If this condition is satisfied, the magnetic field develops regions of low field strength separated by regions of enhanced field

154

strength. Where the field is stronger than the undisturbed field, the particle mirror points shift in such a way that the plasma density decreases. Where the field is weaker than the equilibrium field, the plasma density increases.

For simplicity we consider only two dimensional propagation in the xz plane, i.e., $k_y = 0$. Considering a homogeneous plasma and $d/dx \rightarrow ik_x$, the system (5)-(6) can be reduced to

$$(k_x^2 + m^2)D_A = 0, \quad m^2 = -\frac{(\omega^2 - k_z^2 c_{S\parallel}^2)[\omega^2 - k_z^2 v_A^2(1 - \Gamma)]}{(c_{S\perp}^2 + v_A^2)(\omega^2 - k_z^2 c_T^2)} \tag{9}$$

The Eq. (9) is the dispersion relation for the linear waves in uniform double-polytropic plasma. The relation $D_A = 0$ yields the dispersion relation for the modified Alfvén waves

$$\frac{\omega^2}{k^2} = v_A^{2*} \cos^2 \alpha = v_A^2(1 - \Gamma) \cos^2 \alpha \tag{10}$$

where α is the angle between the magnetic field and the direction of propagation and v_A^{2*} is the modified Alfvén velocity. When $\Gamma = 0$ (i.e., isotropic case, $p_\parallel = p_\perp$) we recover the usual Alfvén velocity. Whereas the isotropic form is positive definite, $(1-\Gamma)$ may be negative ($p_\parallel > p_\perp + B^2/\mu$) (i.e., $\Gamma > 1$). In that case, the Alfvén mode will exhibit non-propagating, pure exponential growth. This instability is called the firehose instability. In general the behaviour of the Alfvén waves can be classified according to the sign of $(1-\Gamma)$. Thus for $(1-\Gamma) > 0$ we have a propagating Alfvén mode (collisional MHD), for $(1 - \Gamma) = 0$ we have a non-propagating, non-growing perfectly inelastic perturbation and for $(1 - \Gamma) < 0$ we have a non-propagating, purely exponential growth, i.e., the firehose instability.

The relation $k_x^2 + m^2 = 0$ yields the dispersion relation for the magnetoacoustic waves, i.e.,

$$\frac{\omega^2}{k^2} = \frac{1}{2}\left[b \pm \sqrt{b^2 - 4c}\right] \tag{11}$$

where

$$b = v_A^2 + c_{S\perp}^2 + \left[(\gamma_\parallel - 1)\frac{c_{S\parallel}^2}{\gamma_\parallel} - (\gamma_\perp - 1)\frac{c_{S\perp}^2}{\gamma_\perp}\right] \cos^2 \alpha$$

$$c = -\left[c_{S\parallel}^2(c_{S\parallel}^2 \cos^2 \alpha - b) + \frac{c_{S\perp}^4}{\gamma_\perp^2}\right] \cos^2 \alpha \tag{12}$$

where the signs $\pm$ in the Eq. (11) correspond to the fast and slow waves respectively. For $\gamma_\perp = 2$ and $\gamma_\parallel = 3$ (the CGL approximation) we recover the result obtained by [4] and the dispersion relations reduce to the usual dispersion relations found in the isotropic MHD for $\alpha = 0$, $c_{S\parallel}^2 = c_\perp^2 = c_S^2$ and $\gamma_\parallel = \gamma_\perp = 1$.

In the isotropic MHD, we find for propagation along the magnetic field ($\alpha = 0$) Eq. (11) the phase speeds $c_{S\parallel}^2$ and $v_A^2(1 - \Gamma)$ while for propagation across the magnetic field ($\alpha = \pi/2$) we get 0 and $c_{S\perp}^2 + v_A^2$.

A   Waves in an isolated slab

Let us consider waves in a magnetically isolated double-polytropic plasma where the plasma is confined in a slab with walls at $x = \pm x_0$.

The plasma inside the slab and the fluid outside the slab are characterized by the following equilibrium configuration

$$\begin{cases} B^{(0)}, p_\perp^{(0)}, p_\parallel^{(0)}, \rho^{(0)}, & |x| < x_0, \\ 0, \quad p^{(e)}, \quad 0, \quad \rho^{(e)}, & |x| > x_0, \end{cases} \tag{13}$$

During the derivation we use the as boundary conditions the continuity of total pressure (magnetic and kinetic) and the normal component of the velocity across the slab. A straightforward calculation shows that the dynamics of a two-dimensional perturbation inside and outside the slab in longitudinal direction relative to the equilibrium magnetic field satisfies the 2nd order ODE for the normal component of the velocity perturbation

$$\frac{d^2 \delta v_x}{dx^2} - m^{(0)2} \delta v_x = 0, \quad |x| < x_0, \quad \frac{d^2 \delta v_x}{dx^2} - m^{(e)2} \delta v_x = 0, \quad |x| > x_0 \tag{14}$$

where

$$m^{(e)2} = k^2 - \frac{\omega^2}{c_S^{(e)2}}, \quad m^{(0)2} = -\frac{(\omega^2 - k^2 c_{S\parallel}^{(0)2})[\omega^2 - k^2 v_A^2(1 - \Gamma)]}{(c_{S\perp}^{(0)2} + v_A^2)(\omega^2 - k^2 c_T^{(0)2})} \tag{15}$$

The general solution δv_x inside and outside the slab can be written as

$$\delta v_x = A^{(0)} \cosh m^{(0)} x + B^{(0)} \sinh m^{(0)} x, \quad |x| < x_0 \tag{16}$$

and

$$\delta v_x = \begin{cases} A^{(e)} e^{-m^{(e)}(x - x_0)}, & x > x_0 \\ B^{(e)} e^{m^{(e)}(x + x_0)}, & x < -x_0. \end{cases} \tag{17}$$

where $A^{(0)}$, $B^{(0)}$, $A^{(e)}$, $B^{(e)}$ are arbitrary constants describing the amplitudes of oscillations.

The waves in magnetically isolated slabs can be divided in kink and sausage waves depending on the fact that δv_x is an even or an uneven function. According to ([9]), δv_x an even function of x corresponds to kink waves, while an odd function of x corresponds to sausage modes.

With the use of the boundary conditions at $x = \pm x_0$ (continuity of δv_x and $P_\perp$) we obtain the dispersion relation in its most general case

$$[k^2 v_A^2 (1 - \Gamma) - \omega^2] = \omega^2 R \begin{Bmatrix} \tanh \\ \coth \end{Bmatrix} m_0 x_0 \tag{18}$$

where
$$R = \frac{\rho_e m_0}{\rho_0 m_e}$$

and the tanh/coth terms correspond to the sausage ($A^{(0)} = 0$) and kink modes respectively ($B^{(0)} = 0$).

Due to their transcendental form we solve these equations in the limit of a slender slab, i.e., $kx_0 \ll 1$. Thus in Eq. (18) $\tanh m^{(0)}x_0 \approx m^{(0)}x_0$ and this equation takes the form

$$(\omega^2 - k^2 c_T^2)m^{(e)} = \frac{\rho_e}{\rho_0} \frac{\omega^2(\omega^2 - k^2 c_{S\|}^2)}{c_{S\perp}^2 + v_A^2} x_0 \tag{19}$$

This equation takes the form of dispersion relation obtained by [9] for isotropic case, i.e., when $c_{S\|}^{(0)} = c_{S\perp}^{(0)} = c_S^{(0)}$. As pointed out in the isotropic case the limit $kx_0 \to 0$ implies the existence of waves, namely, $\omega^2 \to k^2 c_T^2$ and $\omega^2 \to k^2 c_{S\|}^2$. Similar to the isotropic case it can be shown that there are always modes with the phase velocity below the minimum of c_T and $c_{S\|}^{(0)}$. When $\omega^2 \approx k^2 c_T^2$ the phase velocity of wave propagating with near tube velocity is given by

$$\frac{\omega^2}{k^2} \approx c_T^{(0)2} \left[1 - \frac{\rho_e}{\rho_0} \frac{c_S^{(e)}}{(c_S^{(e)2} - c_T^{(0)2})^{1/2}} I^2 kx_0 \right] \tag{20}$$

if $c_T^{(0)} < c_S^{(e)}$ where

$$I = \frac{c_{S\perp}^{(0)2}}{\gamma_\perp(c_{S\perp}^{(0)2} + v_A^2)}$$

The second mode arises when $c_{S\|}^{(0)} < c_{S\|}^{(e)}$, i.e. when ω^2 tends to $k^2 c_S^{(e)2}$. In this case the dispersion equation (18) gives for $\omega^2 \approx k^2 c_S^{(e)2}$

$$\frac{\omega^2}{k^2} \approx c_S^{(e)2} \left[1 - \left(\frac{\rho_e}{\rho_0} \frac{c_S^{(e)2}(c_S^{(e)2} - c_{S\perp}^{(0)2})}{c_{S\perp}^{(0)2} + v_A^2} \right)^2 (kx_0)^2 \right] \tag{21}$$

In the limit of a slender slab, the kink mode leads to the following dispersion relation

$$\omega^2 = k^2 v_A^2 (1 - \Gamma)\frac{\rho_0}{\rho_e}(kx_0) \tag{22}$$

Let us now look at the other extreme of a wide slab ($kx_0 \gg 1$) where we suppose that $m_0 x_0 \to \infty$ for $kx_0 \to \infty$. Thus we approximate $\tanh m_0 x_0$ by unity and we obtain for the sausage and kink waves the dispersion relation

$$[k^2 v_A^2(1 - \Gamma)]m_e = \left(\frac{\rho_e}{\rho_0} \right) \omega^2 m_0 \tag{23}$$

provided that $m_0 > 0$ and $m_e > 0$. This relation coincides with the dispersion relation obtained by [7] for surface waves at a magnetic surface and we can conclude that the propagation of surface non-leaky waves is equivalent to propagation at a single interface. In addition in the approximation of wide slab, there is no any difference between sausage and kink waves.

For body waves $m_0^2 = -n_0^2 < 0$ and the dispersion relation for the sausage and kink waves is

$$[k^2 v_A^2 (1 - \Gamma) - \omega^2] = \omega^2 R' \left\{ \begin{matrix} -\tan \\ \cot \end{matrix} \right\} n_0 x_0 \tag{24}$$

where now

$$R' = \frac{\rho_e n_0}{\rho_0 m_e}$$

and the tan/cot terms correspond to the sausage and kink modes respectively.

IV NONLINEAR WAVES

Waves with large amplitude in dispersive plasmas can form solitons as a result of the competition between the nonlinearity, which steepens the wave amplitude, and dispersion, which tends to broaden the waves by 'smearing' them out.

Following the procedure given by [10] we find that the motion along a magnetic slab is governed by the equation

$$\frac{\partial v_1}{\partial \tau} + \Psi v_1 \frac{\partial v_1}{\partial \xi} + \frac{1}{2} \left(\frac{\rho_e}{\rho_0} \right) I^{3/2} c_T x_0 \frac{\partial^2}{\partial \xi^2} \mathcal{H}(v_1) = 0 \tag{25}$$

where the coefficient Ψ is rather cumbersome and $\mathcal{H}$ is the Hilbert transform. Eq. (25) is the Benjamin-Ono equation for a dilute plasma. This equation is integrable. Its solutions are known as solitons with an algebraic decay and can be written in the original variables as

$$v(z, t) = \frac{N}{1 + [(z - st)/L]^2} \tag{26}$$

where the soliton amplitude N is related to the soliton speed s and length L by the expressions

$$s = c_T + \frac{1}{4} \Psi N, \quad L = 2 \left(\frac{\rho_e}{\rho_0} \right) \frac{I^{3/2} c_T x_0}{\Psi N} \tag{27}$$

The result given by (26) is limited by the condition that c_T has to be real, i.e., this solution is not valid for the region where the mirror instability appears. In the instability region the coefficient of the nonlinear and the dispersive terms become imaginary, the solution is given in form of (26) but the soliton speed becomes imaginary.

V CONCLUSIONS

In this study two themes were pursued in the context of dilute plasmas when the anisotropy in the kinetic pressure and a double-polytropic law is considered. One advantage of using this model is that the wave dispersion relations contain not only the double adiabatic but also the isothermal limit which in space plasmas is sometimes more favorable both observationally and theoretically (see the results obtained by AMPTE spacecraft). First, the linear waves were studied in an isolated slab. The possible propagation modes for slender and wide slabs were given. The anisotropy in the kinetic pressure modifies the dispersion relations of the waves. This model is rather limited by the mirror and firehose instability thresholds.

The evolution of the nonlinear waves have been described by a Benjamin-Ono-type solitary wave equation with algebraic decaying solutions. Again, this result is valid outside the mirror instability region.

VI ACKNOWLEDGMENTS

I. Ballai acknowledges the financial support by 'Onderzoeksfonds K.U. Leuven'. R. Erdélyi acknowledges M. Kéray for patient encouragement. This research was also supported by the Hungarian National Science Fund, Nr. T032462.

REFERENCES

1. Chew, G.F., Goldberger, M.L., Low, F.E.: 1956, *Proc. Roy. Soc. London*, Ser A, 236, 112
2. Song, P., Russel, C.T. & Thomsen, M.F.: 1992, *J. Geophys. Res.*, 97, 8295
3. Phan, T.D., Paschmann, G., Baumjohann, W., Sckopke, N. & Lühr, N.: 1994, *J. Geophys. Res.*, 99, 121
4. Kato, Y., Tajiri, M. & Taniuti, T.: 1966, *J. Phys. Soc. Jpn*, 66, 765
5. Abraham-Shrauner, B.W.: 1973, *J. Plasma Phys.*, 1, 361
6. Hau, L.-N. & Sonnerup, B.U.Ö.: 1993, *Geophys. Res. Lett*, 20(17), 1763
7. Hau, L.-N. & Lin, C.A.: 1995, *Phys. Plasmas*, 2(1), 294
8. Abraham-Shrauner, B.W.: 1973, *Plasma Phys.*, 15, 375
9. Roberts, B.: 1981, *Solar Phys.*, 69, 39
10. Roberts, B.: 1985, *Phys. Fluids*, 28, 3280

Long period oscillations in the polar plumes

D. Banerjee*, E. O'Shea†, J.G. Doyle# and M. Goossens*

*Centre for Plasma Astrophysics, KULeuven, 3001 Heverlee, Belgium
†Dept. of Pure & Applied Physics, Queens University Belfast, BT7 1NN, N.Ireland
#Armagh Observatory, College Hill, Armagh BT61 9DG, N. Ireland
dipu@wis.kuleuven.ac.be, E.Oshea@qub.ac.uk, jgd@star.arm.ac.uk,
Marcel.goossens@wis.kuleuven.ac.be

Abstract.
We examine spectral time series of the transition region line O v 629 Å, observed with the Coronal Diagnostic Spectrometer (CDS) on the SoHO spacecraft. Both Fourier and wavelet transforms have been applied independently to the analysis of plume oscillations in order to find the most reliable periods. The wavelet analysis enables us to derive the duration as well as the periods of the oscillations. Our observations indicate the presence of compressional waves with periods of 10-20 minutes. We have also detected a 10 ± 2 minute periodicity in the network regions of the north polar coronal hole. The waves are produced in short bursts with coherence times of about 20-30 minutes. We interpret these oscillations as outward propagating slow magneto-acoustic waves, which may contribute significantly to the heating of the lower corona by compressive dissipation and which may also provide enough energy flux for the acceleration of the fast solar wind. The data support the idea that the same driver is responsible for the network and plume oscillations with the network providing the magnetic channel through which the waves propagate upwards from the lower atmosphere to the plumes.

INTRODUCTION

Polar plumes are the most prominent features in polar coronal holes where they play an important role in the generation of the high speed wind. SoHO observations have confirmed that plumes are denser and cooler than the surrounding regions. From SUMER/SoHO observations, [1] showed that the plume appears wider and more diffuse in Ne VIII 770 Å, with the base of the plume originating at the intersection of the network boundary. Thus it is important to study not only the dynamics of the network regions in the coronal holes but also the plumes which are believed to originate from the boundaries of these network regions [2]. We investigate the temporal behaviour of polar plumes as observed in the transition region line, O v 629 Å as observed by the Coronal Diagnostic Spectrometer (CDS) onboard the

CP537, *Waves in Dusty, Solar, and Space Plasmas,* edited by F. Verheest, et al.
© 2000 American Institute of Physics 1-56396-962-9/00/$17.00

Solar Heliospheric Observatory (SoHO). In this short contribution we will focus on one of the dataset, s8474r01.

OBSERVATIONS AND DATA REDUCTION

The CDS is a dual extreme ultraviolet spectrometer, covering the wavelength range 150 to 780Å, comprising of a normal incidence and a grazing incidence spectrometer [4]. The observations reported here were performed on 23 July 1997, starting from 00:18 UT and ending at 01:18 UT with pointing (X,Y) as (80, 937) . This dataset, s8474r01 was obtained for the three lines of O v 629 Å (log T_e=5.4 K), Mg IX 368 Å (log T_e=6.0 K) and Fe XVI 335 Å (log T_e=6.3 K). The counts for the Mg IX and Fe XVI lines were too low and so these lines were not used for the power analysis. Details of this observation campaign and the data analysis procedures have been reported in [3]. For the data reported here, the 4x119 arcsec slit was used. To improve the signal-to-noise of this data we binned by three pixels along the slit (i.e. 3×1.68 arcsec), in effect creating new pixels of 5×4 arcsec^2.

The data were checked for periods using two methods, a Fourier analysis (see [5]) and a wavelet analysis (see [6]). For both the intensity and velocity power spectra we use confidence levels of 99.9%. To determine the Doppler shifts, wavelength calibration is needed. We use the *'limb method'*, where we assume that above the limb all (non-radial) wave or mass motions on average cancel out. The localised (in time) nature of the wavelet transform enables us to study the duration of any statistically significant oscillations as well as their period. Hence, to find the most reliable periods, we also performed wavelet analysis on the data. By decomposing a time series into time-frequency space, one is able to determine both the dominant modes of variability and how those modes vary in time. The wavelet software uses the Morlet wavelet and moreover allows the calculation of confidence levels. Again we choose a confidence level of 99.9%. We should point out that wavelet transforms suffer from edge effects at both ends of the time series. The region in which these effects are important are defined by the 'cone of influence' (COI) [6].

RESULTS

The pointing of the slit was positioned at the solar limb in such a way that part of the slit was outside the limb and the remainder on the disk part of the coronal hole. Thus our slit covers network regions in the coronal hole and the plume at the same time. Fig. 1 shows the power spectra in the 0-4 mHz range together with the contrast enhanced intensity map for the dataset s8474r01 for all pixels along the slit. In the upper panel of Fig. 1, black indicates power above the 99.9% detection level. In general we find significant power within the range 0.5-1.75 mHz at different locations across the slit.   To bring out the details of the original intensity map we have filtered out the bright components in the image. For details of this procedure we refer to [5], [3], [7]. In this contrast enhanced image (lower

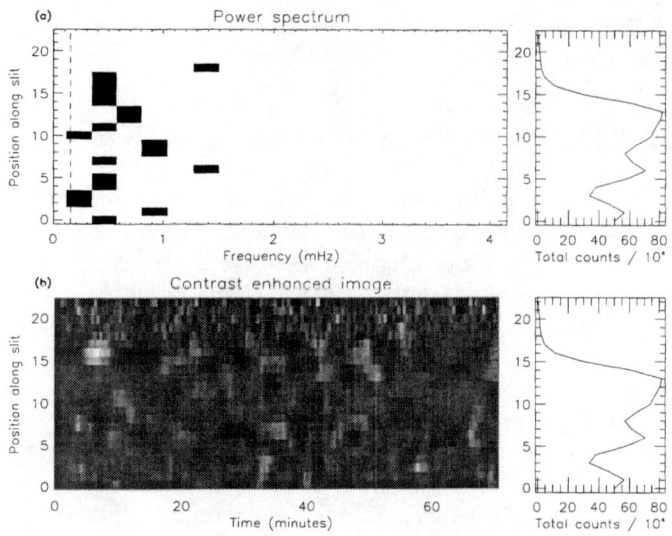

FIGURE 1. The power above the significance level as a function of frequency of oscillation and spatial position along the slit for the s8474r01 dataset (upper panel). The gray scale plot shows the space-time behaviour of the intensity in the O V 629 Å line (lower panel). The gray scale coding has the most intense regions as dark. The right panels show the counts summed over all time against the slit locations. Each pixel is 5 arcsec.

panel of Fig. 1), the solar north-south (SOLAR_Y) direction is on the vertical axis, the horizontal axis is time. Fluctuations in the bright features are clearly visible at several pixel locations and their appearance seems periodic with a periodicity of $\sim$ 10 - 15 minutes. The total number of counts in a pixel (summed counts) during the observation is shown in the right columns of Fig. 1, and is useful in identifying the location of the solar limb and the different network enhancements.

In the following sections we discuss our results for the network regions in the coronal hole and plumes separately and try to find whether there is any connection between these regions.

Network oscillations

We concentrate on pixel 5 (see Fig. 1 for the location), which corresponds to a network boundary. In Fig. 2, the O V intensity and velocity power spectrum for this network boundary is shown in panels (a) and (c) respectively. The lighter and the darker line correspond to the unsmoothed and smoothed power spectra respectively. The solid and dashed horizontal lines represent the 99.9% significance level of the unsmoothed power and the smoothed power respectively. The corresponding intensity and velocity variations with time are plotted in panels (b) and (d)

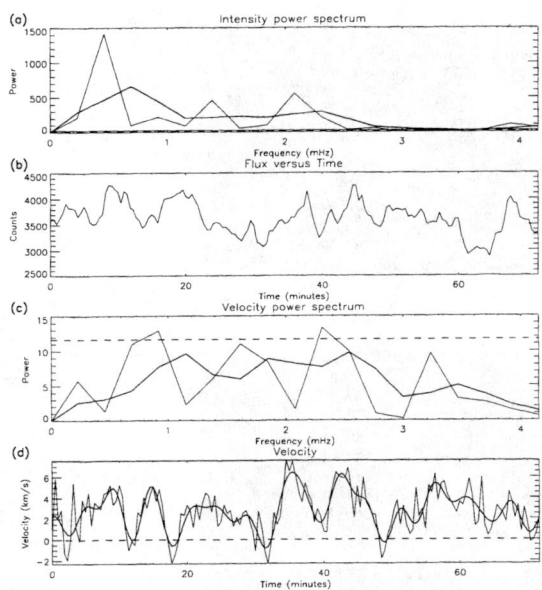

FIGURE 2. A typical network region, corresponding to pixel 5 of s8474r01 for the O v 629 Å line (see Fig. 1), (a) & (c) shows the intensity and velocity power spectra respectively, (b) shows the variation of intensity with time (light curve) and (d) shows the velocity oscillation. In panels (a) & (c) the lighter line corresponds to unsmoothed data and the bold line corresponds to the smoothed data. The solid and the dashed horizontal lines represent 99.9% significance levels for the unsmoothed and smoothed power respectively.

respectively. Since we are looking for long period (low frequency) waves, a low-pass filter of everything above 4 mHz has been applied to the velocity variations and the result of this filtering is shown as the bold line in panel (d). The O v intensity power shows a strong peak around 0.5 mHz and weaker peaks around 1.5 mHz and 2.1 mHz, whereas the velocity power shows peaks around 0.9 mHz, 1.6 mHz and 2.3 mHz, though these peaks are marginally above the significance level. Note that we also find an average redshift of 4 km s^{-1} in this network boundary (see Fig. 2d).

The results from the wavelet analysis for the same pixel are presented in Fig. 3. The thick contours in the phase plot encloses regions of greater than 99.9% confidence. Cross-hatched regions, on either side indicate the 'cone of influence' (COI), where edge effects become important. The time frequency phase plane plot of the intensity (Fig. 3, lower panel) shows significant power between 0.4-0.7 mHz, with a strong peak around 0.5 mHz ($\sim$ 33 min period) for the entire part of the observing sequence. There is also weak power around 1.7 mHz intermittently. The velocity wavelet (Fig. 3, upper panel) shows strong power around 1 mHz for the second half

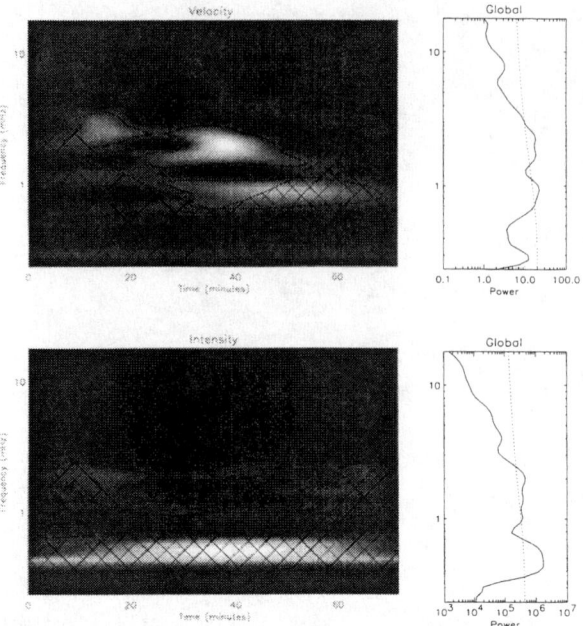

FIGURE 3. The wavelet analysis for O V in the network at the same location as Fig. 2. The left panels show the time frequency phase plot while the right hand panels show the average of the wavelet power spectrum over time.

of the observing sequence, whereas the power around 2 mHz is strong and appears on and off. In the right panel the global wavelet spectrum is plotted, which is just the average of the wavelet power spectrum over time. The dotted line in the global wavelet spectrum is again the 99.9% significance level. The smoothed Fourier transform (see Fig. 2) shows the same behaviour as the global wavelet spectrum. Note that the power around 0.5 mHz is within the 'COI', so it could be affected by edge effects, but the intermittent power around 2 mHz is significant.

Plume Oscillations

Now we turn our attention to a plume location and concentrate on pixel position 16. Fig. 4 shows the velocity and intensity variations plus the resulting power spectra. The intensity power shows a strong peak around 0.5 mHz, and a weaker peak around 1.2 mHz, the velocity power peaks around 0.2 mHz and 0.9 mHz. We find a substantial blue shift for the first twenty minutes and the last twenty minutes of the observing sequence. With the intensity wavelet (Fig. 5 lower panel), we find strong power around 1.2 mHz ($\sim$ 14 min period) for the first half of the observing sequence. These oscillations are also clear in the light curve (see Fig. 4b). There is also strong power around 0.5 mHz for the entire sequence, but we suspect that

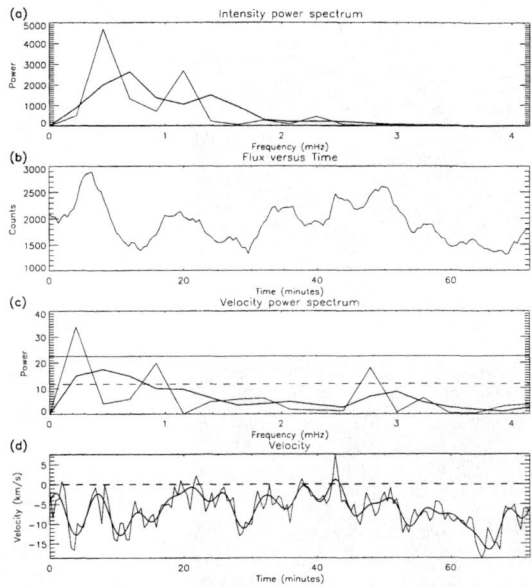

FIGURE 4. A plume region, corresponding to pixel 16 of s8474r01 for the O v 629 Å line (see Fig. 1 for location). Representations are the same as Fig. 2.

it could be due to edge effects. The global spectrum shown in the right panel also depicts the same behaviour. The velocity wavelet results on the other hand (Fig. 5 upper panel) show power around 0.9 mHz ($\sim$ 19 min) for part of the observing sequence and also around 2.8 mHz for very short durations.

DISCUSSION

High-cadence SoHO/EIT observations indicate that quasi periodic fluctuations with periods of 10-15 minutes are present in polar plumes [8] with a filamentary structure within the plume, on a spatial scale of 3-5 arc sec. These authors conclude that the waves are either sound waves or slow magneto-acoustic waves, propagating along the plumes at $\sim$ 75 – 150 km s^{-1}. Recently, [9] detected quasi-periodic variations in the polarization brightness (pB) at 1.9 $R_\odot$, at both plume and inter plume regions. Their Fourier power spectrum shows significant peaks around 1.6-2.5 mHz and additional smaller peaks at longer and shorter time-scales. Their wavelet analysis of the pB time series shows that the coherence time of the fluctuations is about 30 minutes.

Recently, [7] reported on the existence of very long period ($\sim$25 minute) compressional waves in polar plumes, using a 4 hour long time series observed with CDS in July '99. They studied the dynamics of a macro-spicule and reported on

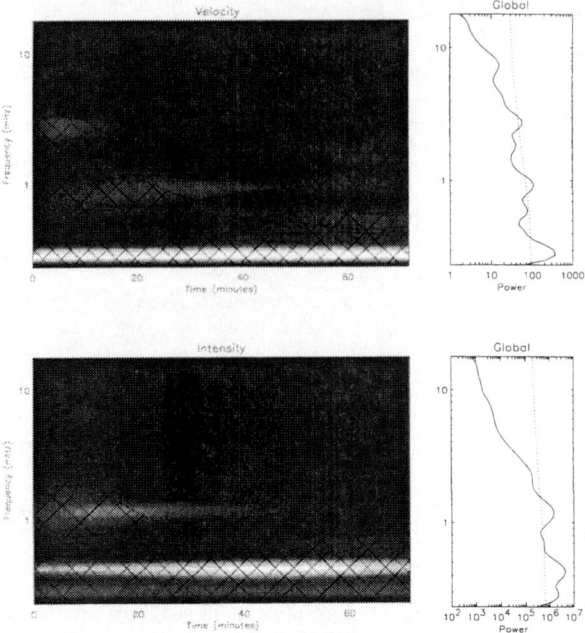

FIGURE 5. The wavelet Results for O V in the plume for dataset s8474r01 at pixel position 16 of Fig. 1 (same location as Fig. 4).

it's effect on the background plume plasma. It was noted that the macro-spicule was probably not connected with the plume. From two one hour time sequences (e.g. s8474r00 & s8474r01) [3] have shown that the dominant periodicity is in the 10-13 minute range, whereas for the longer time sequence s8488r00 ($\sim$ 2 hrs) the dominant period is $\sim$ 25 minutes, although the shorter periods are still present. Even from their contrast enhanced space time plots one can clearly see a 25 minute period. They have reported that the coherence time of the 1.5 mHz fluctuations in both regions (network and plume) is about 30-40 minutes (see their Figs. 4, 5, 7 & 8) . In the present analysis the power spectra obtained by Fourier and wavelet transforms have both established the existence of similar long periods.

From Fourier analysis of our network observations we find a clear periodicity of between 10-12 minutes in intensity. The wavelet analysis indicates that the power is most significant around 1.7 mHz ($\sim$ 10 min period) over a part of the time sequence, which suggests that the driver is working intermittently. The nature of these oscillations may be the same as those observed in the polar plumes (e.g. their intermittent nature). If one compares Fig. 3 (the network) with Fig. 5 (the plume), one can notice that the power is most significant for the same part of the time sequence and also that it peaks around 1.2 mHz. The localised (in time) nature of the wavelet transform has enabled us to quantify the duration of the significant oscillations as well as their periods. The wavelet analysis indicates that these oscillations are not present all the time. Rather it appears that something is driving the oscillations intermittently, this driving also increases the line intensity as we have significant power present during those intervals. The primary cause of the 10-25 minute periodicity is not yet known. The waves are produced in short bursts with coherence times of about 20-30 minutes. This supports the idea that the same driver is responsible for oscillations in both regions. The network boundary probably provides the magnetic channel through which the waves propagate upwards, from the lower atmosphere up to the polar plumes. Thus these waves can be modelled as waves propagating through magnetic tubes or magnetic slabs.

Finally one should note that Alfvénic oscillations are essentially velocity oscillations and do not cause any density fluctuations. The compressional modes may however reveal themselves in the form of intensity oscillations through a variation in the emission measure. This fact, together with the oscillations in intensity, allows us to interpret the waves as slow magneto-acoustic in nature.

Acknowledgements

We would like to thank the CDS team at Goddard Space Flight Center for their help in obtaining the present data.

REFERENCES

1. Hassler D M., Dammasch I. E., Lemaire P., Brekke P., Curdt W., Mason H.E., Vial Jean-Claude., Wilhelm K., 1999, *Science* **283**, 810
2. DeForest C.E., Hoeksema J.T., Gurman J.B., Thompson B.J., Plunkett S.P., Howard R., Harrison R., & Hassler D.M., 1997, *Solar Phys.* **175**, 393
3. Banerjee D., O'Shea E. , & Doyle J.G., 2000b, *Solar Phys.* (in press)
4. Harrison et al. 1995, *Solar Phys.* **162**, 233
5. Doyle J.G., van den Oord G.H.J., O'Shea E., Banerjee D., 1999, *A&A* **347**, 335
6. Torrence C., & Compo G.P., 1998, *Bull. Amer. Meteor. Soc.* **79**, 61
7. Banerjee D., O'Shea E. , & Doyle J.G., 2000a, *A&A* **355**, 1152
8. DeForest C.E., & Gurman J.B., 1998, *ApJ* **501**, L217
9. Ofman L., Romali M., Poletto G., Noci G., & Kohl J.L., 2000, *ApJ* **529**, 592

Observations of solar wave/instability phenomena as imaged by EIT/SOHO, TRACE and Yohkoh/SXT

D. Berghmans* and D. McKenzie[†]

* Royal Observatory of Belgium, Ringlaan 3, 1180 Brussel, Belgium
[†] Montana State University, 264 EPS Building, Bozeman MT 59717, USA

Abstract. On May 13 1998, active region NOAA 8218 was observed in the context of the SOHO/JOP80 campaign by an array of 8 different ground-based and space-born instruments. The emphasis was set on imaging of small-scale dynamics in this relatively small but rapidly evolving AR. In particular, SOHO/EIT (195 Å), TRACE (171 Å) and YOHKOH/SXT produced subfield image sequences at their respective highest possible rates.

We searched for wave and instability phenomena by using an automated recognition scheme. This result in a wide inventory of propagating disturbances and localised transient brightenings. By comparing the soft X-ray signature as recorded by SXT with the EUV-signature as collected by EIT and TRACE, we are able to distinguish between various types of active region transients.

As such we find that the strongest brightenings observed by EIT are indeed the EUV counterparts of the previously reported ARTBs seen by SXT. Weaker brightenings seen by EIT do often not have an X-ray counterpart. Moreover, in an extended system of faint quasi-open loops, we find propagating disturbances, with speeds of the order of 100 km/s, both in EIT and TRACE images. These are interpreted as sonic perturbations. The brightenings will be discussed in this paper while the propagating disturbances are described in the presentation by Eva Robbrecht at this conference.

INTRODUCTION

'Active region' is almost an euphimism for those areas in the solar corona that literally burst of activity on all scales. The very performant Soft X-ray telescope (SXT) revealed the frequent occurence of soft X-ray brightenings in solar active regions. In a series of papers Shimizu and coauthors (eg [8] or [9]) identified a class of events that they baptized 'active region transient brightenings' (ARTBs). They take the form of sudden brightenings of magnetic loops and last from a few minutes to tens of minutes. ARTBs are usually interpreted as dwarf versions of flares (microflares), thought alternative interpretations in terms of global mode

CP537, *Waves in Dusty, Solar, and Space Plasmas*, edited by F. Verheest, et al.
© 2000 American Institute of Physics 1-56396-962-9/00/$17.00

Alfvén wave resonant heating [6] or as unstable coronal loops [7] have also been suggested.

Since the lauch of SOHO, instruments like EIT or CDS observed frequently short-lived explosive brightenings in active regions in the EUV. These were noted in early CDS observations by [5] as strong increases in transition region lines. CDS also observed very rapid time variability in active region loops [4]. The question now arises whether all these observations of transients involve truly different phenomena or merely different aspects of the same recurrent process. With this in mind, SOHO JOP 80 was designed as a simultaneous campaign of the best solar instruments operating at their highest cadence on the same target. This approach allows to probe as much aspects as possible from the observed events.

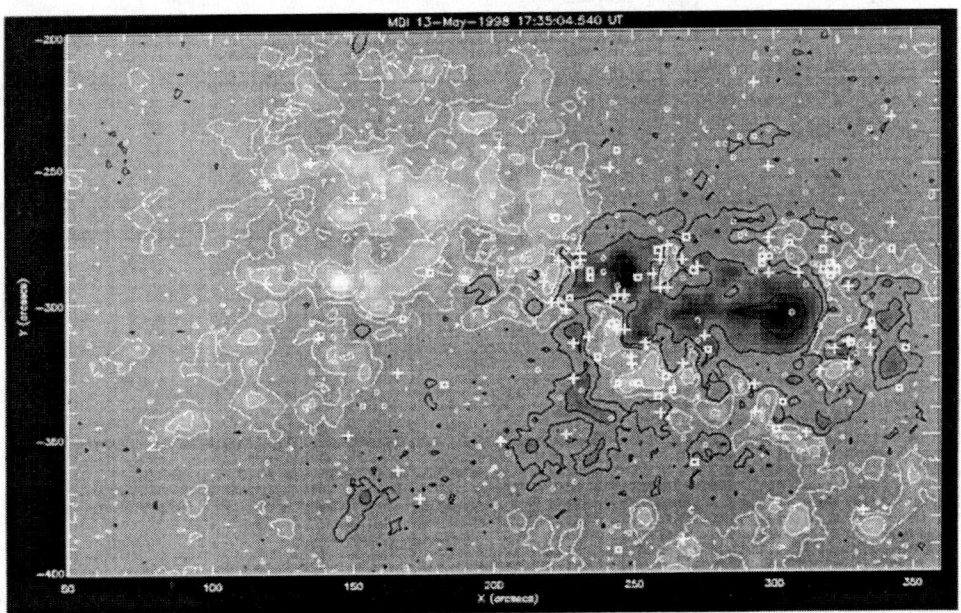

FIGURE 1. *An MDI magnetogram of active region AR8218 with bright contours encircling positive polarity regions (contours at 25, 150 and 900 Gauss) and dark contours encircling negative polarity fluxes (contours at -25, -150 and -900 Gauss). Superimposed we show the location of the events detected in the SXT image sequence (squares) and the events detected in the EIT sequence (high thresholds: crosses, low thresholds: circles).*

EIT [3] was run in 'shutterless mode' to increase the cadence up to 15 sec. AR 8218 (see Fig. 1) was imaged trough the Fe XII (195 Å) bandpass (peak formation temperature: 1.6×10^6 K) from 17h32 up to 18h29. Simultaneously, TRACE observed in the complementary 171 Å bandpass at a 25 sec cadence. SXT imaged the hotter plasma of the same active region. From 17h21, to 17h57 SXT took images at an 8 sec cadence through the Al/Mg/Mn filter, interleaved at a 98 sec cadence with the Al 1265 Å filter. Several other space born instruments (MDI,

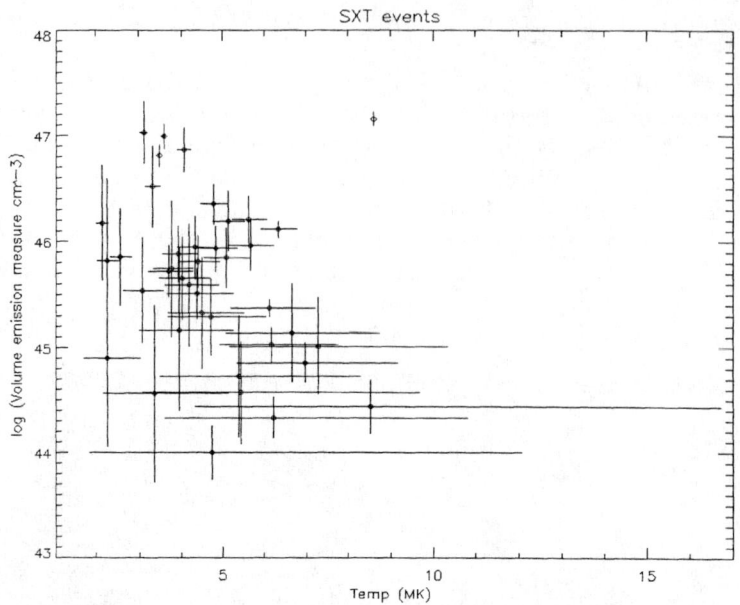

FIGURE 2. *The temperature versus the total emission measure of the brightenings detected by SXT in soft X-rays. The error bars show the 1 sigma uncertainty interval.*

CDS, SUMER), as well as two observatories on the ground (in La Palma and Sac Peak) also participated (see [1] for an overview).

DETECTION OF ACTIVE REGION TRANSIENT BRIGHTENINGS

In order to objectively determine the nature of the localised time-variability, we applied an automated scheme (see [1] and [2] for all technical details) that detects localised transient brightenings. In the SXT data sequence, we looked for events with a peak intensity exceeding a calculated background level at the $\Sigma_P = 5\,\sigma_o$ significance level. Neighbouring pixels are assumed to be part of the same event if their light curve exceeds the $\Sigma_E = 3\,\sigma_o$ significance level simultaneously. The duration of the event is traced backwards and forwards in time until the light curves involved go below the $\Sigma_D = 2\,\sigma_o$ significance level. These settings resulted in 41 events brightening in soft X-rays.

Using the Al 1265 Å images that are interleaved in between the high cadence Al/Mg/Mn images, estimations can be calculated for the peak temperature, the total emission measure and the thermal energy of these events. The variety of all these parameters (see Fig. 2 and Fig. 3) covers the same range as those for the ATRBs described by Shimizu [8]. Therefore we identify the brightenings that our

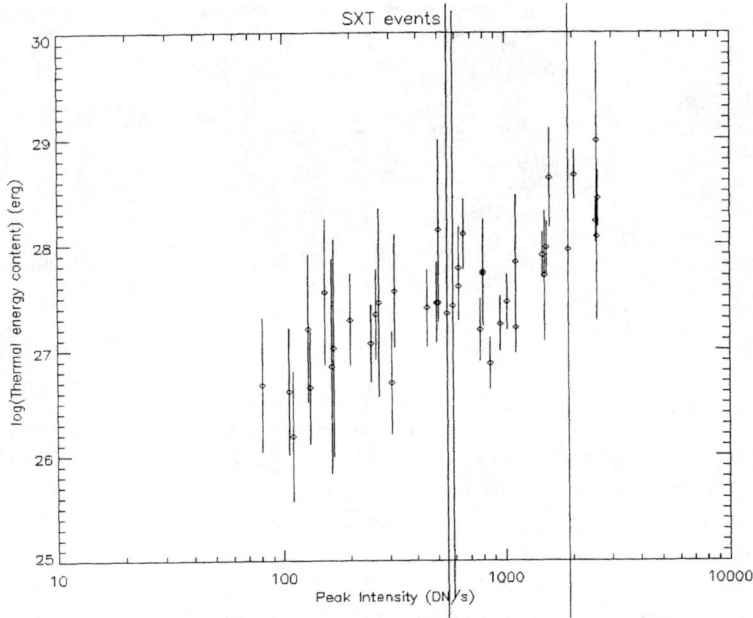

FIGURE 3. *The peak intensity versus the thermal energy content of the brightenings detected by SXT in soft X-rays. The error bars show the 1 sigma uncertainty interval.*

automated detection scheme finds in the SXT data as ARTBs.

Running the automated detection scheme with the same thresholds on the EIT data yields 373 events. We noted that the strongest events in the center of the active region tended to merge with smaller neighbours. We therefore repeated the detection procedure using the higher thresholds ($\Sigma_P = 10\,\sigma_o, \Sigma_E = 6\,\sigma_o, \Sigma_D = 4\,\sigma_o$) which yield a more appropriate estimate of the dimensions of the 68 strongest events. An extensive discussion of the appearance of these events has already been presented in [1].

In Figure 1, we have superimposed all the detected events on an MDI magnetogram. The active region consists of a leading negative flux region containing two well developped sunspots and a more diffuse trailing positive flux region. Around the two main sunspots a few 'islands' exist of the opposite (i.e. positive) polarity. It can cleary be seen that virtually all SXT events (squares) as well as most of the strongest EIT events (crosses) originate at the neutral line between the main sunspots and these nearby island of opposite polarity. The weaker EIT events (circles) however are spread more randomly over the FOV.

On average, the ARTBs observed with SXT last 5.1 min and become 145 Mm² large. In comparison, the 373 brightenings seen with EIT last on average longer (7.8 min) but remain smaller (79 Mm², on average).

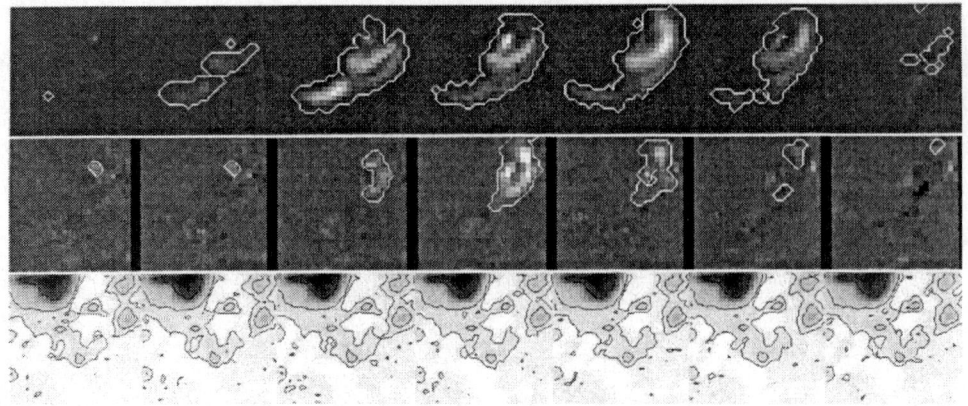

FIGURE 4. *An example of an ARTB-type 1: a microflare-subflare. Time runs from left (17h30m) to right (17h45). Top row: background subtracted SXT images (contour encircling the detected soft X-ray event); middle row: background subtracted EIT images (contour encircling the detected EUV event), bottom row: MDI magnetograms (contours at ± 25, ± 150, ± 900 Gauss)*

CLASSIFICATION OF THE BRIGHTENINGS IN EVENT CLASSES

A brightening seen in the EIT data on the same place and on the same time as a brightening seen in the SXT data, must obviously be related to each other. In such cases we can define an 'event' of which the brightenings seen by EIT and SXT are the EUV and soft X-ray signatures. Given the limited number of SXT ARTBs, our strategy was to start from the list of brightenings found in the SXT image sequence and look for each of those for corresponding brightenings in the EIT data. After that, we took a look at brightenings seen in the EUV without any (clear) soft X-ray counterpart.

At first sight, every single SXT event is unique and different from all others. A closer look however reveals that among this variety, one can find recurrent themes which we have baptized 'event classes'. To give the reader an impression of the appearance of the events, we show below a representative example of each event class and discuss the typical characteristics (morphology and evolution) of the members of each class.

ARTB-type 1, microflare-subflare

The events in this class yield the strongest brightenings in both the EUV and X-rays. Their thermal energy content, as estimated from the SXT data, is above 10^{28} erg and so these events are in the microflare-subflare range. They are all originating near the neutral lines between the main sunspots and the surrounding islands of

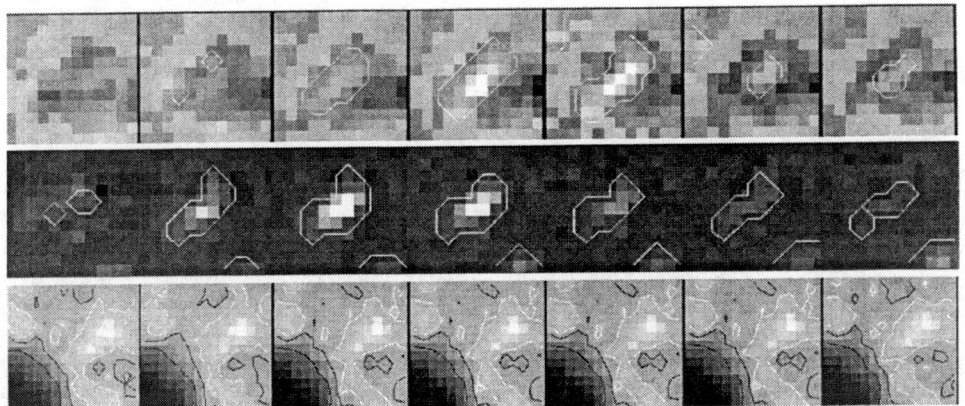

FIGURE 5. *An example of an ARTB-type 2: a penumbral brightening. Time runs from left (17h33) to right (17h37). Top row: background subtracted SXT images (contour encircling the detected soft X-ray event); middle row: background subtracted EIT images (contour encircling the detected EUV event), bottom row: MDI magnetograms (contours at $\pm$ 25, $\pm$ 150, $\pm$ 900 Gauss).*

opposite flux. Often multiple subevents occur recurrently at the same locations. Ussually several EIT brightenings can be attributed to each SXT brightening. Some of these islands of opposite flux look like a cauldron of continuing violent activity both in the EUV and in X-rays. For some other cases the situation is simpler: a brightening X-ray loop preceeded by a smaller brightening EUV-loop with which it shares a common footpoint. An example of such an event is shown in Fig. 4.

ARTB-type 2, penumbral brightenings

These are smaller events (of the order of 50 Mm^2) with the appearance of small loops which brighten both in X-rays and in the EUV. These little preexisting loops bridge the neutral line just around the main sunspot. Several cases were seen for which the X-ray brightening originated from the two footpoints simultaneously. At the moment of peak intensity the EUV brightening and the soft X-ray brightening seem to fill the same volume. The thermal energy contents peaks for these events in the range 27.3-27.7 erg and the peak X-ray flux is ussualy very small ($<$ 300 DN/sec). An example is shown in Fig. 5.

ARTB-type 3, long loop brightenings

In contrast to the more impulsive events in the previous categories, these events consist of the gradual brightening of some larger X-ray loops. No signature at all is found in the EUV. These larger loops arch in between the main preceeding and main following polarity. Whereas for the previous events a clear rise and decay phase is seen, the event in this category consists of a rise phase only: the loop stays at it's higher level of emission. The enhancement is more pronounced towards the loop's footpoints but can be found in the whole body of the loop.

ARTB-type 4, weak EUV brightenings

The events in this class are smaller and weaker EUV brightenings without any X-ray counterpart. They are ussualy detected in the EIT sequence as rather irregularly shaped but the corresponding location in the TRACE images shows clear, brightening loops. This type of brightenings is not restricted spatially to the 'core' of the active region but occurs more uniformily over the field of view.

ARTB-type 5, footpoint brightenings of extended EUV loops

In the North East of the field of view, a bundle of widely opening fieldlines is present (see Fig. 1). At the footpoints of these fieldlines (see top left of Fig. 4), a last type of events is found. These footpoint brightenings are completely invisible in X-rays and even their EUV signature is weak. These brightenings are the origin of propagating disturbances (PDs) that travel along the extended EUV loops. We refer to the presentation of Eva Robbrecht at this conference for an indepth study of these PDs.

DISCUSSION

Based on our ongoing detection and comparison of active region transient brightenings as observed by SXT and EIT, we can already present the following results:

- our detection scheme recovers ARTBs in high cadence SXT data with similar characteristics/dimensions as those described by Shimizu ([8], [9]).

- In a simulateneous & cospatial high cadence EIT sequence we discover the EUV counterparts of the ARTBs. These are much more numerous by about a factor 5.

- On average, the SXT-ARTB last 5.1 min and become 145 Mm^2 large. In comparison, the EIT-ARTB last longer (7.8 min, on average) but remain smaller (79 Mm^2, on average).

- Based on the SXT data, we estimate the radiative losses of the brightenings range from 10^{25} erg to 10^{29} erg.

- Almost all the SXT-ARTB and the bigger EIT-ARTB originate at the neutral lines between the main sunspot and nearby island of opposite polarity.

- By comparing carefully the EUV signature and the soft X-ray signature it is possible to separate different subpopulations within the group of ARTBs.

- Among the weaker events we find a new subpopulation: small scale brightenings at the footpoints of long loops, followed by wave-like disturbances that propagate upward at an apparent speed of the order of 100 km/s. These wave-like disturbances are discussed in a paper by Eva Robbrecht at this conference.

The smaller size and the higher birthrate that we find in EIT as compared to SXT, are both consistent with the hypothesis that most of the Fe XII brightenings detected by EIT, correspond to weaker energy releases for which no plasma is heated beyond $2 \times 10^6 K$. These transients are thus undetectable in (higher-temperature) soft X-rays.

REFERENCES

1. Berghmans, D. and Clette, F., 1999, Sol. Phys., 186, 207
2. Berghmans, D., Clette, F. and Moses, D., 1998, Astron. Astrophys., 336, 1039
3. Delaboudinière, J.-P. et al, 1995, Solar Phys., 162, 291
4. Kjeldseth-Moe, O. and Brekke, P., 1998, Solar Phys., 182, 73
5. Mason, H.E., Young, P.R., Pike, C.D., Harrison, R.A., Fludra, A., Bromage, B.J.I. and Del Zanna, G. 1997, Solar Phys., 170 ,143
6. Ofman, L., Davila, J.M., Shimizu, T., 1996, Astrophys. J. Lett., 459, L39
7. Arber, T.D., Longbottom, A.W, and Van der Linden, R.A.M., Astophys. J., 517, 990
8. Shimizu, T., Tsuneta, S., Acton, L.W., Lemen, J.R. and Uchida, Y., 1992, PASJ, 44, L147
9. Shimizu, T., Tsuneta, S., 1997, Astrophys. J., 486,1045

Convective collapse in a thin flux tube

David Boddie and B. Roberts

School of Mathematics and Statistics
University of St Andrews
St Andrews, KY16 9SS
Scotland

Abstract. The magnetic field in the photospheric layers of the Sun outside sunspots is found to be in the form of concentrated magnetic flux tubes of 1–2 kG field strength. The model of convective collapse in a thin flux tube, where an instability causes "draining" of the tube and concentration of the magnetic field, has proved useful in attempting to explain the inferred field concentrations. Here we explore the mechanism of convective instability allowing for the inflow of matter in the upper reaches of the tube.

INTRODUCTION

The determination of field strengths in the magnetic flux elements of the solar photosphere is a basic problem in solar physics. Both numerical models and stability analyses have been employed to account for the range of field strengths detected. Flux elements in the intranetwork regions have typical field strengths of below 500 G [1], whereas network fields are generally 1000–2000 G [2]. There is no established theory to explain the coexistence of these two field regimes, although the intranetwork fields can be interpreted as belonging to structures with typical equipartition field strengths in the photosphere, where magnetic energy balances the kinetic energy of convection. However, alternative explanations for weak fields in the intranetwork, such as a weak collapse scheme or a description of a shallow flux concentration, exist [3] with this explanation.

The process of convective collapse [4–7], whereby cool material in the upper reaches of a flux tube moves down towards the lower, super-adiabatically stratified regions of the tube, thus allowing a compression of the upper parts under the action of a confining plasma pressure, has been proposed as a mechanism for producing kilogauss field strengths. Such a mechanism has been modelled extensively using stability analyses [5,7,8] and numerical models [10,11], and has been supported by recent observations [1,12].

CP537, *Waves in Dusty, Solar, and Space Plasmas*, edited by F. Verheest, et al.
© 2000 American Institute of Physics 1-56396-962-9/00/$17.00

BACKGROUND THEORY

We describe the dynamics of an intense flux tube as a perfectly conducting ideal gas embedded in a magnetic field using the equations of ideal magnetohydrodynamics (MHD). We ignore the effects of radiative transfer (see [10]). To illustrate the general approach, consider the momentum equation:

$$\rho \left(\frac{\partial \mathbf{v}}{\partial t} + (\mathbf{v} \cdot \nabla) \mathbf{v} \right) = -\nabla \left(p + \frac{\mathbf{B} \cdot \mathbf{B}}{2\mu_0} \right) + \rho \mathbf{g} + \frac{1}{\mu_0} (\mathbf{B} \cdot \nabla) \mathbf{B}. \tag{1}$$

Here p and ρ are the pressure and density of a plasma with permeability μ_0, $\mathbf{v}$ is the flow velocity, $\mathbf{B}$ is the magnetic field, and $\mathbf{g}$ is the gravitational acceleration.

We employ the thin flux tube approximation [9] for a vertical flux tube with untwisted magnetic field located at the solar surface, so that we are able to describe axisymmetric perturbations about the equilibrium configuration of the tube. This is achieved by performing an expansion about the central axis of the tube, writing

$$\mathbf{B} = B(r = 0, z, t)\hat{\mathbf{z}} + O(\epsilon), \qquad \mathbf{v} = v(r = 0, z, t)\hat{\mathbf{z}} + O(\epsilon),$$
$$p = p(r = 0, z, t), \qquad \rho = \rho(r = 0, z, t), \tag{2}$$

where $\epsilon \equiv r_0/L_0$ is a measure of the thinness of a tube with vertical variations occurring over a lengthscale L_0 assumed to be much greater than the variations over a radial lengthscale r_0.

Consider the equilibrium case where there is no plasma flow inside the tube. The momentum equation (1), with the acceleration due to gravity taken as $\mathbf{g} = -g\hat{\mathbf{z}}$, describes a vertically stratified medium. Using the ideal gas law, the equilibrium pressure $p_0(z)$ within the tube is given by

$$p_0(z) = p_0(0) \exp \left[-\int_0^z dz'/\Lambda_0(z') \right] \tag{3}$$

where $\Lambda_0(z) = kT_0/\hat{m}g$ is the pressure scale height inside a tube of temperature $T_0(z)$, assumed to be equal to the temperature of the external hydrostatic atmosphere, see [9] for details. In this case, the plasma beta, $\beta_0(z) = 2c_{s0}^2(z)/\gamma v_{A0}^2(z)$, is a constant, for sound speed $c_{s0}(z)$ and Alfvén speed $v_{A0}(z)$ in the tube.

The thin tube equations for perturbations about this equilibrium are written in the form [9]

$$\rho_0(z) \frac{\partial v}{\partial t} = -\frac{\partial p}{\partial z} - \rho g, \tag{4}$$

$$\frac{\partial p}{\partial t} + v \frac{\partial p_0(z)}{\partial z} = c_{s0}^2(z) \left(\frac{\partial \rho}{\partial t} + v \frac{\partial \rho_0(z)}{\partial z} \right), \tag{5}$$

$$B_0(z) \frac{\partial \rho}{\partial t} - \rho_0(z) \frac{\partial B}{\partial t} + \left(B_0(z) \frac{\partial \rho_0(z)}{\partial z} - \rho_0(z) \frac{\partial B_0(z)}{\partial z} \right) v + \rho_0(z) B_0(z) \frac{\partial v}{\partial z} = 0, \tag{6}$$

where the "0" suffix denotes an equilibrium value and quantities without a suffix describe perturbations about the equilibrium. We consider pressure perturbations in the environment to be negligible. Assuming an exponential form for the time dependence, so that the vertical flow is of the form $v(z,t) = \hat{v}(z)e^{i\omega t}$, for frequency ω (other variables are Fourier analysed similarly), an ordinary differential equation determines the velocity amplitude $\hat{v}$ (see [9]):

$$\frac{d^2\hat{v}}{dz^2} - \frac{1}{2\Lambda_0(z)}\frac{d\hat{v}}{dz} + \left(\frac{\omega^2 - N_0^2(z)}{c_{T0}^2(z)} + \left(1 - \frac{\gamma}{2}\right)\frac{N_0^2(z)}{c_{s0}^2(z)}\right)\hat{v} = 0, \tag{7}$$

where the square of the Brunt-Väisälä frequency, $N_0^2(z)$, is given by

$$N_0^2(z) = \frac{g}{\Lambda_0(z)}\left(\frac{\gamma-1}{\gamma} + \Lambda_0'(z)\right), \tag{8}$$

and the tube speed, c_{T0}, is defined as $c_{T0} = c_{s0}v_{A0}/\sqrt{c_{s0}^2 + v_{A0}^2}$. The dash (') refers to the derivative with respect to z. Our analysis of the problem of convective collapse in the flux tube is concerned with the properties of equation (7).

MODEL AND RESULTS

We model the interior of the flux tube by two sections, the atmosphere ($z > 0$) above $z = 0$ (which corresponds to an optical depth of $\tau_{5000} = 1$) and a convection zone ($z < 0$) lying below. We use a local analysis (see also [5]) for the region $z < 0$, employing the approximation that all quantities are taken to be constant with depth; a WKB analysis is also possible. The atmosphere ($z > 0$) of the tube is taken to be isothermal with $\Lambda_0(z) = \Lambda_i$, a constant.

It is assumed that the perturbed velocity has an evanescent form in the isothermal atmosphere: this is somewhat analogous to numerical approaches [11,13] and earlier analytical work where the upper parts of the tube were closed by setting $\hat{v}(z = 0) = 0$. In a tube of finite depth, the lower boundary condition can be set to either no motion at depth $z = -d$, $\hat{v}(-d) = 0$, or to allow motion through the lower boundary, but taking $d\hat{v}/dz = 0$ at $z = -d$. Although the former condition may be regarded as somewhat artificial [11], it is convenient to consider it so that we can compare results with those obtained earlier.

At the interface $z = 0$ between the convection zone and the isothermal atmosphere, we require that $\hat{v}$ and $d\hat{v}/dz$ are continuous. This leads to a dispersion relation of the form

$$\frac{1}{4\Lambda_0} + \kappa_0\cot(\kappa_0 d) = \frac{1}{4\Lambda_i} - \kappa_i \tag{9}$$

where the effective wavenumbers in the convection zone and isothermal atmosphere, κ_0 and κ_i, are defined as

$$\kappa_0 = \left[\frac{\omega^2}{c_{T0}^2} - \frac{1}{\Lambda_0^2}\left(\frac{9}{16} - \frac{1}{2\gamma} + \left[\frac{\gamma-1}{\gamma^2}\right]\frac{c_{s0}^2}{v_{A0}^2} + \frac{\Lambda_0'}{\gamma}\left[\frac{\gamma}{2} + \frac{c_{s0}^2}{v_{A0}^2}\right]\right)\right]^{1/2},$$

$$\kappa_i = \left[\frac{1}{\Lambda_i^2}\left(\frac{9}{16} - \frac{1}{2\gamma} + \left[\frac{\gamma-1}{\gamma^2}\right]\frac{c_{si}^2}{v_{Ai}^2}\right) - \frac{\omega^2}{c_{Ti}^2}\right]^{1/2}. \qquad (10)$$

The sound and Alfvén speeds in the convection zone ($z < 0$) are denoted by c_{s0} and v_{A0}; their values in the upper atmosphere ($z > 0$) of the tube are c_{si} and v_{Ai}, respectively. The dispersion relation (9) contains the local dispersion relation for a vertical flux tube with closed boundary conditions, $\hat{v}(-d) = 0$ and $\hat{v}(0) = 0$, obtained formally by requiring that $\Lambda_i \to 0$.

We are interested in the photospheric magnetic field strength $B_0(z = 0)$ of the tube, determined as a function of the tube depth d and the gradient in the pressure scale height Λ_0'. It is useful to determine whether there are unstable modes for various choices of the depth d of the tube, its equilibrium field strength $B_0(z = 0)$ and the strength of the super-adiabaticity, as measured by Λ_0'; adiabatic stratification corresponds to $\Lambda_0' = -1/6$ for $\gamma = 1.2$. We allow also for freedom in the choice of the adiabatic index γ. We look for the existence of unstable modes (corresponding to $\omega^2 < 0$) for a range of depths and field strengths with given gradient in pressure scale height.

We require that the velocity by evanescent in the isothermal atmosphere; this implies that κ_i be a real quantity, providing in general an upper limit on ω^2. Similarly, the requirement that the velocity be oscillatory in the convection zone implies that there is a lower limit on ω^2. For unstable modes to be present, we require that this lower limit allow $\omega^2 < 0$ and hence that

$$\frac{B_0^2(0)}{\mu p_0(0)} < -\frac{(\Lambda_0'(0) + (\gamma-1)/\gamma)}{9/16 - 1/(2\gamma) + \Lambda_0'(0)/2}. \qquad (11)$$

Figure 1 gives the stability diagram dividing stable modes from unstable modes. The unstable region lies above the curve for each value of the scale height gradient. The curves were determined for the case where the pressure scale height in the isothermal atmosphere was taken to be the same as that in the convection zone, having an equivalent photospheric temperature of 5000 K. In the case where the pressure scale height in the isothermal regime is adjusted so that $\Lambda_i \to 0$, the dispersion relation reproduces the local results for a closed tube [5]. Small values of the isothermal scale height produce results which are in good agreement with those from this purely local model. Alternatively, setting the scale height Λ_i to large values allows instability to occur at higher field strengths for a given depth. However, this enhanced instability falls some way short of that exhibited by the purely local model with an open boundary at $z = 0$. These comparisons are made in Figures 2 and 3.

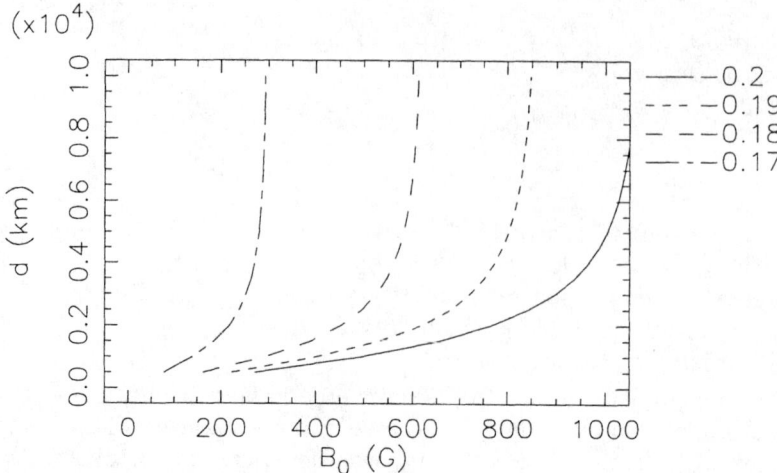

FIGURE 1. Divisions between stable and unstable modes in a thin flux tube. Unstable modes exist above each curve, drawn for various choices of the scale height gradient $\Lambda'_0 = -0.20$ to -0.17, with adiabatic index $\gamma = 1.2$.

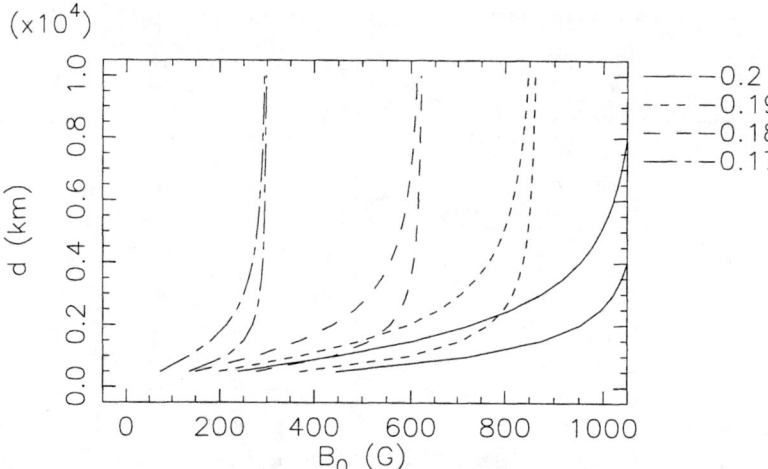

FIGURE 2. The stability boundary for a closed flux tube compared with an open tube (Fig. 1). For given depth, an open tube requires a stronger field strength for stability than a closed tube. The adiabatic index is $\gamma = 1.2$.

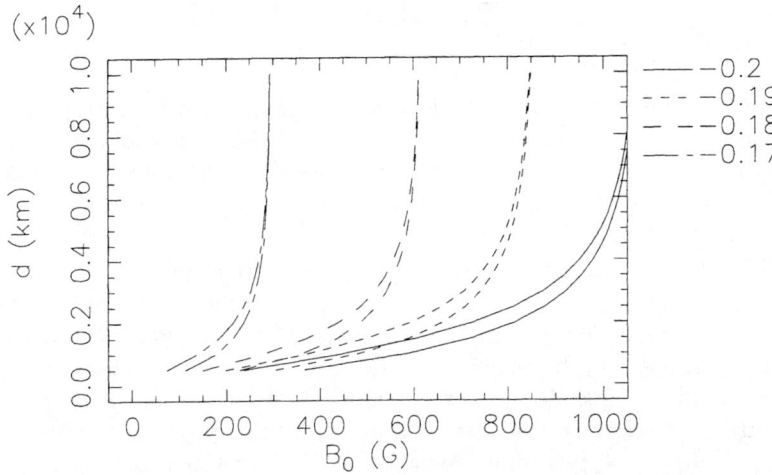

FIGURE 3. Stability curves (as defined in Fig. 1) for the model matching the local and isothermal media for the cases of hot and cold isothermal atmospheres. A hot atmosphere allows instability at higher field strengths. The adiabatic index is $\gamma = 1.2$.

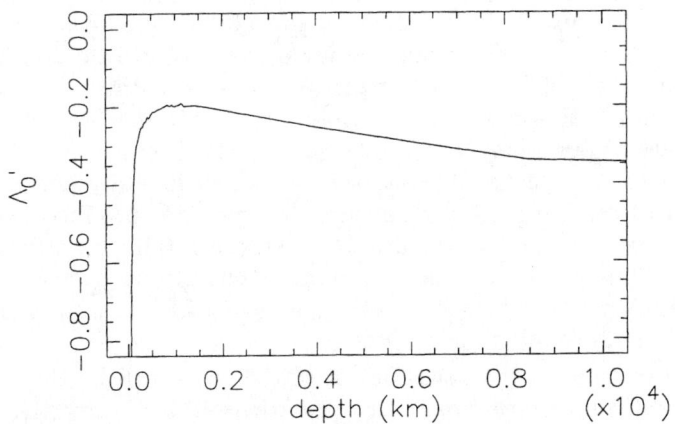

FIGURE 4. The variation in the pressure scale height gradient, Λ_0', with depth in the convection zone, as given by Spruit's [14] model. Note that the maximum gradient is $\Lambda_0' \approx -0.19$.

DISCUSSION

The general form of the curves indicating the maximum field at which unstable modes exist for a tube of given depth is intuitively what we would expect: the tube is susceptible to instability at low field strengths and large depths, and only stable modes exist for the strongest fields. We can use our model to set limits on the field strength of flux tubes.

The form of the scale height gradient is shown in Figure 4, taking data from the convection zone model of Spruit [14]. The gradient becomes approximately constant over a region near its maximum value, $\Lambda'_0 \approx -0.190$, which enables us to choose tubes which may be unstable at fields up to 800 G. Since at fields above this critical field strength for a particular depth there are no unstable modes, to describe intranetwork magnetic elements through this approach requires us to view them as shallow flux tubes. For values of the gradient below the maximum, e.g. at $\Lambda'_0 \approx -0.3$, the tube is more susceptible to unstable modes at higher fields and therefore the model may be used to describe more intense magnetic structures.

For the case where the adiabatic index is chosen to be $\gamma = 5/3$ the allowed values for the scale height gradient at the photosphere are restricted to a range which are lower than those available at $\gamma = 1.2$. However, the profile for Λ'_0 (see Fig. 4) is clearly rapidly varying when it passes through this range of values, leading us to conclude that our model requires a non-local treatment.

A model which includes an isothermal atmosphere exhibits behaviour which lies between the extremes of the purely local model with closed boundaries and one with an open boundary. This may be understood by considering that the scale height is a measure of behaviour of plasma density in the tube. Therefore, the case corresponding to a closed tube involves an isothermal atmosphere with high plasma density, suppressing the downflows associated with instability, whereas an open tube involves a low density atmosphere which is easily perturbed and more susceptible to instability. The parameters we consider produce results which lie between these two extremes, but quite close to those obtained with a cold isothermal atmosphere. Thus, the case of closed boundary conditions gives results that are in reasonable agreement with those obtained allowing for an atmosphere, at least as far as a local analysis allows us to judge.

An important step in the development of this work will be the inclusion of the effects of a realistic variation in super-adiabaticity in the convective layer.

ACKNOWLEDGMENTS

David Boddie would like to acknowledge financial support from the Particle Physics and Astronomy Research Council (PPARC) of the UK.

REFERENCES

1. Solanki, S.K., Zufferey, D., Lin, H., Rüedi, L., Kuhn, J.R., *Astron. Astrophys.* **310**, L33–L36 (1996).
2. Solanki, S.K., "Dynamics of flux tubes in the solar atmosphere" in *Solar and Heliospheric Plasma Physics*, edited by G.M. Simnett, C.E. Alissandrakis, L. Vlahos, Springer, p. 49 (1997).
3. Rutten, R.J., "(Inter-) Network Structure and Dynamics" in *Third Advances in Solar Physics Euroconference: Magnetic Fields and Oscillations*, edited by B. Schmieder, A. Hofmann, J. Staude, ASP Conference Series 184, pp. 181–200 (1999).
4. Parker, E.N., *Astrophys. J.* **221**, 368–377 (1978).
5. Webb, A.R. and Roberts, B., *Solar Phys.* **59**, 249–274 (1978).
6. Spruit, H.C., *Solar Phys.* **61**, 363–378 (1979).
7. Spruit, H.C. and Zweibel, E.G., *Solar Phys.* **62**, 15–22 (1979).
8. Hasan, S.S., *Mon. Not. R. Astron. Soc.* **219**, 357–372 (1986).
9. Roberts B. and Webb, A.R., *Solar Phys.* **56**, 5–35 (1978).
10. Hasan, S.S., *Astron. Astrophys.* **143**, 39–45 (1985).
11. Takeuchi, A., *Astrophys. J.* **522**, 518–523 (1999).
12. Lin, H. and Rimmele, T., *Astrophys. J.* **514**, 448–455 (1999).
13. Steiner, O., "Flux Tube Dynamics" in *Third Advances in Solar Physics Euroconference: Magnetic Fields and Oscillations*, edited by B. Schmieder, A. Hofmann, J. Staude, ASP Conference Series 184, pp. 38–54 (1999).
14. Spruit, H. C., *Solar Phys.* **34**, 277–290 (1974).

Alfvén waves in a plane parallel flowing atmosphere

L. M. B. C. Campos, N. L. Isaeva and P. J. S. Gil

Secção de Mecânica Aeroespacial, ISR,
Instituto Superior Técnico, 1049-001 Lisboa, Portugal.

Abstract. The present paper considers Alfvén waves in an inhomogeneous medium, under a non-uniform external magnetic field and in the presence of a non-uniform flow, to obtain the wave equations for the velocity and magnetic field perturbations in the case of propagation along a vertical uniform external magnetic field in an atmosphere with a vertical mean flow velocity and satisfying mass conservation. The wave equation has a critical layer where the mean flow velocity equals the Alfvén speed, and in the case of an isothermal atmosphere, the wave equation for the magnetic field perturbation can be reduced to a Bessel type with complex order and imaginary variable. Thus it is possible to obtain exact solutions for all frequencies and distances, discussing the cases in which there are propagating or evanescent waves far below or far above the critical layer i.e. the latter acts as a reflecting layer. The solution in the vicinity of the critical layer allows the amplitude and phase of the wave field to be plotted, as a function of distance from the critical layer (measured in scale heights), for different values of the dimensionless frequency and initial Alfvén number (ratio of flow velocity to Alfvén speed). One application is the solar atmosphere, for which the MHD equations are appropriate, and the wavelengths are large compared with the density scale height, thus excluding the ray approximation.

INTRODUCTION

The observation of Alfvén waves in the solar wind, call attention to several effects, namely stratification, the background flow and the change in strength and direction of the external magnetic field among others. These have been considered for Alfvén waves in the solar wind, mostly using the JWKB, or ray approximation, i.e., that the wavelength is short compared with the lengthscale of changes in background properties (e.g. mass density, mean flow velocity, and magnetic field strength and direction); the problem has also been addressed without making the ray approximation. It is perhaps useful conceptually to look first separately at the effects of (i) the non-uniform external magnetic field of varying direction and (ii) the mean flow. The latter is the purpose of the present paper.

The effects of mean flow and non-uniform magnetic field on Alfvén waves in the solar wind are indeed separated in the initial stages of coronal expansion [1]. The

CP537, *Waves in Dusty, Solar, and Space Plasmas*, edited by F. Verheest, et al.
© 2000 American Institute of Physics 1-56396-962-9/00/$17.00

atmosphere may then be treated as plane parallel, under a uniform vertical magnetic field; the flow velocity must vary with altitude inversely to the mass density, to conserve the mass flux. A model of Alfvén waves with vertical external magnetic field and background flow is thus relevant to the initial stages of solar wind expansion in the corona. It is also relevant to the solar photosphere and chromosphere [2,3], since it has been shown that Alfvén waves can propagate in magnetic flux tubes [4], and in magnetic slabs [5]; besides, it has been shown that both the magnetic flux tubes in the chromosphere and photosphere, and magnetic arcades in the corona, can support internal flows [6]. In this case of vertical magnetic flux tube, with upward flow, the magnetic field would be internal.

ALFVÉN WAVE EQUATIONS WITH FLOW

The fundamental equations of non-dissipative (or ideal) MHD in the case of non-uniform, unsteady perturbations $\mathbf{v}, \mathbf{h}$, superimposed on a steady, non-uniform mean flow $\mathbf{U}$ and background magnetic field $\mathbf{B}$ can be linearized for small perturbations. If these are assumed to be incompressible then the linearized equations for small perturbations take a relative simple form.

Alfvén waves with non-uniform flow and mass density

For the present case of a plane parallel atmosphere under a uniform vertical magnetic field $\mathbf{B} = B\mathbf{e}_z$ the geometry is simpler than in the three-dimensional case. In addition, Maxwell's equation $0 = \nabla \cdot \mathbf{B} = B'$ implies that $\mathbf{B}$ must be uniform. It is necessary to consider a non-uniform vertical mean flow $\mathbf{U} = U(z)\mathbf{e}_z$ so that the mass flux is conserved $\rho(z) U(z) = const$, in a 'stratified' medium for which the mass density $\rho(z)$ depends on altitude z. The velocity $\mathbf{v}$ and magnetic field $\mathbf{h}$ perturbations also depend on altitude and time t i.e. $\mathbf{v}, \mathbf{h} = v, h(z,t)\mathbf{e}_x$, and since they are parallel, they may be taken along the x-axis. The linearized equations for the perturbations take then the simpler form:

$$\dot{h} + (Uh)' = Bv', \qquad \dot{v} + Uv' = (\mu B/4\pi\rho)\, h', \tag{1}$$

where dot and prime denotes derivative respectively with regard to time and altitude. The mean state momentum equation simplifies to

$$\rho U'U + p' = -\rho g. \tag{2}$$

In the case presented here of Alfvén waves in a flowing atmosphere the equation for the magnetic field perturbation, deduced from (1), is

$$\ddot{h} + 2\left(U\dot{h}\right)' + \left[U\left(Uh\right)'\right]' - \left(A^2 h'\right)' = 0, \tag{3}$$

185

where A is the Alfvén speed, defined by $A^2 \equiv \mu B^2/4\pi\rho$. The equation for the magnetic field perturbation is simpler than the one for the velocity perturbation (not shown here).

Since the mean state does not depend on time, it is convenient to use a Fourier decomposition. The equation that the magnetic field perturbation spectrum $H(z; w)$ must satisfy has two singularities when $U = \pm A$, and since $U, A > 0$ the physically relevant one is the critical layer at an altitude z_* such that the mean flow velocity and Alfvén speed are equal $U(z_*) = A(z_*)$. The exact location of the critical layer, and the altitude dependence of the coefficients of the equation that describes the perturbations, are determined by the background state, specified below. Most of the literature is concerned with Alfvén waves in an atmosphere at rest, under an uniform magnetic field, for which hydrostatic equilibrium applies. In the present case, of Alfvén waves in a flowing atmosphere, the background conditions are modified, thus justifying a discussion of the mean state.

Isothermal atmosphere with vertical flow as mean state

The mean state is determined by the equation of state for a perfect gas, equation (2) and the equation of continuity $\rho U = \rho_0 u$ where $u \equiv U(0)$ its value at zero altitude. There are three equations and four variables, viz. pressure $p(z)$, density $\rho(z)$, temperature $T(z)$ and velocity $U(z)$, so one of them can be chosen arbitrarily. In the case of an isothermal atmosphere, this leads to $u/U(z) = \rho(z)/\rho_0 = p(z)/p_0$ and to an implicit equation specifying pressure profile

$$p(z)/p_0 = e^{-z/L} \exp\{(u^2/2RT_0)[1 - (p_0/p(z))^2]\} \tag{4}$$

where $L \equiv RT_0/g$ is the scale height. In the absence of mean flow u = 0, the pressure profile is explicit and simplifies to

$$p(z)/p_0 = e^{-z/L} = \rho(z)/\rho_0 \tag{5}$$

The same result will hold, at least for small altitude, provided that

$$1 \gg u^2/2RT_0 = (\gamma/2)(u/c)^2 = (\gamma/2)M^2 \tag{6}$$

where $c = \sqrt{\gamma RT_0}$ is the sound speed, and $M \equiv u/c$ the initial Mach number of the mean flow. In these conditions, of mean flow with low initial Mach number

$$M^2 \ll 2/\gamma: \qquad p(z)/p_0 = \rho(z)/\rho_0 = e^{-z/L} = u/U(z) \tag{7}$$

specifies the pressure, density and velocity profiles for the isothermal atmosphere, subject to the restriction [see (4)]

$$[p(z)/p_0]^2 \gg u^2/(2RT_0) = (\gamma/2)M^2 \tag{8}$$

which puts an upper bound on the range of altitudes over which (7) may be used. Note that the latter imply for the Alfvén speed the profile

$$A(z) = ae^{z/2L}, \qquad a \equiv B\sqrt{\mu/4\pi\rho_0} \tag{9}$$

where a is the value at altitude zero.

Using (7) at the critical layer, where the mean flow velocity equals the Alfvén speed and noting that $p(z)^2 \leq p_0$, it follows that (8) is satisfied at the critical layer if any of the following three equivalent conditions is met:

$$[A(z)]^2 \ll 2RT_0 \Leftrightarrow [A(z)]^2 \ll (2/\gamma)c^2 \Leftrightarrow p_0 \gg \mu B^2/8\pi \equiv P, \tag{10}$$

the latter stating that the gas pressure at altitude zero is much larger than the magnetic pressure, which is constant for an uniform external magnetic field. In conclusion, a self-consistent background or mean state is specified by the stratification laws, for substitution in the wave equation, over an altitude range extending beyond the critical layer.

Solution of the wave equation

Substituting the mean state stratification laws into the wave equation (3) for the magnetic field perturbation and performing the Fourier decomposition one gets the equation for $H(z;\omega)$

$$\left(1 - N^2 e^{z/L}\right) L^2 H'' + \left(2i\Omega N + 1 - 3N^2 e^{z/L}\right) LH'$$
$$+ \left(\Omega^2 e^{-z/L} + 2i\Omega N - 2N^2 e^{z/L}\right) H = 0, \tag{11}$$

which involves two dimensionless quantities, the dimensionless frequency $\Omega \equiv \omega L/a$ and the Alfvén number $N = u/a$.

The dimensionless frequency is also $\Omega = \omega L/a = kL = 2\pi L/\lambda$ a measure of the ratio of the scale height to the local wavelength $\lambda = 2\pi/k = 2\pi a/\omega$; the JWKB or ray approximation assumes that the medium is nearly homogeneous on a wavelength scale $\lambda^2 \ll L^2$, i.e. $\Omega^2 \gg 1$. In the present paper all values of Ω are allowed, including the limit opposite to ray theory, of long wavelengths $\Omega \ll 1$. The Alfvén number (or Alfvén Mach number), i.e. the ratio of initial flow to Alfvén speed is the hydromagnetic analogue of the hydrodynamic Mach number, defined as the ratio of flow velocity to sound speed $M \equiv u/c$. The second condition in (10) implies that

$$N^2 \equiv (u/a)^2 \gg (\gamma/2)(u/c)^2 = (\gamma/2)M^2, \tag{12}$$

the initial Alfvén number is much larger than the initial Mach number, which is a modest restriction, since the latter is small by (6). This non-intrusive constraint follows from the condition of initial gas pressure much larger than the magnetic pressure. Note that the gas pressure decays with altitude in an isothermal atmosphere, whereas the magnetic pressure remains constant, for a uniform external magnetic field. For example, in the solar photosphere, where Alfvén waves are generated [7–10], the gas pressure dominates the magnetic pressure except at granulation boundaries and sunspots, where they can be comparable; farther into the corona the magnetic pressure dominates the gas pressure [11]. Even starting at an altitude z = 0 with a small Alfvén number, the Alfvén number will increase with altitude $U(z)/A(z) = (u/a)e^{z/2L}$, becoming unity at the critical layer:

$$U(z_*) = A(z_*) : \qquad z_* = 2L \log(a/u) = -2L \log N, \tag{13}$$

which is a singularity of the wave equation (11).

The wave equation can be transformed into a Bessel equation of nonintegral order that only involves one single dimensionless parameter $K \equiv \Omega N = \omega L u/a^2$. Its solution, specifying the wave field at all altitudes, is

$$H(z; \omega) = s^{-1/2+iK} (1 - s) Z_{1-2iK} \left(2iK\sqrt{s}\right) \tag{14}$$

where $s = 1 - N^{-2}e^{-z/L} = 1 - e^{-(z-z_*)/L}$ vanishes at the critical layer $z = z_*$ and Z_ν is a linear combination of Bessel J or Hankel H functions

$$Z_\nu(r) = C_1 J_\nu(r) + C_2 J_{-\nu}(r) = C_+ H_\nu^{(1)}(r) + C_- H_\nu^{(2)}(r). \tag{15}$$

Boundary conditions, transmission coefficient at the critical layer and wave reflection

Asymptotic analysis of the solution shows that in the deep layers $H^{(1)}$ represents an upward propagating wave and $H^{(2)}$ a downward propagating wave. Thus application of a 'radiation condition' far below the critical layer, implies choosing $C_+ = 0$ for downward, and $C_- = 0$ for upward, propagating waves. The identification of upward and downward propagating waves concerns only the general behaviour. The exact solution takes into account gradual reflection of the waves by the stratified medium. Another type of boundary condition is to specify the magnetic field perturbation at some altitude e.g. $H(z_1; \omega) = H_1(\omega), H(z_2; \omega) = H_2(\omega)$. Two boundary conditions of this type, or one boundary condition and one radiation condition, specify completely the two independent constants of integration in the solution. Note that at high altitude, far above the critical layer, the wave field vanishes so that this yields no useful boundary condition.

To study the wave field near the critical layer $z \to z_*$ or $s \to 0$, it is preferable to use Bessel rather than Hankel functions since the general solution can then be written as a linear combination of solutions, one finite at the critical layer

$$H^+(z; \omega) \sim [\Gamma(2 - 2iK)]^{-1} \exp\left[(1 - 2iK) \log(iK)\right] \tag{16}$$

and the other divergent that can be put on the form

$$H^-(z;\omega) \sim [(z - z_*)/L]^{-1} \exp\{2iK \log[|z - z_*|/L]\} \times \begin{cases} 1 & \text{if } z > z_* \\ e^{2K\pi} & \text{if } z < z_* \end{cases} \quad (17)$$

This result corresponds to a transmission coefficient

$$T = e^{-\phi} = e^{-2\pi K} = e^{-2\pi\omega Lu/a^2} \quad (18)$$

which depends only on the scattering parameter K. For low-frequency waves in the sense $\phi \ll 1$, the transmission coefficient is about unity $T \sim 1$, whereas for high-frequency waves $\phi \gg 1$, it is much smaller than unity, showing that waves are almost totally reflected. The transmission coefficient is much smaller than unity for frequencies above $\omega_* \equiv a^2/2\pi Lu$. Note that what is in question here is wave reflection localized at the critical layer, in addition to the gradual wave reflection by the stratification of the medium. These effects are illustrated by plotting the wave fields (see figure 1).

Results are presented using the dimensionless variable $Z \equiv (z - z_*)/L$ which has two advantages: the critical layer lies at the origin and there is only one dimensionless parameter, namely the scattering parameter $\phi = \omega/\omega_*$, for which is given a range of values from nearly total reflection to total transmission.

The phase is not plotted in the range $(-\pi, +\pi)$ because this would give rise to many 2π 'phase jumps' without physical meaning. Instead, they are plotted as monotonic functions from 0 to $N\pi$, with N as large as necessary.

Far above the critical layer $(z \gg 1)$ all wave fields vanish i.e., the amplitude decays and the phase becomes constant; thus wave reflection at the critical layer renders the waves evanescent above it. Below the critical layer the wave fields are oscillatory, and the way in which the transition to a decaying field above the critical level occurs, is different for the divergent H^- and finite H^+ wave field components.

DISCUSSION

Questions can be raised about the extent to which the present results would be affected by instabilities, non-linearity, dissipation and other phenomena. The present problem is a linear perturbation of the MHD equations in a stratified, flowing background, and thus can be seen as a stability analysis or a wave theory. The former interpretation would identify the critical layer of the wave as an instability, since the wave field is singular there. Since the background medium is steady, i.e. does not depend on time, the frequency is conserved, i.e. there are no modes growing in time.

The non-linear effects are clearly important near the critical layer, where linear theory predicts unbounded amplitude; as the Alfvén waves grow in amplitude the threshold for modulation [12] or parametric [13–15] instabilities may be exceeded. The additional question of whether non-linear effects are important at all altitudes,

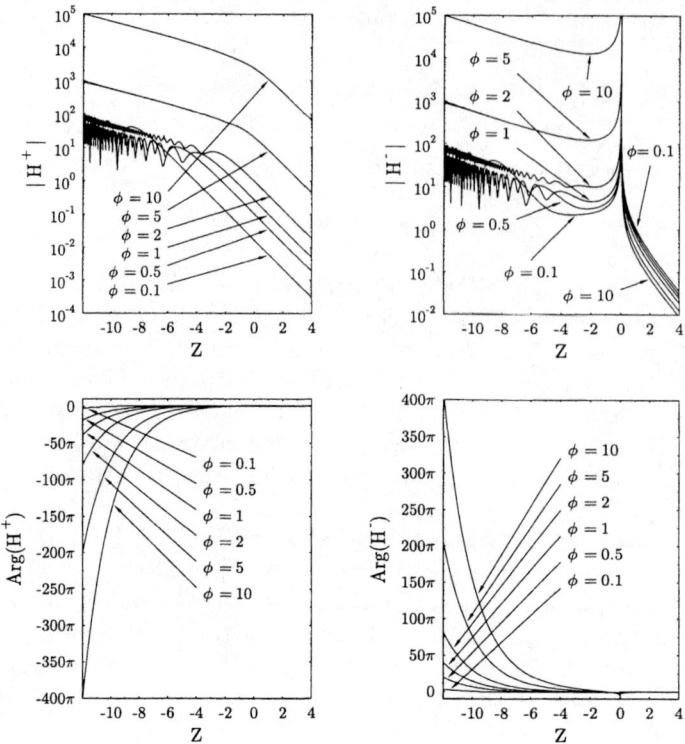

FIGURE 1. Modulus (top) and phase (bottom) of magnetic field perturbation spectrum for vertical Alfvén waves in a flowing isothermal atmosphere; component of the wave field finite H^+ (l.h.s.) and singular H^- (r.h.s.) at the critical layer, versus dimensionless distance from critical layer, for several values of dimensionless frequency ϕ.

and should have been considered from the outset, can be answered in the negative. As can be seen from the figure (top), the magnetic field perturbation H decays with altitude (except near the critical layer), and since the background magnetic field B is uniform, the ratio H/B decays with altitude, i.e. Alfvén waves become 'more linear', and do not form shocks. The fact that Alfvén waves become 'more linear' with altitude, and do not form shocks [16,17], relates to their transversal, incompressible nature, and contrasts with compressive modes, e.g. sound waves.

Concerning dissipation, it is important near the critical layer, where gradients become large, and it could be significant everywhere, because it changes the form of the Alfvén wave equation. An Alfvén wave, being transversal, and hence incompressible, can be dissipated by Ohmic electrical resistance and shear viscosity; in an atmosphere these can lead to the appearance of a critical layer [17–20]. Also, the oscillatory behaviour of the amplitude below the critical layer could give rise to large gradients and so dissipation could act like a filter since it would be more important for lower frequencies.

The particle density in the solar atmosphere is sufficiently high for the MHD description to remain valid. The Alfvén in the solar atmosphere exceeds $a > 10$ km s^{-1} and for the typical periods $\tau \sim 1 - 5$ min, the local reference wavelength is hundreds of thousands of km, i.e., larger than the scale height for the atmospheric mass density. Under these conditions MHD remains an adequate model, provided that the JWKB approximation is not made, because the wavelengths are long compared to the scale height.

Due to the effects discussed above, this model should be seen as a 'prototype' problem, allowing the study of Alfvén waves in a flowing medium. With this in mind it should nevertheless be an indication of the existence in general of a critical layer for Alfvén waves in a moving medium, leading to a gradual wave reflection process. As can be seen in the figure the 'evanescent' Alfvén waves beyond the critical layer decay slowly for lower frequencies. Thus the critical layer acts more like a partial filter favoring the transmission of lower-frequency waves and the spectrum of Alfvén perturbations tends to steepen at higher frequencies, as is indeed observed in the solar wind. Details of all the calculations can be found in [21].

REFERENCES

1. Foukal, P., *Solar Astrophysics*, Wiley, 1990.
2. Bray, R.G., and Loughhead, R.E, *The Solar Chromosphere*, Chapmann & Hall, 1974.
3. Stix, M., *The Sun*, Springer-Verlag, 1989.
4. Spruit, H.C., *Solar Phys.* **75**, 3 (1982).
5. Roberts, B., *Solar Phys.* **69**, 39 (1981).
6. Van Ballegooijen, A.A., *Astron. & Astrophys.* **106**, 43 (1982).
7. Kulsrud, R.M., *Astrophys. J.* **121**, 461 (1955).
8. Campos, L.M.B.C., *J. Fluid Mech.* **81**, 529 (1977).
9. Campos, L.M.B.C., *Month. Not. Roy. Astron. Soc.* **241**, 215 (1989).
10. Stein, R.F., *Astrophys. J.* **246**, 966 (1981).
11. Parker, E.N., *Cosmical magnetic fields*, Oxford Univ. Press, 1979.
12. Lashmore-Davies, C. N., *Phys. Fluids* **19**, 587 (1976).
13. Wong, H.K., and Goldstein, M.L., *J. Geophys. Res.* **91**, 5617 (1986).
14. Viñas, A.F., and Goldstein, M.L., *J. Plasma Phys.* **46**, 107 (1991).
15. Viñas, A.F., and Goldstein, M.L., *J. Plasma Phys.* **46**, 129 (1991).
16. Campos, L.M.B.C., *J. Phys.* **A16**, 417 (1983).
17. Campos, L.M.B.C., *Rev. Mod. Phys.* **59**, 363 (1987).
18. Campos, L.M.B.C., *J. Mec. Theor. Appl.* **2**, 861 (1983).
19. Campos, L.M.B.C., *Wave Motion* **17**, 101 (1993).
20. Campos, L.M.B.C., *Europ. J. Mech.* **B12**, 187 (1993).
21. Campos, L.M.B.C., and Gil, P.J.S., *Phys. Plasmas* **6**, 3345 (1999).

On Alfvén waves in an atmosphere with viscosity and anisotropic resistivity

L.M.B.C. Campos and P.M.V.M. Mendes

Instituto Superior Técnico
1096 Lisboa, Codex – Portugal
fax: (351) 1 841 72 74, email: lbc@isr.ist.utl.pt

Abstract. We consider dissipative Alfvén waves in a viscous atmosphere with a magnetic field, so that the electrical conductivity is considered as a non-diagonal tensor. Under a vertical uniform magnetic field we consider transverse magnetic and velocity perturbations. We obtain two decoupled second-order systems to right and left polarised Alfvén waves. The first electrical diffusivity χ_1 and the viscous diffusivity η are associated with "real" damping; the second electrical diffusivity χ_2 is associated with "imaginary damping". Boundary conditions are imposed in order to obtain the velocity perturbation spectrum of an Alfvén wave. The velocity perturbation is plotted versus altitude over a distance of five scale heights, for different values of the parameters.

INTRODUCTION

Alfvén waves are important in the stellar atmosphere heating study. In an atmosphere the Alfvén speed may be a function of altitude (Ferraro & Plumpton 1958; Hollweg 1978; Leroy 1983; Campos 1983; Muzielak & Moore 1995), as well as the Ohmic resistivity and shear viscosity (Heyvaerts & Priest 1983; Steinholfson 1985; Campos 1988). In the present study the external magnetic field is assumed to be uniform allowing for non-uniform Alfvén speed, viscous and resistive diffusivities. The anisotropic resistivity (Shkarofsky 1960; Oster 1967) is considered as a generalisation of the ohmic diffusivity and together with shear viscosity, they are the dissipation mechanisms of Alfvén waves. The presence of an external magnetic field breaks the isotropy of the medium, implying that the electric current and the electric field are no longer parallel. There are two independent resistive diffusivities (Campos & Mendes 2000). The second resistive diffusivity, or *anisotropic resistive diffusivity*, χ_2, in a homogeneous medium, implies that the velocity and magnetic field perturbation of an Alfvén wave are not collinear and they rotate in the plan orthogonal to the external magnetic field (Campos & Mendes 2000). The anisotropic resistivity introduces also a change in the velocity of the wave.

In the present paper we obtain the Alfvén wave equation to the velocity perturbation (section 3 and 4) in an isothermal atmosphere. With the assumption that χ_1 = const and $\chi_2 \propto e^{z/L}$, this wave equation reduces to a gaussian hypergeometric equation (section 5).

CP537, *Waves in Dusty, Solar, and Space Plasmas*, edited by F. Verheest, et al.
© 2000 American Institute of Physics 1-56396-962-9/00/$17.00

ANISOTROPIC DIFFUSIVITY TENSOR

The derivation of the Alfvén wave equation requires only two of the fundamental equations of MHD, namely the equations of induction and momentum. The equation of momentum is:

$$\rho\left[\partial_t \vec{V} + (\vec{V}.\vec{\nabla})\vec{V}\right] + \vec{\nabla}p = \frac{\mu}{c}\vec{J} \times \vec{H} + \partial_j\left(\rho\eta\partial_j V_i\right) + \nabla\left(\rho\eta_* \vec{\nabla}.\vec{V}\right) \tag{1}$$

where $\partial_t \equiv \partial/\partial t$, $\partial_j \equiv \partial/\partial x_j$, ρ is the mass density, $\vec{V}$ the velocity, $\vec{H}$ the magnetic field, $\vec{J}$ the electric current, p the gas pressure, μ the magnetic permeability, c the speed of light in vacuum, and η, η_*, respectively the shear and bulk kinematic viscosities, which may be non-uniform. On the r.h.s. of (1) appears the magnetic force, involving the electric current, given by:

$$\vec{J} = \frac{c}{4\pi}\vec{\nabla} \times \vec{H} \tag{2a}$$

The magnetic field satisfies the induction equation:

$$\partial_t \vec{H} + \vec{\nabla} \times (\vec{H} \times \vec{V}) = -\frac{c}{\mu}\vec{\nabla} \times \vec{E} \tag{2b}$$

where the electric field is related to the current

$$J_i = \sigma_{ij}E_j \tag{3}$$

through the conductivity tensor, σ_{ij}, whose inverse σ_{ij}^{-1} is the resistivity tensor. Inverting (3) and substituting together with (2a) in the induction equation (2b), leads to:

$$\partial_t \vec{H} + \vec{\nabla} \times (\vec{H} \times \vec{V}) = -\vec{\nabla} \times \left[\chi_{ij}\left(\vec{\nabla} \times \vec{H}\right)_j\right] \tag{4a}$$

where the electrical diffusivity tensor is used:

$$\chi_{ij} = \frac{c^2}{4\pi\mu}\sigma_{ij}^{-1}, \tag{4b}$$

The isotropic case corresponds to the ohmic electrical conductivity σ and diffusivity χ:

$$\sigma_{ij} = \sigma\delta_{ij}, \quad \chi_{ij} = \chi\delta_{ij}, \quad \chi = c^2/4\pi\mu\sigma \tag{5a,b,c}$$

In the more general case of anisotropic electrical conductivity (5a) is replaced (Shkarofsky 1960; Oster 1967) by:

$$\sigma_{ij} = \begin{bmatrix} \sigma_{11} & \sigma_{12} & 0 \\ -\sigma_{12} & \sigma_{11} & 0 \\ 0 & 0 & \sigma_{33} \end{bmatrix}, \quad \text{and} \quad \chi_{ij} = \begin{bmatrix} \chi_1 & \chi_2 & 0 \\ -\chi_2 & \chi_1 & 0 \\ 0 & 0 & \chi_3 \end{bmatrix} \tag{6,7}$$

where the electrical diffusivity (7) involves three scalar electrical diffusivities, one analogous to the ohmic case (5c) and two distinct (8b,c):

$$\chi_3 \equiv c^2 / 4\pi\mu\sigma_{33} , \tag{8a}$$

$$\chi_1 \equiv (c^2 / 4\pi\mu)\sigma_{11} / (\sigma_{11}^2 + \sigma_{12}^2) , \tag{8b}$$

$$\chi_2 \equiv -(c^2 / 4\pi\mu)\sigma_{12} / (\sigma_{11}^2 + \sigma_{12}^2) . \tag{8c}$$

Comparison of (6) with (5a), or of (5b) with (7; 8a,b,c) shows that the ohmic or isotropic case corresponds to $\sigma_{12} = 0$, $\sigma_{11} = \sigma_{33} = \sigma$, or equivalently $\chi_2 = 0$, $\chi_1 = \chi_3 = \chi$. The induction equation (4a) can be written in index notation:

$$\partial_t \vec{H} + \vec{\nabla} \times (\vec{H} \times \vec{V}) = -e_{ijk}\partial_j (\chi_{kl} e_{lmn} \partial_m H_n), \tag{9a}$$

and the momentum equation (1):

$$\rho\left[\partial_t \vec{V} + (\vec{V}.\vec{\nabla})\vec{V}\right] + \vec{\nabla}p = \frac{\mu}{4\pi}\left(\vec{\nabla} \times \vec{H}\right) \times \vec{H} + \partial_j\left(\rho\eta\partial_j V_i\right) + \partial_i\left(\rho\eta.\partial_j v_j\right) \tag{9b}$$

where (2a) was used in the magnetic force.

LEFT AND RIGHT-POLARISED ALFVEN WAVE EQUATIONS

The total state of the fluid is assumed to consist of a mean state of rest under a uniform magnetic field $\vec{B}$, whose direction is taken as the x_3 axis, and transverse magnetic field $\vec{h}$ and velocity $\vec{v}$, perturbations:

$$\vec{v}(\vec{x},t) = v_1(z,t)\vec{e}_1 + v_2(z,t)\vec{e}_2 , \tag{10a}$$

$$\vec{H}(\vec{x},t) = h_1(z,t)\vec{e}_1 + h_2(z,t)\vec{e}_2 + B\vec{e}_3 , \tag{10b}$$

$\vec{H}$ and $\vec{v}$ are unsteady and non-uniform in the $x_3 \equiv z$ direction. Substitution of (11a,b) in the induction (9a) and momentum (9b) equations yields:

$$\dot{h}_1 - Bv_1' = \left(\chi_1 h_1' + \chi_2 h_2'\right)' , \quad \dot{h}_2 - Bv_2' = \left(\chi_1 h_2' - \chi_2 h_1'\right)' \tag{11a,b}$$

$$\dot{v}_1 - (A^2 / B)h_1' = \rho^{-1}\left(\rho\eta v_1'\right)' , \quad \dot{v}_2 - (A^2 / B)h_2' = \rho^{-1}\left(\rho\eta v_2'\right)' \tag{12a,b}$$

where A is the Alfvén speed (13a):

$$A^2 \equiv \mu B^2 / 4\pi , \tag{13}$$

and dot and prime denotes the derivatives (13b,c) with regard to time t and altitude z. Note that velocity and magnetic field perturbation of an Alfvén wave, in the presence of viscosity and anisotropic resistivity, form a fourth-order coupled system (11a,b; 12a,b).

The introduction of right (v_+, h_+) and left (v_-, h_-) polarised waves:

$$v_\pm(z,t) = v_1(z,t) \pm iv_2(z,t), \quad h_\pm(z,t) = h_1(z,t) \pm ih_2(z,t) \tag{14a,b}$$

leads to two decoupled second-order systems:

$$\dot{v}_\pm - (A^2 / B)h_\pm' = \rho^{-1}\left(\rho\eta v_\pm'\right)' , \quad \dot{h}_\pm - Bv_\pm' = \left[(\chi_1 \mp i\chi_2)h_\pm'\right]' . \tag{15a,b}$$

194

Note that the viscous diffusivity η and the first electrical diffusivity χ_1 , which corresponds to the Ohm effect, act on the same way on left and right polarised waves, i.e. are associated with *real* damping; the second electrical diffusivity χ_2 , which represents anisotropy, is associated with *imaginary* damping $i\chi_2$, and has opposite signs for right and left polarised waves.

The magnetic field perturbation can be determined by (15a) from the velocity perturbation:

$$\dot{h}_{\pm} = \left(B / A^2 \right) \left[\dot{v}_{\pm} - \rho^{-1} (\rho \eta v_{\pm}')' \right] . \tag{16}$$

The velocity perturbation satisfies the Alfvén wave equation obtained by taking the time derivative of (15a), and then substituting in turn (15b) and (16), leading to

$$\ddot{v}_{\pm} - A^2 v_{\pm}'' = \rho^{-1} \left(\rho \eta \dot{v}_{\pm}' \right)' + A^2 \left\{ A^{-2} (\chi_1 \mp i\chi_2) \left[\dot{v}_{\pm} - \rho^{-1} (\rho \eta v_{\pm}')' \right] \right\}'' \tag{17}$$

The exact Alfvén wave equation (17) is of the fourth-order in space, due to the presence of the last term on the r.h.s., involving the product of diffusivities. If the viscous and resistive diffusivities are small, in the sense that their product is negligible compared with the combination of wave frequency ω and Alfvén speed with the same dimensions:

$$\chi_1 \eta , \ \chi_2 \eta \ll A^4 / \omega^2 \tag{18}$$

The Alfvén wave equation simplifies to the second-order in space:

$$\ddot{v}_{\pm} - A^2 v_{\pm}'' = (\eta + \chi_1 \mp i\chi_2) \dot{v}_{\pm}'' + \left\{ \frac{(\rho \eta)'}{\rho} + 2A^2 \left[\frac{\chi_1 \mp i\chi_2}{A^2} \right]' \right\} \dot{v}_{\pm}' + A^2 \left[\frac{\chi_1 \mp i\chi_2}{A^2} \right]'' \dot{v}_{\pm} \tag{19}$$

DISSIPATIVE ISOTHERMAL ATMOSPHERE

Next we considerer the case of an atmosphere at rest, under a uniform vertical magnetic field with the z direction of stratification parallel to the external magnetic field. Since the mean state depends on z but not on t, it is convenient to make a Fourier transform in time:

$$v_{\pm}(z,t) = \int_{-\infty}^{+\infty} V_{\pm}(z;\omega) e^{-i\omega t} d\omega , \tag{20}$$

since the frequency is conserved. Substitution of (20) into the Alfvén wave equation (19) leads to linear ordinary differential equations with variable coefficients:

$$\left[(A^2 \mp \chi_2 \omega) - i\omega(\eta + \chi_1) \right] V_{\pm}'' - i\omega \left\{ \rho^{-1} (\rho \eta)' + 2A^2 \left[A^{-2} (\chi_1 \mp i\chi_2) \right]' \right\} V_{\pm}' +$$

$$+\left\{\omega^2 - i\omega A^2\left[A^{-2}(\chi_1 \mp i\chi_2)\right]''\right\}V_\pm = 0. \tag{21}$$

The coefficient of $V_\pm''$ shows the distinction between the first resistive diffusivity χ_1, which adds to the viscous diffusivity η, and thus represents a damping effect, and the second resistive diffusivity χ_2, which adds to or subtracts from the Alfvén speed, and thus represents a propagation effect. The coefficients of the differential equation (21) simplify somewhat in the case: $\rho\eta$, χ_1, $\chi_2 A^{-2} \approx$ const:

$$\left[(A^2 \mp \chi_2\omega) - i\omega(\eta + \chi_1)\right]V_\pm'' - 2i\omega\chi_1 A^2\left[A^{-2}\right]'V_\pm' + \left[\omega^2 - i\omega\chi_1 A^2\left(A^{-2}\right)''\right]V_\pm = 0. \tag{22}$$

which will be justified next.

Consider an isothermal atmosphere, for which the mass density decays exponentially with altitude on the scale height

$$\rho(z) = \rho_0 e^{-z/L}, \quad L \equiv RT/g \tag{23a,b}$$

For an uniform external magnetic field, this implies that the Alfvén speed (13) increases exponentially on twice the scale height:

$$A(z) = ae^{z/2L}, \quad a^2 \equiv \mu B^2/4\pi\rho_0 \tag{24a,b}$$

The static viscosity $\rho\eta$ is a function of temperature, implying that it is constant for an isothermal atmosphere, and thus the kinematics viscosity or viscous diffusivity varies inversely with the mass density $\eta \sim \rho^{-1}$, i.e. (23a) leads to (25a):

$$\eta(z) = \eta_0 e^{z/L}, \quad \chi_2(z) = \chi_0 e^{z/L} \tag{25a,b}$$

and a similar scaling is adopted for the second resistive diffusivity, which varies rapidly with altitude in the low chromosphere of the sun. The first resistive diffusivity varies more slowly with altitude, and is taken as a constant (26a):

$$\text{const} \approx \chi_1, \quad \rho(z)\,\eta(z), \quad \chi_2(z)[A(z)]^{-2}, \tag{26a,b,c}$$

and the remaining constants (26b,c) follow from (23a, 24a, 25a,b). The condition (26a,b,c) justifies the simplified form (22) of the differential equation specifying dissipative Alfvén waves.

The present problem involves six quantities ω, a, L, η_0, χ_0, χ_1, with which can be formed four dimensionless parameters:

$$\Omega = \omega L/a, \quad \varphi = a^2/\omega\chi_1, \quad \varepsilon = \eta_0/\chi_1, \quad \delta = \chi_0/\chi_1, \tag{27a,b,c,d}$$

namely the dimensionless frequency (27a), inverse ohmic damping (27b), and ratio of the viscous (27c) and second resistive diffusivity (27d) to the first resistive diffusivity. Substituting (23a, 24a; 25a,b; 26a) in the differential equation (22) leads to:

$$[1 + (\varepsilon \mp i\delta + i\varphi)e^{z/L}]L^2V_\pm'' - 2LV_\pm' + (1 + i\Omega^2\varphi)V_\pm = 0 \tag{28}$$

where the dimensionless parameters (27a-d) were used. The change of independent variable (29a)

$$1/u = -(\varepsilon \mp i\delta + i\varphi)e^{z/L}, \quad V_{\pm}(z,t) \equiv \Phi(u) \tag{29a,b}$$

which measures altitude on the inverse of the column mass, leads from (28) to a differential equation for (29b):

$$(1-u)u\Phi'' - (1-3u)\Phi' - (1+i\Omega^2\varphi)\Phi = 0, \tag{30}$$

REDUCTION TO A HYPERGEOMETRIC EQUATION

The differential equation (31) is of the gaussian hypergeometric type

$$(1-u)u\Phi'' + [\gamma - (\alpha + \beta + 1)u]\Phi' - \alpha\beta\Phi = 0, \tag{31}$$

with parameters:

$$\gamma = 1, \quad \alpha = 1 + \Omega\sqrt{-i\varphi}, \quad \beta = 1 - \Omega\sqrt{-i\varphi} \tag{32a,b,c}$$

determined solely by the parameters (27a,b).
The wave equation (30) has a singularity when the coefficient of the highest derivative vanishes, i.e. the coefficient of $V''_{\pm}$ in (21), of Φ'' in (30), corresponding to a critical layer at $u_* = 1$ (and $u_* = 0$ for $z \to \infty$). This critical layer occurs at the complex altitude $z_* = -L\log(-\varepsilon \pm i\delta - i\varphi)$, and hence is of transition level type.

The implication is that the wave field is finite at all altitudes $-\infty < z < +\infty$, and the critical layer limits the radius of convergence of the solutions of the differential equation (31), specifying different forms of the wave field for $|u| > 1$ and $|u| < 1$. The assumption of weak diffusivities (18) implies that the parameter φ is large (27b), whereas ε, δ would be typically of order unity, and so from (29a):

$$|u| = |\varepsilon^2 + (\varphi \pm \delta)^2|^{-1/2} e^{-z/L} \tag{33}$$

is small, at least for positive altitudes. Thus the solution of the hypergeometric equation is chosen for $|u| < 1$.

Since $\gamma = 1$ in (32a), the solution of the hypergeometric equation involves functions of the first F and second G kinds:

$$\Phi(u) = C_1 F(\alpha, \beta; 1; u) + C_2 G(\alpha, \beta; 1; u) \tag{34}$$

where C_1 and C_2 are arbitrary constants of integration. Asymptotically at high altitude $z \to \infty$, as $u \to 0$ in (30a), the hypergeometric function of the first kind is finite but that of second kind has a logarithmic singularity leading to a velocity perturbation that is a linear function of altitude; thus the rate-of-strain tends asymptotically to a constant, $dV_{\pm}/dz \sim const$, and the damping condition, that the total rate of dissipation by viscosity be finite

$$D \equiv \tfrac{1}{2}\rho_0\eta_0 \int_{-\infty}^{+\infty} |dV_{\pm}/dz|^2 dz < \infty, \tag{35}$$

is not met, unless $C_2 = 0$. Thus the damping condition (37) specifies one constant of integration (36a);

$$C_2 = 0, \quad \Phi(u_0) = C_1 F(\alpha, \beta; 1; u_0), \tag{36a,b}$$

and the other constant of integration C_1 is determined by an initial condition (36b), specifying the wave field at a given altitude, e.g. $\rho = \rho_0$ corresponding to $z = 0$.

Thus the velocity perturbation spectrum of an Alfvén wave (30a,b) is specified by:

$$V_\pm(z; \omega) = V_\pm(0; \omega)[F_\pm(z/L)/F_\pm(0)], \tag{37}$$

where:

$$F_\pm(z/L) \equiv F(\alpha, \beta; 1; -e^{-z/L}/(\varepsilon \mp i\delta + i\varphi)), \tag{38}$$

The difference between right and left polarised waves lies only in the sign in $\varphi \pm \delta$ which appears in the variable (29a) of the hypergeometric function (38). It is clear from (27a,d) that if $\varphi \gg \delta$ or $a^2 \gg \chi_0 \omega$ the second resistive diffusivity has little effect on propagation; this condition is equivalent to $A^2 \gg \omega\chi_2$ and could have been predicted from coefficient of $V_\pm''$ in (21).

DISCUSSION

The velocity perturbation is plotted normalised to the initial velocity (39):

$$W_\pm(Z) \equiv V_\pm(z; \omega)/V_\pm(z; 0) = F_\pm(Z)/F_\pm(0), \tag{39}$$

where $0 \le Z \equiv z/L \le 5$, versus altitude z made dimensionless by dividing by the scale height L, over a distance of five scale heights. The four parameters of the problem (27a-d) are given by the baseline values: $\Omega = 1$, $\varphi = 100$, $\varepsilon = 1 = \delta$, corresponding to a local wavelength larger than the scale *height* $\lambda = 2\pi/k = 2\pi a/\omega = 2\pi L/\Omega = 2\pi L > L$, weak damping $a^2 = 100 \chi_1 \omega$, and equal diffusivities $\chi_1 = \chi_0 = \eta_0$. Variations of these parameters are considered in turn:

(i) in Figure 1 for wavelengths smaller than the scale height or larger (ray approximation), $\Omega = 0.1$, 1, 10;

For small dimensionless frequency Ω ($\lambda > L$) we have a standing wave, while for $\Omega = 10$ ($\lambda < L$) the wave is progressive.

(ii) in Figure 2 for zero or strong second resistive diffusivity, compared to the first resistive diffusivity, $\delta = \chi_0/\chi_1 = 0$, 1, 10, 100.

We can see that the increase of the second resistive diffusivity in comparison with the first one favours the propagation until the amplitude attenuation becomes very strong ($\delta \gg 1$). The latter seems to be the case of the solar transition region.

It could be seen that the increase of the viscous diffusivity to the first resistive diffusivity favours the propagation but increase the damping of the wave (it is the situation of the high chromosphere and transition region of the sun). We have a standing wave when the first resistive diffusivity predominates largely over the viscous diffusivity $\chi_1 \gg \eta_0$ (like in the solar low chromosphere).

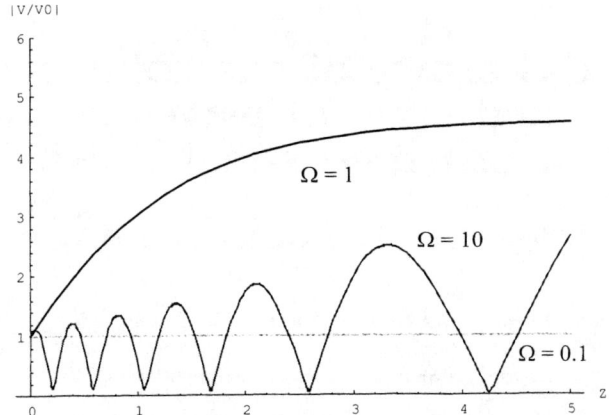

Fig.1 – Modulus of the normalised velocity perturbation. Variation of the dimensionless frequency.

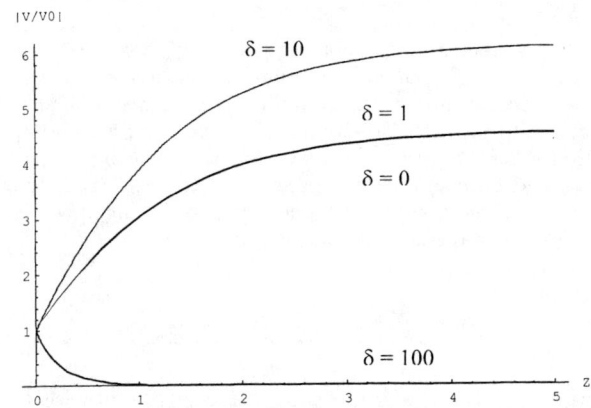

Fig.2 – Modulus of the normalised velocity perturbation. Variation of δ.

REFERENCES

1. Ferraro, C. and Plumpton, V.C.A., *Astrophys. J.* **127**, 459 (1958).
2. Hollweg., J.V., *Solar Phys.* **62**, 307 (1978).
3. Leroy, B. *Astron. Astrophys.* **125**, 371 (1983).
4. Campos, L.M.B.C., *J.Phys.* **A16**, 417 (1983).
5. Muzielak, Z.E. and Moore, R.L., *Astrophys. J.* **452**, 434 (1995).
6. Heyvaerts, J. and Priest, E.R., *Astron. Astrophys.* **117**, 220 (1983).
7. Steinholfson, R.S., *Astrophys. J.* **270**, 304 (1985).
8. Campos, L.M.B.C., *J.Phys.* **A21**, 2911 (1988).
9. Shkarofsky, I.P., *Can. J. Phys.* **39**, 1619 (1960).
10. Oster, L., *Solar Phys.* **3**, 543 (1967).
11. Campos, L.M.B.C. and Mendes, P.M.V.M., J. Plasma Phys. (to appear).

Linear dynamics of the solar convection zone: excitation of waves in unstably stratified shear flows

G. D. Chagelishvili[*,†], A. G. Tevzadze[*,¶] M. Goossens[¶]

* Abastumani Astrophysical Observatory, 2ᵃ Ave. Kazbegi 380060 Tbilisi, Georgia
† Space Research Institute, 84/32 Str. Profsoyuznaya, 117810 Moscow, Russia
¶ Center for Plasma Astrophysics, K.U.Leuven, Celestijnenlaan 200B, 3001, Heverlee, Belgium

Abstract.
   In this paper we report on the nonresonant conversion of convectively unstable linear gravity modes into acoustic oscillation modes in shear flows. The convectively unstable linear gravity modes can excite acoustic modes with similar wave-numbers. The frequencies of the excited oscillations may be qualitatively higher than the temporal variation scales of the source flow, while the frequency spectra of the generated oscillations should be intrinsically correlated to the velocity field of the source flow. We anticipate that this nonresonant phenomenon can significantly contribute to the production of sound waves in the solar convection zone.

INTRODUCTION

   The excitation and propagation of waves are important for understanding the dynamics of the sun and stars. It is believed that most of the solar mechanical energy is accumulated in the turbulent motions in its convection zone. In the convection zone the gravitational stratification drives the convective instability providing the dynamical activity of this relatively thin region. The dynamics of the solar convection is studied to explain many observational features of the Sun. Notably, it is thought that the solar acoustic oscillations are excited by the turbulence in the convection zone [1-5].

   Lighthill's ideas of aerodynamic sound generation form the basis of the theoretical investigation of the wave excitation in a hydrodynamic medium [6,7]. This theory of wave excitation by a free turbulence has been generalized for stratified fluids by Stein [8]. From a physical point of view, Lighthill's theory of wave generation employs the concept of stochastic excitation of oscillations (waves). In Lighthill's theory perturbations are described by an inhomogeneous wave equation, with linear terms forming the oscillatory part and the inhomogeneous terms standing for the source function. The source terms, which may be classified by their multipole order,

CP537, *Waves in Dusty, Solar, and Space Plasmas*, edited by F. Verheest, et al.
© 2000 American Institute of Physics 1-56396-962-9/00/$17.00

are stochastically created by the turbulent perturbations. The amplifying effect of a sheared mean flow on the fluctuations of the Reynolds stress (nonlinear source term) and thus on the wave production has been noted by Lighthill [7]. However, this effect has not received further attention within the context of stochastic excitation.

Significant advances in the investigation of the dynamics of flows with velocity shear have been achieved together with the disclosure of specific features of shear flow phenomena [9,10]. Operators arising in the mathematical formalism of the canonical modal analysis in the study of the linear dynamics of shear flows are not self-adjoint. Consequently eigenmode interference introduces principal complications. The nonmodal approach has proved to be an alternative successful route for exploring the dynamics of shear flows. This approach employs the study of temporal evolution of the spatial Fourier harmonics of perturbations.

Impressive progress has been made by use of the nonmodal analysis (see e.g., [11-16]). This approach has led to the discovery of new channels of energy exchange between different modes in shear flows. Resonant phenomena of wave transformations have been studied in [17-23]. The nonresonant phenomenon of the conversion of vortices into acoustic waves has been described in [24]. The same mechanism is found to operate for magnetosonic [25] as well as for plasma Langmuir oscillations [21].

In this report we introduce a new dynamical source of acoustic waves in unstably stratified shear flows. Namely, the *linear* nonresonant conversion of convective into acoustic wave modes in a stratified shear flows. Convectively unstable exponentially growing buoyancy perturbations generate acoustic wave oscillations in presence of a sheared mean flow. We identify this linear conversion of modes in shear flows as a new excitation mechanism of the solar oscillations and waves. It differs in principle from the stochastic excitation mechanism and should significantly contribute to the process of acoustic wave generation in the solar convection zone.

PHYSICAL APPROACH

The equations governing the dynamics of a compressible stratified flow are:

$$[\partial_t + (\mathbf{V}\nabla)]\,\rho + \rho(\nabla\mathbf{V}) = 0, \tag{1.a}$$

$$[\partial_t + (\mathbf{V}\nabla)]\,\mathbf{V} = -\nabla P/\rho + \mathbf{g}, \tag{1.b}$$

$$[\partial_t + (\mathbf{V}\nabla)]\,P = (\gamma P/\rho)\,[\partial_t + (\mathbf{V}\nabla)]\,\rho. \tag{1.c}$$

We consider the hydrodynamic situation where a horizontal shear flow $\mathbf{V}_0 = (Ay, 0, 0)$ occurs in a vertically stratified medium $\mathbf{g} = (0, 0, -g)$. For simplicity we assume that $A = const$ and $g = const$. This yields the stratified equilibrium state:

$$P_0(z)/P_0(0) = \rho_0(z)/\rho_0(0) = \exp(-zk_H), \tag{2}$$

where $k_H \equiv \gamma g/c_s^2$ and $c_s^2 \equiv \gamma P_0/\rho_0$. We introduce the linear perturbations in the following way:

$$\mathbf{V} = \mathbf{V}_0 + \mathbf{V}'\rho_0(0)/\rho_0(z), \quad P = P_0 + P', \quad \rho = \rho_0 + \rho'. \tag{3}$$

Here the velocity perturbations are normalized to exclude the exponential height dependence due to the vertical stratification of the background flow. We use the Cowling approximation [26] and neglect the perturbations of the gravitational acceleration. Following the standard method of nonmodal analysis (see [27] for a rigorous mathematical interpretation) we introduce the spatial Fourier harmonics (SFH) of the perturbations with time dependent phases:

$$\Psi(\mathbf{r}, t) = \psi(\mathbf{k}(t), t) \exp(ik_x x + ik_y(t)y + i\tilde{k}_z z), \tag{4.a}$$

$$k_y(t) = k_y(0) - Ak_x t, \tag{4.b}$$

where $\tilde{k}_z \equiv k_z + ik_H/2$. For compactness of notation we introduce the generalized vector of perturbations and their SFHs as follows: $\Psi \equiv (\mathbf{V}', p', \rho')$ and $\psi \equiv (\mathbf{u}, p, \rho)$. To avoid complex coefficients in the dynamical equations, we construct the normalized entropy and vertical velocity perturbation SFHs in the following way:

$$s \equiv (ic_s^2 \tilde{k}_z^*/g - 1)(p - c_s^2 \rho)/(\gamma - 1), \tag{5.a}$$

$$v \equiv (c_s^2 \tilde{k}_z^* + ig)u_z, \tag{5.b}$$

where $\tilde{k}_z^* = k_z - ik_H/2$. From Eqs. (1-5) we obtain, by the use of straightforward manipulations, the following set of differential equations that govern the SFH of the linear perturbations in stratified shear flow:

$$\dot{p}(t) = c_s^2(k_x u_x + k_y(t)u_y) + v, \tag{6.a}$$

$$\dot{u}_x(t) = -Au_y - k_x p, \tag{6.b}$$

$$\dot{u}_y(t) = -k_y(t)p, \tag{6.c}$$

$$\dot{v}(t) = (N_B^2 - c_s^2 \bar{k}_z^2)p - N_B^2 s, \tag{6.d}$$

$$\dot{s}(t) = v. \tag{6.e}$$

N_B^2 is the square of the frequency of the Brunt-Väisälä: $N_B^2 \equiv gk_H(\gamma - 1)/\gamma$ and $\bar{k}_z^2 = |\tilde{k}_z|^2 = k_z^2 + k_H^2/4$. In an unstably stratified flow negative buoyancy ($N_B^2 < 0$) requires that the adiabatic index $\gamma < 1$. Such an effective value may be assigned to this parameter under a certain thermodynamic approach [28]. However, in Eqs (6.a-e) we retain only N_B^2 and argue that these equations are more general than the underling γ prescription.

Further we note that vorticity is conserved in the wave-number space: $I = k_x u_y - k_y(t)u_x - (A/c_s^2)(p - s)$. The spectral energy of the perturbations can be defined as follows:

$$E = \rho_0/(2c_s^2)\left(E_K + E_P + E_T\right), \tag{7.a}$$

202

$$E_K = c_s^2(u_x^2 + u_y^2) + v^2/(c_s^2 \bar{k}_z^2 - N_B^2), \tag{7.b}$$

$$E_P = p^2, \quad E_T = N_B^2 s^2/(c_s^2 \bar{k}_z^2 - N_B^2). \tag{7.c, d}$$

where E_K, E_P and E_T correspond to the kinetic, elastic and thermobaric energies of the perturbations, respectively. Formally the perturbation energy is conserved in the shearless limit: $\dot{E} = A c_s^2 u_x u_y$. The instability of the convective eddies corresponds to a negative value of the thermobaric energy.

Linear modes in the shearless limit

The linear modes may be classified explicitly in the shearless limit ($A = 0$). In this case the full Fourier expansion of the linear perturbations $\Psi(r, t) \propto \tilde{\psi}(k, \omega)$ yields the dispersion equation:

$$\omega(\omega^4 - c_s^2 k^2 \omega^2 + N_B^2 c_s^2 k_\perp^2) = 0, \tag{8}$$

where $k_\perp^2 \equiv k_x^2 + k_y^2$ and $k^2 = k_\perp^2 + \bar{k}_z^2$. The solutions of Eq. (8) describe the stability and characteristic temporal variation scales of the existing modes:

$$\omega_v = 0, \tag{9.a}$$

$$\omega_{s,c}^2 = \frac{1}{2} c_s^2 k^2 \left\{ 1 \pm \left(1 - \frac{4 N_B^2 k_\perp^2}{c_s^2 k^4} \right)^{1/2} \right\}, \tag{9.b}$$

where the subscripts v, s, c define the frequencies of the vortex, acoustic and convective modes, respectively. In an unstably stratified flow, i. e., when $N_B^2 < 0$, $i\omega_c$ defines the growth rate of the buoyancy perturbations.

Obviously the $I = constant$ law demonstrates the existence of the stationary ($\omega = 0$) vortex mode in the linear spectum. The conserved vorticity I may be considered as the vortex mode measure. The physical eigenfunctions of the acoustic $\Phi_s(t)$ and convective $\Phi_c(t)$ modes may be rigorously defined in this limit:

$$\Phi_s(t) \equiv p(t) + N_B^2 \frac{\Omega_s^2 - \omega_s^2}{\Omega_c^4} \left(s(t) - \frac{\bar{k}_z^2}{k^2} p(t) \right) \tag{10.a}$$

$$\Phi_c(t) \equiv s(t) - \frac{\bar{k}_z^2}{k^2} p(t) - \frac{\Omega_c^2 - \omega_c^2}{N_B^2} p(t) \tag{10.b}$$

where $\Omega_s^2 \equiv c_s^2 k^2$ and $\Omega_c^2 \equiv N_B^2 k_\perp^2/k^2$. Hence the equations governing the dynamics of the perturbations of the different modes may be decoupled as follows:

$$\ddot{\Phi}_s(t) + \omega_s^2 \Phi_s(t) = 0, \tag{11.a}$$

$$\ddot{\Phi}_c(t) + \omega_c^2 \Phi_c(t) = 0. \tag{11.b}$$

Starting from this simple situation we study the velocity shear effects on the perturbation modes.

Effects of a sheared flow

To study the effects of the velocity shear on the linear modes we introduce the small-scale approximation: $k_z^2 \gg k_H^2$. This approximation strongly simplifies the mathematical formulation and is justified for the following two reasons. Firstly, our analysis needs constant vertical gravity, an assumption that may be adopted for perturbations with vertical height scales shorter than the stratification scale. Secondly, this approximation is necessary for our assumption of a constant linear shear of the flow velocity, especially in the turbulent flows. Using Eq. (7) this approximation may be represented by the following condition: $c_s^2 k_z^2 \gg N_B^2$. In terms of the frequencies it yields $(\Omega_s^2 - \omega_s^2) \approx (\Omega_c^2 - \omega_c^2) \approx 0$, which strongly simplifies the characteristic physical quantities of the perturbation modes: $\Phi_s(t) \approx p(t)$ and $\Phi_c(t) \approx (s(t) - \bar{k}_z^2 p(t)/k^2(t))$. To analyze the dynamics of acoustic oscillations in the shear flow we rewrite Eqs. (6.a-e) in the form of coupled second order differential equations for the variables $p(t)$ and $y(t)$:

$$\ddot{p}(t) + f(t)\dot{p}(t) + \Omega_1^2(t)p(t) = \lambda_1(t)\dot{y}(t) + \lambda_2(t)y(t), \tag{12.a}$$

$$\ddot{y}(t) + \Omega_2^2(t)y(t) = 0, \tag{12.b}$$

where the convection variable $y(t)$ is introduced as follows:

$$y(t) \equiv \frac{k_\perp(t)}{k(t)} \left(s(t) - \frac{\bar{k}_z^2}{k^2(t)} p(t) \right) \tag{13}$$

and

$$\Omega_1^2(t) = c_s^2 k^2(t) + 2A^2 \frac{k_x^2}{k^2(t)} - 4A^2 \frac{k_x^2 k_y^2(t)\bar{k}_z^2}{k_\perp^2(t)k^4(t)}, \tag{14.a}$$

$$\Omega_2^2(t) = N_B^2 \frac{k_\perp^2(t)}{k^2(t)} + 2A^2 \frac{k_x^2 k_z^2}{k_\perp^4(t)k^4(t)} \left[3k_\perp^2(t)k^2(t) - 4k_y^2(t)k_\perp^2(t) - k_y^2(t)\bar{k}_z^2 \right], \tag{14.b}$$

$$f(t) = 2A \frac{k_x k_y(t)}{k^2(t)}, \tag{14.c}$$

$$\lambda_1(t) = -2A \frac{k_x k_y(t)}{k_\perp(t)k(t)}, \tag{14.d}$$

$$\lambda_2(t) = -2A^2 \frac{k_x^2 k_\perp(t)}{k^3(t)} \left(1 - \frac{k_y^2(t)\bar{k}_z^2}{k_\perp^4(t)} \right). \tag{14.e}$$

In deriving Eqs. (12.a-b) we have used the following two simplifications. Firstly, we have retained only the terms describing the effect of the buoyancy perturbations on the acoustic waves, and we have neglected the effect of the acoustic pressure perturbations on the evolution (exponential amplification) of the buoyancy perturbations in the right hand side (rhs) of Eq. (12.b). Secondly, we have neglected the source terms in the rhs of the two dynamical equations that describe the shear induced coupling between the vortex and acoustic wave modes (in Eq. 12.a) and vortex and buoyancy modes (in Eq. 12.b). In fact, the coupling of the vortex and acoustic wave is a process that has been studied to reveal the mean flow shear induced nonresonant mode conversion phenomenon in [24]. However, in the present case, the source terms of the acoustic waves that are proportional to the vortex mode measure, conserved quantity I, are dominated by the source terms, associated with the exponentially amplifying convective modes: $y(t)$ and $\dot{y}(t)$. It should be emphasized that the present approach is justified only for a convectively unstable medium with $N_B^2 < 0$, so that the buoyancy modes undergo exponential amplification in the linear regime.

The dynamics of the acoustic waves in the absence of the buoyancy perturbations is described by the homogeneous part of Eq. (12.a). The acoustic wave frequency and amplitude variations are described by the parameters $\Omega_1^2(t)$ and $f(t)$ (see [19] for a detailed study). The dynamics of the convective mode is described by Eq. (12.b). Eq. (16.b) shows the transient stabilization effect of the sheared mean flow in an unstably stratified medium. The stabilization occurs at times, when $|k_y(t)/k_x| < 1$ and reaches its maximum at $t = t^*$, when $k_y(t^*) = 0$ (see Eq. 16.b).

The terms $\lambda_1(t)\dot{y}(t)$ and $\lambda_2(t)y(t)$ in the rhs of Eq. (14.a) describe the coupling between the convective and acoustic waves modes. The shear flow origin of these source terms is obvious from Eqs. (16.d,e). Hence, Eqs. (14.a,b) describe the mean flow shear induced buoyancy – acoustic wave mode conversion in a convectively unstable medium. Some specific features of this phenomenon are due to its linear nature; SFH of the exponentially growing buoyancy perturbations are able to generate SFH of the acoustic waves with the same wave-numbers. The amplitude of the excited wave mode depends on the values of the source terms $\lambda_1(t)$ and $\lambda_2(t)$. So, convective modes with $k_x = 0$ can not generate acoustic waves at all ($\lambda_1 = \lambda_2 = 0$). While maximal efficiency of the mode conversion phenomenon should occur at $k_z = 0$, or in a realistic physical approximation (see Eq. 12) at $k_z^2 \geq N_B^2/c_s^2$. Naturally, acoustic wave emission from convection should generally increase when the mean flow shear parameter A increases.

We numerically analyze Eqs. (6.a-e) to verify the analytical results obtained from the approximate equations (12.a,b). We select the initial perturbations in a specific manner, which enables us to excite the convective and acoustic wave modes individually at the initial moment of time. It appears that exponentially growing buoyancy perturbations instantly excite the acoustic wave mode harmonics at a given point in time, when the perturbation wave-number along the flow velocity shear is zero: $t = t^*$, $k_y(t^*) = 0$. The generation of acoustic waves is clearly traced from the pressure variation, as well as the compression of the perturbations. Numerical analysis shows that the efficiency of this mode conversion phenomenon increases with the flow shear parameter.

DISCUSSION AND CONCLUSIONS

We have presented a study of compressible convection in shear flows. In particular we have focused on linear small-scale perturbations in unstably stratified flows with constant shear of velocity. The linear character of the system enables us to identify the perturbation modes and to study their dynamics individually. We find a mode conversion that originates from the velocity shear of the flow: exponentially growing perturbations of convection are able to excite acoustic waves. This process offers a novel approach to the hydrodynamic problem of the acoustic wave generation.

This wave excitation phenomenon can be important for the acoustic oscillations of the sun. Being responsible for the wave generation in high shear regions of a stratified turbulent flow, this nonresonant phenomenon can contribute to the production of sound in the solar convection zone. Moreover, the process of the wave excitation should be triggered by a weak vertical magnetic field. In this case we anticipate the production of high frequency compressional MHD waves. The latter process will considerably increase the extraction of the mechanical energy of the convection by waves.

Specific to this phenomenon is that perturbations of buoyancy are able to excite acoustic waves with similar wave-numbers. This property makes it clearly distinct from stochastic excitation, where the generated frequencies are similar to the life-times of the source perturbations. In contrast, frequencies of the oscillations generated by the mean flow velocity shear induced mode conversion may be qualitatively higher than the temporal variation scales of the perturbations in the source flow of a compressible convection. The frequency spectrum of the excited acoustic waves should be intrinsically correlated to the velocity field of the turbulent source flow. Shear flow induced wave excitation in stratified flows offers a natural explanation of the fact, that the solar acoustic oscillation are mainly excited in the high shear regions of the convection, intergranular dark lanes [29]. It also explains the puzzling wave-number dependence of the observed mode energies at fixed frequencies (see [5] and references therein). A detailed comparison with observational data requires a more realistic physical model. The simplicity of our model is used to demonstrate the basic features of this excitation phenomenon.

Finally we note that in the present formalism we have focused on the waves with frequencies higher than the characteristic cut-off frequency for the acoustic waves in the convection zone. Shear flow initiates the qualitative change of the temporal variation scales of perturbations and the excitation of the waves that are not trapped in the convective envelope. Hence, this mode conversion presents a new significant contribution into the channel of energy transfer from the dynamically active interior to the atmosphere of the Sun.

ACKNOWLEDGMENTS

A. G. Tevzadze would like to acknowledge the financial support as "bursaal" of the "FWO Vlaanderen", project G.0335.98. This work was supported in part by the INTAS grant GE97-0504.

REFERENCES

1. Stein, R. F. 1968, ApJ 154, 297
2. Goldreich, P. & Keele, D. A. 1977, ApJ 212, 243
3. Blamforth, N. J. 1992, MNRAS 255, 639
4. Goldreich, P. & Kumar, P. 1990, ApJ 363, 694
5. Goldreich, P. Murray, N. & Kumar, P. 1994, ApJ 424, 466
6. Lighthill M. J. 1952, Proc. Roy. Soc. A 211, 564
7. Lighthill M. J. 1954, Proc. Roy. Soc. A 222, 1
8. Stein, R. F. 1967, Solar Physics 2, 385
9. Reddy S. C., Schmidt P. J., & Henningson D. S., 1993, SIAM J. Appl. Math. 53, 15
10. Trefethen L. N., Trefethen A. E., Reddy S. C., & Driscoll T. A. 1993, Science 261, 578
11. Goldreich, P. & Linden-Bell, D. 1965, MNRAS 130, 125
12. Batler, K. M. & Farrell, B. F. 1992, Phys. Fluids A 4, 1637
13. Balbus, S. A. & Hawley J. F. 1992, ApJ 400, 610
14. Lubow S. H. & Spruit H. C., 1995, ApJ 445, 337
15. Chagelishvili, G. D., Chanishvili, R. G., Lominadze, J. G. & Tevzadze, A. G. 1997a, Phys. Plasmas 4, 259
16. Zaqarashvili T. V., 1999, A&A 341, 617
17. Chagelishvili, G. D., & Chkhetiani O. G. 1995, JETP Letters 62, 301
18. Chagelishvili, G. D., Rogava, A. D. & Tsiklauri, D. G., 1996a, Phys. Rev. E 53, 6028
19. Chagelishvili, G. D., Khujadze G. R., Lominadze, J. G. & Rogava, A. D. 1997b, Phys. Fluids 9, 1955
20. Rogava, A. D. & Mahajan, S. M. 1997, Phys. Rev. E 55, 1185
21. Rogava, A. D., Chagelishvili, G. D. & Mahajan, S. M. 1998, Phys. Rev. E 57, 7103
22. Poedts S., Rogava, A. D., & Mahajan, S. M, 1998, ApJ 505, 369
23. Poedts S., Rogava, A. D., & Mahajan, S. M, 1999, Space Science Reviews 87, 295
24. Chagelishvili, G. D., Tevzadze, A. G., Bodo, G. & Moiseev, S. S. 1997c, Phys. Rev. Letters 79, 3178
25. Tevzadze, A. G. 1998, Phys. Plasmas 5, 1557
26. Cowling, T. G., 1941, MNRAS, 101, 367
27. Criminale, W. O. & Drazin, P. G. 1990, Stud. Appl. Math. 83, 123
28. Ryu, D. & Goodman, J. 1992, ApJ 388, 438
29. Rimmele, T. R., Goode, P. R., Harold, E., and Stebbins, R. T. 1995, ApJ, 444, L119

Resonant absorption in randomly driven coronal loops

Anik De Groof and Marcel Goossens

Centre for Plasma-Astrophysics, K.U. Leuven, Celestijnenlaan 200B, 3001 Leuven, Belgium

Abstract. De Groof et al. '98 [1] and '00 [2] studied the time evolution of fast magnetosonic and Alfvén waves in a coronal loop driven by radially polarised footpoint motions in linear ideal MHD. Footpoint driving seems to be an efficient way of generating resonant absorption since the input energy is mainly stored in body modes which keep the energy in the loop. The most important feature in this study is the stochastic driving of the loop. While in earlier models with a periodic driver or a single pulse, the loop is only heated at one single layer, we now find multiple resonance layers which results in a more globally heated loop. Moreover, these resonances (created on a realistic time scale) have lengthscales which are small enough to explain energy dissipation. An important aspect to take into account is the mass transfer between corona and chromosphere since the density becomes time dependent and consequently, the resonant surfaces shift throughout the loop [3]. Combined with the multiple resonances we found in the previous study, this result can lead to the globally heated coronal loops we observe.

INTRODUCTION

The several-million-degree plasma in the solar corona is contained in a large number of discrete magnetic loops, clustered in active regions. X-ray images show that these coronal loops have the largest heating requirements. As the footpoints of the magnetic field lines are anchored in the photosphere, they are forced to follow the convective motions. Footpoint motions which are 'fast' in comparison with the Alfvénic transit time, generate magnetosonic waves and Alfvén waves reflecting back and forth along the length of the loop. The loop is then expected to act as a leaking, resonant cavity for MHD waves, in which dissipation is enhanced by means of turbulence, resonant absorption and/or phase mixing (see [2] for references). Besides, reconnection events higher in the solar atmosphere are also potential sources for waves in coronal loops. Observational evidence of MHD waves propagating in coronal loops has indeed been reported [4]. Tataronis & Grossmann [5] were the first to give a basic theory on plasma heating by the Alfvén continuum waves. In 1978, Ionson [6] suggested that resonant absorption might effectively heat the solar corona. Since the original suggestion, resonant absorption has remained a popular mechanism for explaining the heating of the solar corona. A lot of work was done

CP537, *Waves in Dusty, Solar, and Space Plasmas*, edited by F. Verheest, et al.
© 2000 American Institute of Physics 1-56396-962-9/00/$17.00

on sideways excitation of these resonant Alfvén waves ([2] for references). Later studies focussed on different kinds of footpoint motions in order to excite resonant Alfvén waves either directly by azimuthally polarised footpoint motions or indirectly by radially polarised footpoint motions. Berghmans et al. [7] and Tirry et al. [8] revealed the importance of the presence of quasi-modes in both cases. Recently, more attention is paid to the fact that coronal loops are finite and bounded by the underlying atmospheric layers. Goedbloed et al. [9] showed that line-tying leads to the presence of MHD waves of mixed nature which have the right signature for effective energy transfer from the photosphere to the corona and subsequent resonant Alfvén wave heating. Moreover, the coupling with the chromosphere needed some investigation. Beliën et al. [10] concentrated on the influence of varying loop lengths and the higher (but finite) density in the chromosphere. They restricted to $k_y = 0$ and consequently they removed the linear coupling of Alfvén waves to fast waves and the quasi-modes out of their analysis. On the other hand, Ofman et al. [3] studied the effect of mass flows between the corona and the chromosphere.

So far, most of the modelled footpoint motions were harmonic or consisted of one pulse. In the present paper we want to figure out how the results alter for quasi-randomly driven loops. In [1], we studied the behaviour of the fast waves within the loop without coupling to Alfvén waves and found that driving at the loop's feet forms a good basis for resonant absorption as heating mechanism. [2] investigated the efficiency of resonant absorption in the coupled case ($k_y \neq 0$), where the excited body waves can couple to Alfvén waves. In the present paper, we analyse the results of [1] and [2] and present some future plans.

I   PHYSICAL AND MATHEMATICAL MODEL

In [1] and [2], the coronal loop is modelled as a static, straight, gravitationless plasma slab with thickness b, obeying the standard set of ideal MHD equations. In our Cartesian coordinate system, the x-coordinate corresponds to the direction of the inhomogeneity in the equilibrium, the y-coordinate is the (ignorable) azimuthal coordinate and the z-coordinate represents the direction along the loop. At $z = 0$ we impose a given footpoint motion whereas at $z = L$ we assume the loop to be line-tied (see [1] for more details). The plasma is permeated by a uniform magnetic field ($\mathbf{B_0} = B_0\mathbf{e_z}$) and has a uniform pressure p_0 which we neglect in comparison with the magnetic pressure. The inhomogeneity of the plasma is introduced by a continuously varying density $\rho_0(x) = \rho_A + \rho_B\cos(\frac{\pi}{b}x)$ with $\rho_B < \rho_A$, which models the higher density inside the loop. The plasma is being shaken by small-amplitude perturbations at the footpoints of the magnetic field lines on the ($z = 0$)-plane. This plane can be viewed as the photosphere or as the position where the magnetic reconnection takes place. As long as non-linear and non-ideal effects are negligible we can follow the temporal evolution of the excited MHD waves inside the loop by solving the linear ideal MHD equations. Since the equilibrium quantities are constant in the y-coordinate which runs over an infinite interval, we can Fourier analyse with respect to y. For the Fourier component corresponding to wave number k_y, the time evolution and the spatial variation in x and z are described by

$$\{\frac{1}{v_A^2}\frac{\partial^2}{\partial t^2} - \frac{\partial^2}{\partial z^2} - \frac{\partial^2}{\partial x^2}\}\xi_x = ik_y\frac{\partial\xi_y}{\partial x}, \tag{1}$$

$$\{\frac{1}{v_A^2}\frac{\partial^2}{\partial t^2} - \frac{\partial^2}{\partial z^2} + k_y^2\}\xi_y = ik_y\frac{\partial\xi_x}{\partial x}. \tag{2}$$

with the Lagrangian displacement $\boldsymbol{\xi}$ and the Alfvén speed $v_A(x) = \sqrt{\frac{B_0^2}{\rho_0(x)}}$.
This coupled system of partial differential equations in ξ_x and ξ_y describes the coupled fast-Alfvén waves. Slow waves are absent ($\xi_z = 0$) because the plasma pressure is neglected. In this study we focus on radially polarised footpoint motions and consequently, only fast waves will be driven directly. Since $k_y \neq 0$, these fast MHD waves couple to Alfvén waves at the resonant surfaces where the ideal Alfvén wave resonance condition is satisfied. Length, speed, magnetic field strength and density are non-dimensionalised with respect to b, $v_A(0)$, B_0 and $\rho(0)$ respectively.

The mathematical approach in this paper is based on the method used in [8]. We represent the footpoint motions by inhomogeneous boundary conditions for equations (1-2) at the $z = 0$ and $z = L$ boundary planes:

$$\begin{aligned} \xi_x(x, z = 0, t) &= R(x)T(t), \\ \xi_x(x, z = L, t) &= 0, \\ \xi_y(x, z = 0, t) &= \xi_y(x, z = L, t) = 0. \end{aligned} \tag{3}$$

We have assumed for mathematical simplicity that the dependencies on x and t of the footpoint motions are separable. In order to avoid complications with initial conditions we assume in addition that at $t = 0$, ξ_x, ξ_y and both their time derivatives are zero and as a consequence: $T(t = 0) = \frac{\partial T(t=0)}{\partial t} = 0$. Apart from these restrictions the functions $R(x)$ and $T(t)$ can be chosen completely arbitrarily. A convenient way to solve the coupled partial differential equations (1-2) is to get rid of as many derivatives as possible, as described in detail by [8]. This approach enables us to obtain an expression which describes the generation of linear MHD waves (coupled fast-Alfvén waves) by radially polarised footpoint motions. The solution can be written as a superposition of eigenmodes (see [1]).

II   CLASSIFICATION OF EIGENMODES

For $k_y = 0$ and $R = 0$, expressions (1-2) form two separate eigenvalue problems for fast and Alfvén eigenmodes; for $k_y \neq 0$ the equations are coupled.
Fig. 1 shows the eigenfrequencies of the first 3 fast eigenmodes together with the upper and lower bound of the Alfvén continuous spectrum as function of k_z. As described in [1], only the fast modes with frequency within the range of the continuous spectrum, called 'body modes', keep the energy inside the loop.

The same paper showed that in the uncoupled case most of the input energy is converted to body mode energy (around 95%) so that the energy leakage is minimal. In the present case where body modes couple to Alfvén waves, this means that a **good basis is formed for resonant absorption as dissipation mechanism**. In the following sections we study whether resonant absorption is efficient as heating mechanism.

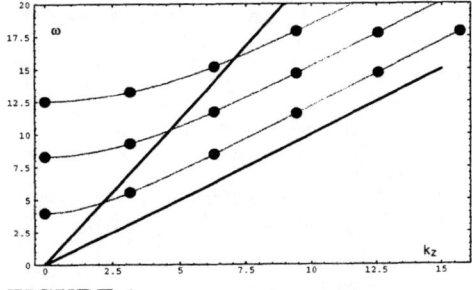

FIGURE 1.

Since, for small values of k_y, the global compressional oscillations are dominated by poloidal perturbations, and the Alfvénic disturbances are dominated by a toroidal polarisation, we will use the term *fast mode energy* for the combined poloidal and compressional energy and the term *Alfvén mode energy* for the toroidal energy. Moreover in what follows, the term *fast mode energy* only refers to the contribution of the body modes.

III   EFFICIENCY OF RESONANT ABSORPTION

To check the efficiency of resonant absorption as heating mechanism, we have to check several criteria. First, we need an efficient energy transfer from the footpoint motions to the Alfvén waves (via the fast waves) and secondly Alfvén resonances have to form at particular magnetic surfaces in the coronal loop. Moreover, because of the extremely high Reynolds numbers in the solar corona, dissipation can only occur if the extremely small lengthscales are produced within the lifetime of the coronal loop. As a third criterion we would like to check whether resonant absorption can heat the loop globally.

A   Time evolution of fast and Alfvén waves

In [2], we studied a loop with dimensions $L = 1$ and $b = 1$. The density parameters ρ_A and ρ_B are taken to be 0.6 and 0.4 respectively. We modelled the footpoint motions as a succession of pulses with the following time and x dependencies:

$$T(t) = \begin{cases} \sin(at - \frac{\pi}{2}) + 1 & \text{as } 0 \leq t \leq \frac{2\pi}{a}, \\ 0 & \text{as } t > \frac{2\pi}{a} \end{cases} \qquad R(x) = \sin(\pi x)$$

which are chosen in order to simulate an instant 'kick' at the loop's feet.

From Fig. 2, we conclude that after the driving has stopped, the fast mode energy is transferred to Alfvén mode energy. The efficiency of this transfer depends on several parameters. First the azimuthal wavenumber k_y acts as a coupling parameter between the fast modes and the Alfvén modes (larger k_y correspond to a stronger

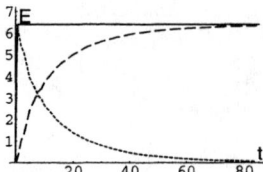 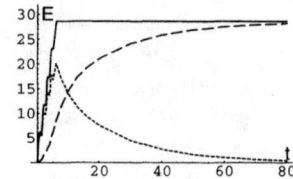

FIGURE 2. *Time evolution of fast mode energy (dotted), Alfvén mode energy (dashed) and total energy (solid) after a driving with respectively one pulse and a succession of 5 identical pulses.*

coupling for values of k_y smaller than 1.6). On the other hand, short pulses bring more Alfvén mode energy into the loop but the energy grows and saturates slower.

B   Creation of small length scales

For the resonant absorption to be valid and to be efficient, resonances have to be formed in the coronal loop. Moreover, the resonant absorption has to produce these extremely small length scales fast enough in order to achieve the observed energy losses. Tirry et al. [8] demonstrated that when the loop is driven harmonically, resonance peaks are formed at the magnetic surfaces where the local Alfvén frequency equals the quasi-mode frequencies which are present in the driving.

As a first step to a real random driving, we drive the loop by a series of 30 pulses with randomly distributed widths and randomly distributed time intervals in between the pulses. The individual

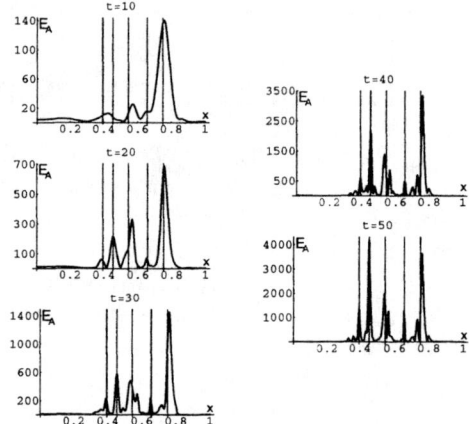

FIGURE 3. *Alfvén energy as function of x for $t=10$ to $t=50$ after driving with a randomly varying pulse train.*

pulses have a time-dependence $\sin(at - \frac{\pi}{2}) + 1$ with pulse widths varying between 0.4 and 2.1 ($3 < a < 16$) and the time intervals are taken to be smaller than 1. As seen on Fig. 3, indeed resonances are built up at the magnetic surfaces corresponding to the quasi-modes which are excited by the driving. The modes corresponding to the first eigenfunctions and small values of k_z are the most dominant (see Fig. 1). In the case presented in Fig. 3, these are the following: $\omega = 5.58, 8.51, 11.61, 14.71, 14.73$, respectively corresponding to $x_A = 0.75, 0.54, 0.45, 0.66, 0.39$.

As expected resonant peaks are growing in time at the magnetic surfaces listed above and indicated by vertical lines on the figure. We remark that with this kind of driving, the peaks are packed rather closely together. When we would vary the pulse widths of the driving pulses more or when we would take a longer loop (see [2]), more quasi-modes will couple to Alfvén modes and consequently the loop can be heated even more globally.

Resonant absorption can only be a viable coronal heating mechanism if the generation of the small scale dissipative features found above takes place on time scales shorter than the life time of coronal loops, which varies from 6 tot 24 hours. As calculated by Tirry et al. [8], this means that lengthscales of about 100 meter are to be generated within half of the loop's life time. Beaufumé et al. [11] observed that a medium coronal loop has typically $L = 3 \times 10^7 m$ and $b = 1.2 \times 10^6 m$.

Fig. 4 shows the reduction in resonance length scale for a footpoint driving consisting of a single pulse at the main resonant points. As expected the length scales of the resonant peaks are reduced proportionally to the inverse of time [12]. (The saturation in Fig. 4 is caused by the fact that we only drive the loop for a small time and because phase-mixing is not taken into account). We see that even for the slowest decrease, the time needed to generate a length scale of about $100m$ lies around 3 hours, estimated using the fit $l = \frac{0.84}{t}$ (where $l = \frac{L}{b}$).

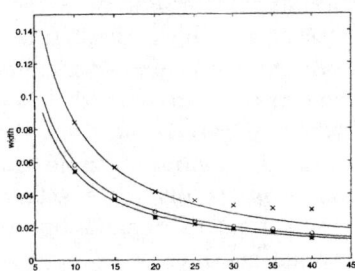

FIGURE 4. *Time evolution of FWHM of 3 main resonant peaks in the Alfvén mode energy after driving the loop by 1 pulse ($\omega = 8.51$).*

We realise that the linear MHD model only enables us to make a rather rough estimation of reality but it is certainly not an overestimation. Nonlinearity would lead to even smaller time scales by means of turbulence [13]. Hence we can conclude that in our model resonant absorption is able to produce small scale dissipative features on a realistic time scale.

IV   COMPARISON WITH PREVIOUS STUDIES

Most of the previous studies on the efficiency of resonant absorption concluded that the heating is highly concentrated in a thin layer of the loop. Our results show multiple resonant peaks and consequently a more globally heated loop. What are the main differences that cause this new result?

First, most of the earlier studies assumed a monochromatic or harmonic driver at the footpoints of the loop. We point out the importance of a more realistic driver by comparing our results with those for an harmonic driver in the same model. We drive the loop by a sine-function with period $p = 2\pi/9$, which is approximately the average of the pulse widths present in the pulse train of section III B. In Fig. 5 we plot the Alfvén mode energy at $t = 20$, this means just before 30 periods have passed such that the graph is comparable to Fig. 3 ($t = 40$).

213

Only one resonant peak is built up resulting from an energy transfer between an initially excited fundamental fast body wave and an Alfvén wave, namely at $x_A = 0.54$, corresponding to the quasi-mode with frequency $\omega = 8.51$ which is the closest to the driving frequency $\omega_d = 9$. The second resonant peak, built up at the magnetic surface $x = 0.59$ where the local Alfvén frequency equals the driving frequency, is forced by the harmonic driving and will not appear in more realistic, random motions. Since we did not take into account the difference in total kinetic energy in the drivers, it only makes sense to compare the figures qualitatively.

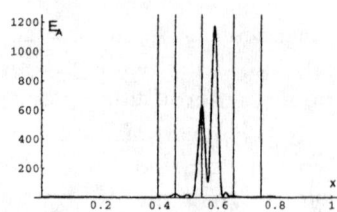

FIGURE 5. *Alfvén mode energy as a function of x after an harmonic driving of 30 sine-pulses with period $2\pi/9$.*

Another very important difference with previous studies which included a broadband driver but nevertheless had only a single heated magnetic shell, is the fact that in our study, all the quasi-modes are taking into account. Assuming k_z to be constant, as done in many papers, excludes all but the first harmonic of the fast modes in the driving spectrum. As seen in Fig. 3, higher harmonics also lead to resonant peaks in the loop and consequently should not be excluded in studies of coronal heating.

V OUTLOOK FOR FUTURE WORK

As an extension to our previous work, we plan to study loops driven by azimuthally polarised footpoint motions. Alfvén waves will be driven directly and when $k_y \neq 0$, these excite fast waves, including quasi-modes, which in turn resonantly couple to Alfvén waves. These direct and indirect resonant excitations, can have both positive and negative effects on the heating [7].

A second important effect to take into account is the coupling with the chromosphere which leads to a mass flow from corona to chromosphere and vice versa. Consequently, the heating of the loop results in a time dependent coronal density. First, at the location of the resonances, the plasma will be underdense relative to the large heating rate and the field lines will fill with chromospheric plasma. On the other hand, all the other field lines will be overdense so that coronal plasma will cool down on the chromosphere. As a result, the density profile of the coronal loop will evolve to resemble the heating profile more closely! This modified density profile has two primary effects: it changes the quasi-mode frequency and secondly, the location where the ideal Alfvén resonance condition is satisfied, will change.

Ofman et al. [3] presented one of the first studies on the effect of this varying density on the heating of the loop. They incorporated a broadband driver but kept k_z constant and consequently started from the idea of one resonant magnetic surface. The dynamic density they modelled resulted in a profile with peak density regions shifting around and varying in magnitude as the simulations evolved. The effect on the heating was very promising: **several resonant surfaces** were found,

shifting throughout the loop.

This result drives us to incorporate the idea of a varying density in our model, which resulted in multiple resonance layers. In that way, we hope to find multiple resonance peaks, shifting throughout the loop. This effect could be consistent with the globally heated loops we observe.

VI   SUMMARY

We studied resonant absorption in a linear, ideal MHD model of a coronal loop, driven by quasi-random footpoint motions polarised normal to the magnetic flux surfaces. We found before that driving at the loop's feet forms a good basis for resonant absorption as heating mechanism. Afterwards we considered the coupled case in order to study the energy transfer from the body modes to the resonant Alfvén waves and the subsequent creation of small lengthscales. We find that the energy is efficiently transferred and Alfvén resonances are built up at the magnetic surfaces corresponding to the quasi-modes of the system. The implementation of a random driver and a varying k_z turns out to be very important since only in that case, multiple resonant peaks are built up on realistic time scales. In future work this analysis will be extended by taking into account the chromospheric coupling. Density profiles will become time dependent and we expect the resonant surfaces to shift in time what will result in an (optically) globally heated loop!

ACKNOWLEDGEMENTS

We gratefully acknowledge James A. Klimchuk for useful discussions and suggestions.

REFERENCES

1. De Groof, A., Tirry, W.J., and Goossens, M., *A&A* **335**, 329-340 (1998)
2. De Groof, A., and Goossens, M., *A&A* **356**, 724-734(2000)
3. Ofman, L., Klimchuk, J.A., and Davila, J.M., *ApJ* **493**, 474-479 (1998)
4. Nakariakov, V.M., Ofman, L., DeLuca, E.E., Roberts, B., and Davila, J.M., *Science* **285**, 862-864 (1999)
5. Tataronis, J.A., and Grossmann, W., *J. Plasma Phys.* **13**, 87 (1975)
6. Ionson, J.A. *ApJ* **226**, 650-673 (1978)
7. Berghmans, D., and Tirry, W.J., *A &A* **325**, 318-328 (1997).
8. Tirry, W.J., Berghmans, D., and Goossens, M., *A&A* **322**, 329 (1997)
9. Goedbloed, J.P., and Halberstadt, G., *A&A* **286**, 275-301 (1994)
10. Beliën, A.J.C., Martens, P.C.H., and Keppens, R., *A&A* , (1998)
11. Beaufumé, P., Coppi, B., and Golub, L. , *ApJ* **393**, 396-408 (1992)
12. Mann, I.R., Wright, A.N., and Cally, P.S., *J.Geophys. Res.* **100**, 19441-19456 (1995)
13. Ofman, L., and Davila, J.M., *J. Geophys. Res.* **100**, 23427 (1995)

Observation of oscillations in coronal loops

I.De Moortel, R.W.Walsh, J.Ireland

School of Mathematical and Computational Sciences, University of St Andrews, North Haugh, St Andrews, Fife KY16 9SS, Scotland
ESA at NASA Goddard Spaceflight Center, Room G-1, Building 26, Mail Code 682.3, Greenbelt, Maryland 20771, USA

Abstract. High cadence TRACE data (JOP 83) in the 171 Å bandpass are used to report on several examples of outward propagating oscillations in the footpoints of large diffuse coronal loop structures close to active regions. The disturbances travel outward with a propagation speed between 70 and 160 km s^{-1}. The variations in intensity are of the order of 2%-4%, compared to the background brightness and these get weaker as the disturbance propagates along the structure. From a wavelet analysis at different positions along the structures, periods in the 200–400 seconds range are found. It is suggested that these oscillations are slow magneto-acoustic waves propagating along the loop, carrying an estimated energy flux of 4×10^2 ergs cm^{-2} s^{-1}.

I INTRODUCTION

Despite the widespread occurrence of hot coronae in the Sun and other stars, the coronal heating mechanism is still not fully understood. Over the past decades, several mechanisms have been suggested ([1]–[3]). The study of coronal oscillations is important since these oscillations could be associated with dissipating wave motions that could be heating of the solar corona. The detection of such oscillations in the corona is a crucial step in determining the presence and relevance of these wave heating mechanisms.

DeForest & Gurman [4] report on quasi-periodic compressive waves in solar polar plumes which Ofman et al. [5] consider to be slow magneto-acoustic waves. Aschwanden et al. [6] reported the detection of spatial displacement oscillations of coronal loops, observed for the first time due to the high spatial resolution of TRACE. Nakariakov et al.[7] analysed these transverse loop oscillations induced by a flare and estimated that the coronal dissipation coefficient could be as much as eight or nine orders of magnitude larger than the theoretically predicted classical value. This larger dissipation coefficient may solve some of the existing difficulties with wave heating and reconnection theories. Berghmans & Clette [8] found

CP537, *Waves in Dusty, Solar, and Space Plasmas*, edited by F. Verheest, et al.
© 2000 American Institute of Physics 1-56396-962-9/00/$17.00

propagating disturbances in coronal loops, observed in the EIT (SOHO) 195 Å bandpass.

In this paper we report on the detection of propagating oscillations out from the footpoints of large diffuse coronal loop structures close to active regions as observed in 171 Å by TRACE. A first report on small scale EUV brightenings in the same high cadence (9 s) TRACE data was given in [8]. As in Ireland et al. [9]–[10], we employ a wavelet analysis to investigate significant periodicity in the observed oscillations.

II   OBSERVATIONS - 23 MARCH 1999

The analysis in this paper is based upon TRACE 171 Å (Fe IX) observations of active region AR 8496, taken as part of JOP 83 - *High cadence activity studies and the heating of coronal loops*. The aim of this study was to reduce the cadence of the TRACE observations as low as possible while still retaining enough counts that events of interest would not be lost in the noise. The 23 March 1999 sequence started at 0647 UT and consists of 157 512×512 pixel images with a 9 s cadence and with a pixel size of $1''$. For more details see De Moortel et al. [11]. Figure 1 (left) shows an image of the footpoint of a large coronal loop near AR 8496. The tube-like area that is indicated shows the region we will be looking at in detail. The bright footpoint is about 40 arcsec long.

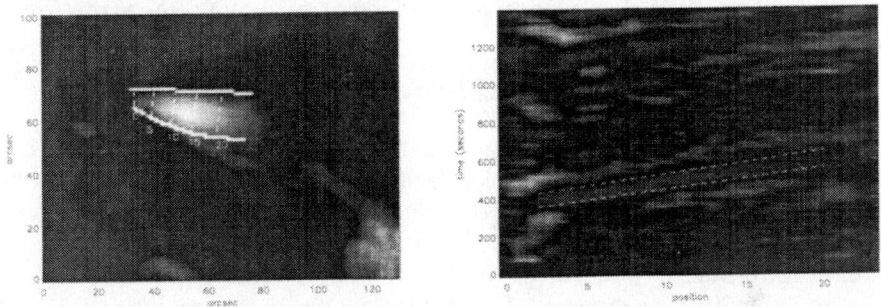

FIGURE 1. (left) Typical image in the 23 March 1999, 0651 UT, 171 Å dataset of the area supporting the oscillatory signal. (right) A plot of the running difference of the average time series for each position along the structure.

III   ANALYSIS AND RESULTS

To obtain results above a satisfactory confidence level, it is necessary to sacrifice some of the spatial and temporal resolution of this dataset. To increase the signal-to-noise ratio, a datacube with an 18 second cadence was created by summing over consecutive images. To emphasise the time-variable aspect of the behaviour of the region, a running-difference image of the average time series is created for each position. From each frame we subtracted the frame taken 90 seconds earlier. In order to obtain a uniform normalisation, we added all the unique datacounts along two consecutive cross-sections and divided by the number of counts. Note that the datacube now has half of the time and spatial resolution of the original.

In the running-difference image (Figure 1), some bright and dark features can be seen clearly running across the image with positive gradients. These diagonal ridges represent outward travelling regions of slightly higher (bright) or lower (dark) intensity. This indicates that a disturbance is travelling along the structure. The propagation speed of these features can be calculated by measuring their slope in the running-difference image. This speed has been calculated to be $\sim 70 - 165$ km s^{-1} throughout the time sequence. Taking into account line-of-sight projections, this range gives us a lower limit for the propagation speed. There is no significant deceleration or acceleration of the features as they propagate. These oscillations are real time variations in intensity since motions due to solar rotation would result in much slower variations. For the first half of the region, the amplitude of the disturbance is a variation of 2%-4% compared to the background brightness, which is more that twice the amplitude of the expected noise level. The signal gets weaker as it propagates along the structure; there is about a 0.6%-2% variation in the background brightness which is only just above the noise level when the signal reaches the end of the analysed region. So there appears to be evidence that the amplitude of the disturbances decreases as they travel along the structure.

Wavelet analysis

To determine an oscillation timescale in the data, we introduce a wavelet analysis at different positions along the tube. Wavelet analysis is an important extension of Fourier analysis as it provides the time localisation of the frequency components. This makes the wavelet analysis ideal for analysing time series where one expects localised variations of power. For further details we refer the reader to [12].

Let us assume a time series x_n of N observations with sample interval δt. The continuous wavelet transform is then defined as the convolution of x_n with an analysis (or mother) wavelet $\psi(\eta)$. It is assumed that ψ is normalised, i.e. $\int_\infty^\infty \psi \psi^* d\eta = 1$. For $\eta = (n' - n)\delta t/s$ we have

$$W_n(s) = \sum_{n'=0}^{N-1} x_{n'} \sqrt{\frac{\delta t}{s}} \psi^* \left[\frac{(n' - n)\delta t}{s} \right] , \qquad (1)$$

where s is the wavelet scale and n allows us to translate the analysing wavelet in time. By varying s and n, one can build up a picture of any features in the time series as a function of the scale s and the localised time index n.

To analyse our oscillatory signal, we will use the Morlet wavelet,

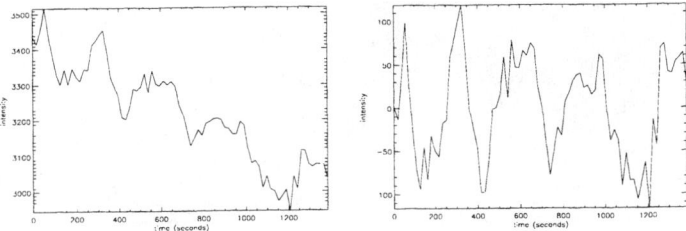

FIGURE 2. An overview of the raw (left) and processed (right) intensity variations at position 4 along the region shown in Figure 1

$$\psi(\eta) = \pi^{-1/4} \exp(6i\eta) \exp\left(-\eta^2/2\right) , \tag{2}$$

consisting of a plane wave modulated by a Gaussian, as our mother wavelet. The wavelet power spectrum is defined as $|W_n(s)|^2$. Ranging through s and n will build up a two-dimensional time-frequency transform of the original time series. The wavelet transform suffers from edge effects at both ends of the time series; this results in a *cone of influence* in the transform, indicated as a dashed line in Fig. 3. Portions of the transform outside the cone of influence are subject to these edge effects and cannot be trusted. To put a confidence level in the analysis, we assume that the noise in our TRACE data is Poissonian, that is, if x counts are observed then the Poisson noise in such measurement is $\sigma_{\text{noise}}(x) = \sqrt{x}$. A 99.0% confidence level is chosen in the analysis above which any wavelet power is considered as real. The same running difference and wavelet method was applied to Quiet Sun regions in the same dataset but no significant periodicity was found.

The left plot of Figure 2 shows the original data at position 4 along the tube like region as defined in Figure 1. The right plot shows the "processed" data at this position; that is, a linear polynomial is fitted to the original data and subtracted subsequently from the original data. Just from a visual inspection, we already see that at this position, there is a signal with a period of roughly 300 seconds for the entire duration of the observation.

The results from the wavelet analysis for position 4 are shown in Figure 3. A clear band of strong wavelet power, situated between 250 and 350 seconds, above the 99.0% confidence level, is detected running throughout the entire time interval. This analysis was repeated for all the positions along the tube and an overview of the results is given in Figure 3 (right). This figure shows the range of periods that are picked up for each position, above the 99.0 % confidence level. From this we see that the periods are situated between 140 s and 500 s. The diamonds in Figure 3 indicate where the strongest wavelet power is situated for each position, which ranges from 180 s to 420 s.

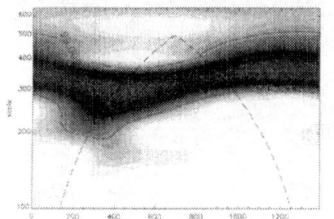 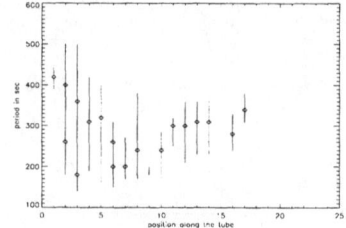

FIGURE 3. (left) Wavelet analysis at position 4 along the tube (confidence level=99.0 %). Darker colours correspond to higher wavelet power. The dashed line indicates the cone of influence. (right) An overview of the results for all positions along the tube (confidence level=99.0 %)

IV   OBSERVATIONS - APRIL 2000

The observations of active region AR 8939, presented in this section were taken on 5 April 2000 (0230 UT), 7 April 2000 (1328 UT) and 9 April 2000 (0904 UT), again as part of JOP 83. The sequences consist of 512×512 pixel images with a 9 s cadence and with a pixel size of $1''$. The data is prepared in the same way as in [10]. We here give a brief overview of the properties of the oscillations found in this data.

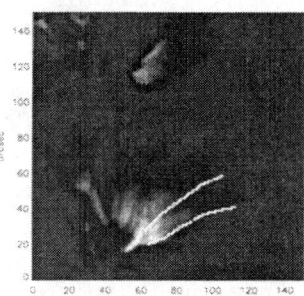 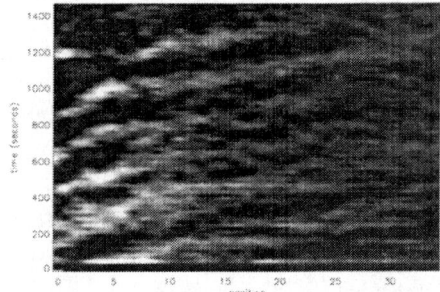

FIGURE 4. (left) Typical image in the 5 April 2000, 0230 UT, 171 Å dataset of the area supporting the oscillatory signal. (right) A plot of the running difference of the average time series for each position along the structure.

In the running difference image (Figure 4), we again see bright and dark diagonal features running across the image, which indicates that disturbances are travelling along the structure. From the slopes in this running-difference image, the propagation speed has been calculated to be between 70 and 185 km s^{-1}. The amplitude of the disturbance is a variation of 2%-3.5% compared to the background brightness and gets weaker as the disturbance propagates along the structure. From a wavelet analysis at different positions along this region, we found periods in the 200–350 seconds range.

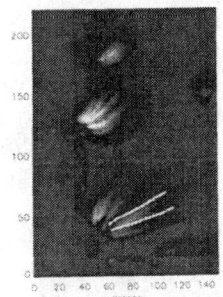 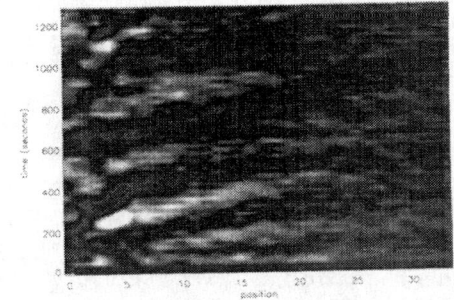

FIGURE 5. (left) Typical image in the 7 April 2000, 1328 UT, 171 Å dataset of the area supporting the oscillatory signal. (right) A plot of the running difference of the average time series for each position along the structure.

The 7 April 2000, 1328 UT dataset (Figure 5), displays very similar results. The propagation speed is of the order 70–150 km s^{-1} and the amplitudes of the disturbance are of the order 2%-4.5% in the lower half of the region. A wavelet analysis indicated periods between 250 and 450 seconds, which is slightly longer than the periods we found on 5 April 2000.

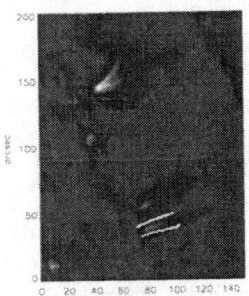 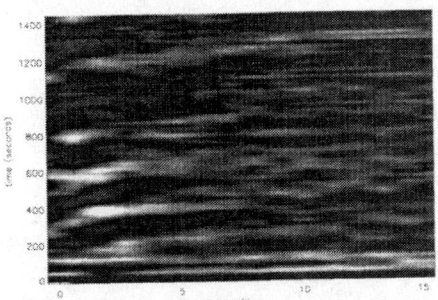

FIGURE 6. (left) Typical image in the 9 April 2000, 0904 UT, 171 Å dataset of the area supporting the oscillatory signal. (right) A plot of the running difference of the average time series for each position along the structure.

Finally, on 9 April 2000, 0904 UT (Figure 6), progation speeds of the order 60–160 km s^{-1} are found and amplitude variations of the order 2%-5% in the lower half of the region. From a wavelet analysis at different positions, we found periods in the 200–400 seconds range.

V DISCUSSION AND CONCLUSIONS

In the magnetically dominated corona, structures are expected to undergo three types of oscillations, driven by different restoring forces: non-compressional Alfvén waves, where the restoring force is provided by the magnetic tension, and slow and fast magneto-acoustic waves, where the magnetic and kinetic pressures are the restoring forces ([13]–[15]). While Alfvénic oscillations are essentially velocity oscillations and do not cause any variation in the intensity, the compressional waves cause density changes. Therefore, they could be observed in the form of intensity variations, provided there are no associated temperature changes big enough to bring the material outside the 171 Å bandpass.

All four examples of oscillations in the footpoints of large coronal loops discussed in this paper display a very similar range of parameters. The disturbances travel outward with a propagation speed between 70 and 160 km s^{-1}. The variations in intensity are of the order of 2%-4%, compared to the background brightness and these get weaker as the disturbance propagates along the structure. From a wavelet analysis at different positions along the structures, periods in the 200–400 seconds range are found.

Since the data displays propagating variations in the intensity (Figure 1), this could be interpreted as magneto-acoustic modes. The moving features that are visible in the loop footpoint region travel at a constant speed throughout our field of view, i.e. there appears to be no significant deceleration or acceleration. As they are travelling at the order of the coronal sound speed ($c_s \sim 150$ km s^{-1}), these propagating oscillations could be considered to be good candidates for being propagating slow magneto-acoustic waves, similar to the propagating features found in polar plumes by DeForest et al. (1998). Ofman et al. [5] calculated the acoustic cutoff frequency to be $\omega_{\text{cutoff}} \approx 1.5 \times 10^{-3}$ rad s^{-1} and consequently, waves with periods over 70 minutes would be evanescent. As the oscillations observed here have much shorter periods, the waves are travelling into the corona. Following [5], it is possible to estimate the energy flux carried by the slow magneto-acoustic waves as $\rho[(\delta v)^2/2]v_s$, where δv is the wave velocity amplitude and $v_s \approx c_s = 150$ km s^{-1} in the corona. Using $\rho = 5 \times 10^{-16}$ g cm^{-3} and $\delta v = 3$ km s^{-1}, an upper bound for the wave energy flux is $\sim 3.5 \times 10^2$ ergs cm^{-2} s^{-1}. This energy is only a very small fraction of the total energy required to heat coronal loops. The above interpretation is only valid if we assume the loop to be linear and homogeneous. For a non-linear, non-homogeneous medium, there could be several other possible interpretations such as nonlinear waves and mode coupling. Unlike the observations done by Aschwanden et al. [6], no flare occurred during or just before these observations.

Due to the coincidence with the photospheric 5-minute period, there could be some form of coupling between the photosphere and the coronal loop, e.g. in the form of photospheric driving of the loop footpoints. The observation of coronal oscillations has important consequences as it could provide us with useful information for many numerical or analytical models of wave heating in the solar corona. Other possible interpretations of the propagating disturbances and the damping of their amplitude as they travel along the structure, will be addressed in De Moortel et al. [16]. More details of the April 2000 datasets will also be discussed.

Acknowledgements

The authors would like to thank A.W. Hood and D.S. Brown for useful comments and critical reading of the manuscript. Wavelet software was provided by C. Torrence and G. Compo, and is available at URL: http://paos.colorado.edu/research/wavelets. I. De Moortel is supported by E.U. grant ERBFMBICT982880.

REFERENCES

1. Narain, U., Ulmschneider, P., *SSR* **75**, 453 (1996)
2. Browning, P.K., *Plasma Phys. and Controlled Fusion* **33**, 539 (1991)
3. Zirker, J.B., *Solar Physics* **148**, 43 (1993)
4. DeForest, C.E., Gurman, J.B., *ApJ* **501**, L217 (1998)
5. Ofman, L., Nakariakov, V.M., DeForest, C.E., *ApJ* **514**, 441 (1999)
6. Aschwanden, M.J., Fletcher, L., Schrijver, C.J., Alexander, D.,*ApJ* **520**, 880 (1999)
7. Nakariakov, V.M., Ofman, L., DeLuca, E.E., Roberts, B., Davila, J.M., *Science* **285**, 862 (1999)
8. Berghmans, D., Clette, F., *Solar Physics* **186**, 207 (1999)
9. Ireland, J., Walsh, R.W., Harrison, R.A., Priest, E.R., *ESA-SP* **446** (1999)
10. Ireland, J., Wills-Davey, M., Walsh, R.W., *Solar Physics, in press* (1999)
11. De Moortel, I., Ireland, J., Walsh R.W., *A&A* **355**, L23 (2000)
12. Torrence, C., Compo, G.P., *Bull. Amer. Meteor. Soc.* **79**, 61 (1998)
13. Porter, L.J., Klimchuk, J.A., Sturrock, P.A., *ApJ* **435**, 482 (1994)
14. Porter, L.J., Klimchuk, J.A., Sturrock, P.A., *ApJ* **435**, 502 (1994)
15. Roberts, B., Edwin, P.M., Benz, A.O., *ApJ* **279**, 857 (1984)
16. De Moortel, I., Ireland, J., Walsh R.W., 2000, *in preparation*

Phase mixing of Alfvén waves in an open and stratified atmosphere

I.De Moortel, A.W.Hood, T.D.Arber

School of Mathematical and Computational Sciences, University of St Andrews, St Andrews, Fife, KY16 9SS, Scotland

Abstract. Phase mixing was introduced by Heyvaerts and Priest [1] as a mechanism for heating plasma in open magnetic field regions. Here we include a stratified density and a diverging background magnetic field. We present numerical and WKB solutions to describe the effect of stratification and divergence on phase mixing of Alfvén waves. It is shown that the decrease in density lengthens the oscillation wavelengths and thereby reduces the generation of transverse gradients. However, the divergence of the field lines shortens the wavelengths and thus enhances the generation of gradients. Furthermore we found that in a stratified atmosphere, ohmic heating is spread out over a greater height range whereas viscous heating is not strongly influenced by the stratification. A wavelet analysis indicated that the wavelet transform could povide us with information about the medium the waves are travelling through.

I INTRODUCTION

The coronal heating mechanism remains one of the major unsolved problems in solar physics. The possibility of heating by MHD waves has been investigated intensively and a prime candidate for transferring energy up to coronal levels is a flux of Alfvén waves. Heyvaerts and Priest [1] suggested damping of Alfvén waves due to phase mixing as a possible source of coronal heating. Since then, the propagation and damping of shear Alfvén waves in an inhomogeneous medium has been studied in more detail ([2]–[6]) by relaxing the limits of weak damping and strong phase mixing. Possible observational evidence of coronal heating by phase mixing is discussed in [7]. More recently, Ruderman et al. [8] considered phase mixing of Alfvén waves in planar two-dimensional open magnetic configurations, using a WKB method.

In this paper we study the effect of both vertical and horizontal density stratifications and divergence of the background magnetic field on the phase mixing of Alfvén waves in an open magnetic atmosphere. We restrict ourselves to a study of travelling waves, generated by photospheric motions that cause disturbances to propagate outwards from the Sun without total reflection. For a more detailed description of this study we refer the reader to [9]–[10].

CP537, *Waves in Dusty, Solar, and Space Plasmas*, edited by F. Verheest, et al.
© 2000 American Institute of Physics 1-56396-962-9/00/$17.00

II EQUILIBRIUM AND LINEARISED EQUATIONS

To study the effect of gravitational stratification of the density and divergence of the background magnetic field on phase mixing, we set up the equilibrium in spherical coordinates. Assuming a low-β-plasma and an isothermal atmosphere, the leading order solution is a radially diverging potential magnetic field, $\mathbf{B}_0 = B_0 \frac{r_0^2}{r^2}\hat{\mathbf{r}}$ and $p_0 = p_0(\theta)e^{-\frac{r_0}{H}\left(1-\frac{r_0}{r}\right)}$ where H is the pressure scale height. The linearised MHD equations can then be combined to give either an equation for the perturbed magnetic field b,

$$b + \frac{v_A^2(\theta)}{\Omega^2}\frac{r_0^4}{r}\frac{\partial}{\partial r}\left(e^{\frac{r_0}{H}\left(1-\frac{r_0}{r}\right)}\frac{1}{r^4}\frac{\partial(rb)}{\partial r}\right) + i\frac{\eta}{\Omega}\nabla^2 b = 0\,, \tag{1}$$

where we neglect the dynamic viscosity $\rho_0\nu$, or to give an equation for the perturbed velocity v,

$$\frac{\Omega^2}{v_A^2(\theta)}e^{-\frac{r_0}{H}\left(1-\frac{r_0}{r}\right)}v + \frac{r_0^4}{r^3}\frac{\partial^2}{\partial r^2}\left(\frac{v}{r}\right) + i\frac{\rho_0\nu\Omega\mu}{B_0^2}\nabla^2 v = 0\,, \tag{2}$$

neglecting the resistivity η and where $v_A^2(\theta) = B_0^2/\mu\rho_0(\theta)$ and $\rho_0 = \rho_0(\theta)e^{-\frac{r_0}{H}\left(1-\frac{r_0}{r}\right)}$. Including both dissipation terms increases the order of the equations and obscures the physical effects of each term (see [9]). Working with dimensionless variables ($r = r_0\bar{r}$ and $\theta = \theta_0\bar{\theta}$ and dropping the barred variables) and only including the dominant second order derivatives in the damping terms, WKB solutions to the above equations are given by,

$$b = \sin(\pi\theta)e^{-\frac{1}{4H}\left(1-\frac{1}{r}\right)}\exp\left(-i\frac{2\pi r_0}{\lambda_0}k(\theta)R\right)\exp\left[-\frac{1}{2}\frac{\Lambda_\eta^2}{\Omega}\left(\frac{2\pi r_0}{\lambda_0}\right)^3 k^3\right.$$
$$\left.\int_1^r\left(\theta_0^2 u^6 e^{-\frac{3}{2H}\left(1-\frac{1}{u}\right)} + \frac{k'^2}{k^2}R^2 e^{-\frac{1}{2H}\left(1-\frac{1}{u}\right)}\right)du\right]\,, \tag{3}$$

for the perturbed magnetic field when we consider resistivity, with $\Lambda_\eta^2 = \frac{\eta}{\Omega_0 r_0^2\theta_0^2}$, $k(\theta) = (1+\delta\cos(m\pi\theta))^{-1/2}$ and $R = \int_1^r u^2 e^{-\frac{1}{2H}\left(1-\frac{1}{u}\right)}du$ and,

$$v = \sin(\pi\theta)e^{\frac{1}{4H}\left(1-\frac{1}{r}\right)}\exp\left(-i\frac{2\pi r_0}{\lambda_0}k(\theta)R\right)\exp\left[-\frac{1}{2}\Lambda_\nu^2\Omega\left(\frac{2\pi r_0}{\lambda_0}\right)^3 k\right.$$
$$\left.\int_1^r\left(\theta_0^2 u^6 e^{-\frac{1}{2H}\left(1-\frac{1}{u}\right)} + \frac{k'^2}{k^2}R^2 e^{\frac{1}{2H}\left(1-\frac{1}{u}\right)}\right)du\right]\,, \tag{4}$$

for the perturbed velocity when we consider viscosity, with $\Lambda_\nu^2 = \frac{\rho_0\nu\Omega_0\mu}{B_0^2}\left(\frac{\lambda_0}{2\pi r_0\theta_0}\right)^2$. For the derivation of these equations, see [10]. The results in this paper are obtained by using a numerical code but in all cases, the WKB solutions give very good agreement with the numerical results.

III VERTICAL STRATIFICATION OF THE DENSITY

In this section, we study the effect of the gravitational stratification of the density on phase mixing of Alfvén waves in a non-diverging magnetic field. By assuming $r = 1 + \frac{\lambda_0}{2\pi r_0} z$ and $x = r\theta$ in Eqs. (3) and (4), we see that at low heights, i.e. $\bar{r} \approx 1$, and for small initial wavelengths λ_0 we return to a Cartesian coordinate system.

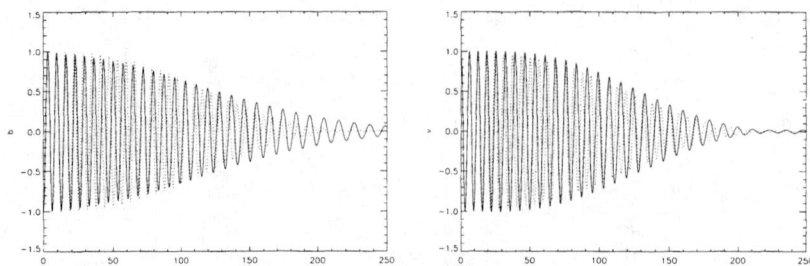

FIGURE 1. *A cross-section of (top) the perturbed field with $\Lambda_\eta^2 = 10^{-4}$ and (bottom) the velocity with $\Lambda_\nu^2 = 10^{-4}$ at $x = 0.5$ for $H = 200$ and $\delta = 0.1$. The dotted line is the corresponding solution for an unstratified plasma.*

As the Alfvén speed $v_A \sim e^{z/2H}$, the wavelength $\lambda \sim e^{z/2H}$ as well so the wavelengths will get longer when the scale height H gets smaller, as we can see from Fig. 1. Therefore phase mixing will be less efficient when the plasma is stratified since it will take longer for waves on neighbouring field lines to get out of phase and hence, to create the necessary short lengthscales for dissipation to become important. On the other hand, we see a slight initial increase in the amplitude of the velocity which will enhance the viscous heating of the plasma. Furthermore, we see that the perturbed velocity decays faster when we consider dynamic viscosity than the magnetic field when we consider resistivity.

The contour plots of j^2 (Fig. 2) show that the ohmic heating is spread out over a wider area and builds up less high when the scale height is smaller. Because the process of phase mixing is slower than in an unstratified atmosphere, for a similar height, less energy will have been transferred into heat through ohmic dissipation in the stratified case. We see, however, that the vorticity is only slightly affected by the vertical stratification. This different behaviour is due to the initial increase in the amplitude of the perturbed velocity and the fact that we considered the *dynamic* viscosity $\rho_0 \nu$ to be constant, rather than the kinematic viscosity ν. We see that in the stratified case, the maximum of the vorticity becomes larger and occurs at lower height than the maximum of the current density. By calculating the total amount of ohmic and viscous heating, we find that these remain constant as the scale height decreases. So although the viscous heating will dominate the ohmic heating at lower heights, the total amount of heat deposited into the plasma by either mechanism will remain the same.

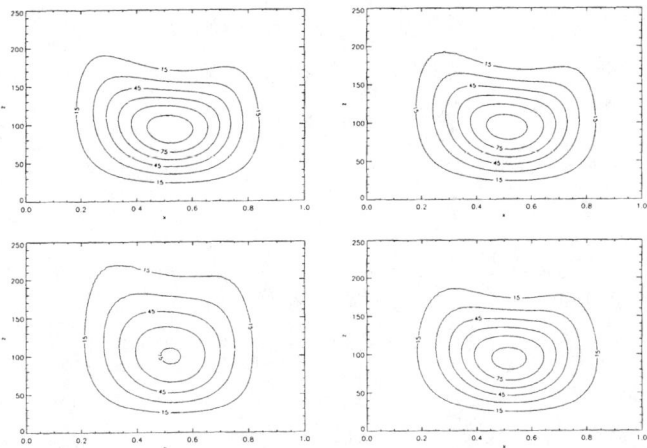

FIGURE 2. *Contour plots of (left) j^2 and (right) ω^2 with $\Lambda^2 = 10^{-4}$ and $\delta = 0.1$ for (top) $H = \infty$ and (bottom) $H = 200$.*

IV   RADIAL DIVERGENCE

We now investigate the effect of a radially diverging background magnetic field but eliminate the vertical stratification from the model by letting $H \to \infty$ in Eqs. (3) and (4). We see that, unlike the stratification case, the perturbed magnetic field and the perturbed velocity behave similarly. We will therefore concentrate on the magnetic field to see the effect of varying the plasma parameters.

From Fig. 3, we see that the divergence of the background magnetic field causes wavelengths to become shorter, as the Alfvén speed and the wavelength λ now behave like $\frac{1}{r^2}$. This suggests that phase mixing will be more efficient in a diverging medium as the short length scales, necessary for efficient dissipation, will be created much faster. The overall damping of the wave amplitudes is enhanced and therefore we expect heat to be deposited into the plasma at lower heights.

Fig. 4 (a) and (b) are contour plots of the current density with a diverging and non-diverging background magnetic field respectively. This figure shows that j^2 is concentrated at lower heights in the radially diverging geometry. However, although the maximum of the current density occurs at a lower height in the spherical case, its value is less than in the Cartesian case. Due to the combination of strong phase mixing and the shortening of the length scales caused by the divergence of the background magnetic field, the perturbed magnetic field is damped more quickly.

By comparing the results for the total ohmic heating with the Cartesian case, we found that the amount of heat deposited through ohmic dissipation does not depend on the geometry of the background magnetic field.

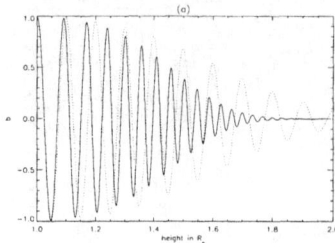

FIGURE 3. *A cross-section of the perturbed magnetic field for a radially diverging background magnetic field at $\theta = 0.5$ with $\Lambda^2 = 10^{-4}$, $\delta = 0.5$ and $\lambda_0 = 0.1$. The dotted lines represent the corresponding solutions in Cartesian coordinates at $x = 0.5$ for a non-diverging background magnetic field.*

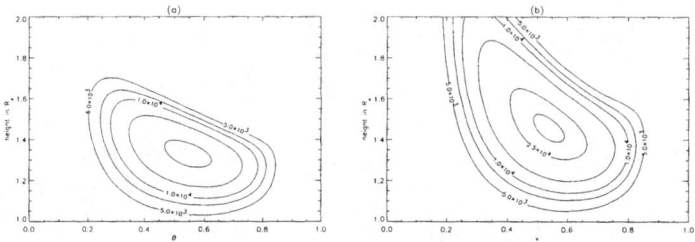

FIGURE 4. *A contour plot of the current density j^2 with $\Lambda^2 = 10^{-4}$, $\delta = 0.5$, $\lambda_0 = 0.1$ and $\theta_0 = 0.1$ for (a) a diverging background magnetic field and (b) a non-diverging background magnetic field.*

V   STRATIFICATION AND DIVERGENCE

In the previous sections we saw that stratification and divergence basically have opposite effects on phase mixing of Alfvén waves. We now want to find out what the combined effect is of gravitational stratification and radial divergence.

As v_A and λ will behave as $\frac{1}{r^2}e^{\frac{1}{2H}\left(1-\frac{1}{r}\right)}$, the behaviour of the wavelength will depend on the value of H and will only increase for $r < \frac{1}{4H}$. For $H = 0.2$, λ initially increases till $r = 0.125$ and then decreases. The magnetic field initially decays faster when we include stratification but overall, it will take longer for the waves to be completely decayed. Analysing the amplitude of the velocity in the stratified plasma, we notice a slight initial increase. The difference between the velocity results for the stratified and the unstratified atmosphere are considerably smaller than the magnetic field results. When we consider viscous dissipation, the perturbed velocity decays faster than the perturbed magnetic field with ohmic dissipation. However, in general we still find that the wave amplitudes decay faster in the unstratified atmosphere. For both the perturbed magnetic field and velocity

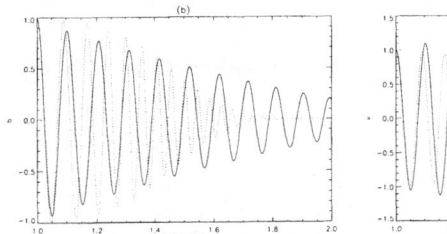
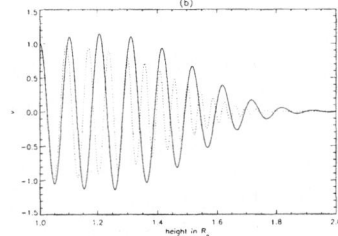

FIGURE 5. *A cross-section of the perturbed (top) magnetic field and (bottom) velocity for a radially diverging background magnetic field at $\theta = 0.5$ with $\Lambda^2 = 10^{-4}$, $\delta = 0.5$ and $\lambda_0 = 0.1$ for $H = 0.2$. The dotted lines represent the corresponding solution for $H = \infty$.*

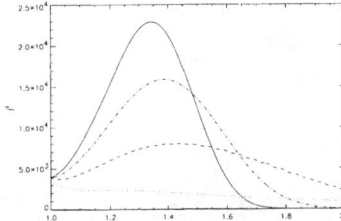
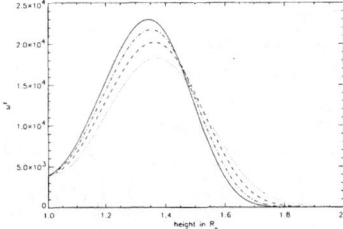

FIGURE 6. *A cross-section of (left) the current density and (right) the vorticity for a radially diverging background magnetic field at $\theta = 0.5$ with $\lambda_0 = 0.1$ for different values of the scale-height (solid line: $H = \infty$, dot-dashed line: $H = 0.5$, dashed line: $H = 0.2$, dotted line: $H = 0.1$).*

we mainly recover the results we found when including (only) stratification. The effect of a radially diverging background magnetic field does not seem to be strong enough to compensate for the stratification of the density.

The cross sections (Fig. 6) of the current density and the vorticity confirm the dominant effect of the stratified density. The current density is spread out over a wider area when the scale height is smaller, i.e. the maximum of j^2 is less high and situated lower down. The vorticity is only spread out slightly due to the lengthening of the wavelengths in the stratified atmosphere. When comparing corresponding spherical and Cartesian results we find that the maximum of the current density and the vorticity is situated at a lower height in the spherical case but also builds up less high than the Cartesian case.

Including both stratification and divergence shows that phase mixing can be more or less efficient than the Heyvaerts and Priest [1] solution, depending on the value of the scale height. A similar conclusion can be found in [8], but a direct comparison cannot be made. In this study we have a different Alfvén speed and we considered a truly open atmosphere, i.e. at no point do the magnetic field lines connect back to the solar surface.

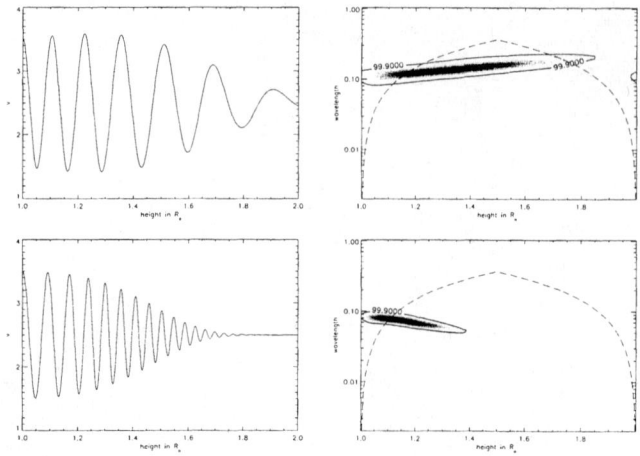

FIGURE 7. Velocity (left) and corresponding wavelet transform with $\Lambda^2 = 10^{-4}$, $\delta = 0.5$ and $\lambda_0 = 0.1$, for (top) a non-diverging magnetic field, with H=0.5 and (bottom) a diverging magnetic field, no stratification.

VI   WAVELET ANALYSIS

Finally, we perform a wavelet analysis of phase mixed Alfvén waves, to see whether there is a distinct signature of phase mixing in the wavelet transform. Wavelet analysis provides the spatial localisation of different wavelengths present in an oscillation. We will use the Morlet wavelet, $\psi(\eta) = \pi^{-1/4} \exp(6i\eta) \exp(-\eta^2/2)$, as our basic wavelet. For further details we refer the reader to [11]. We chose a 99.9% confidence level in our analysis above which we consider any wavelet power as real. The wavelet transform also suffers from edge effects, which results in a *cone of influence* in the transform (dashed line in Figure 7) and portions of the transform outside this cone of influence cannot be trusted.

From Figure 7, we see that the wavelet transform exhibits clear signatures when either a diverging magnetic field or gravitational stratification are present. Figure 7 (top) shows the velocity and corresponding wavelet transform in a non-diverging but stratified atmosphere with dissipation. From the previous sections, we know that wavelengths lengthen when Alfvén waves propagate through a stratified plasma. The wavelength of the wavelet transform indeed starts of at 0.1, but then increases as we go up in height. The wavelet power drops below the 99.9% confidence level when the amplitudes damp due to phase mixing. In Figure 7 (bottom), we consider radial divergence, with no vertical stratification. A diverging background magnetic field shortens the wavelengths when the waves propagate outwards, which again gives a very clear signature in the wavelet transform.

The wavelet transform provides us with some information about the medium the

230

waves are travelling through. A decrease in the wavelength could imply a diverging background magnetic field, whereas an increasing wavelength could indicate gravitational stratification. By doing a wavelet analysis of oscillations found in solar observations, we could not only determine which frequencies or wavelengths are present in the oscillations but also get some information on the part of the solar atmosphere in which the oscillations were found.

VII   DISCUSSION AND CONCLUSIONS

Let us put the results obtained in this paper into typical solar coronal conditions. A coronal hole plasma is strongly inhomogeneous due to the presence of, e.g., plumes, so phase mixing will occur. If we, for example, assume that the plasma density inside a coronal plume is a factor 4 higher than the surrounding plasma, we find $\delta \approx 0.5$. If we further assume a driver with a 1-minute period and assume v_A = 1000 km/s, we find that $\lambda_0 \approx 0.1$ R_o. For $T = 2 \times 10^6$K, the pressure scale height $H = 0.2$ R_o. With these values, we expect the maximum of the ohmic heating to occur at 1.4 R_o and the maximum of the viscous heating at 1.35 R_o(Fig. 6). A 1-minute oscillation could therefore be a candidate for heating the coronal holes, while e.g. a 5-minute oscillation might deposit heat into the solar wind.

From this study of the effect of stratification and radial divergence on phase mixing of Alfvén waves, we found that stratification lengthens the wavelengths, whereas the wavelengths shorten when the waves travel through a diverging atmosphere. Investigating the combined effect of stratification and divergence on phase mixing of Alfvén waves, we found that the efficiency of phase mixing depends on the particular geometry of the configuration. Depending on the value of the scale height the wave amplitudes can damp either slower or faster than in the uniform non-diverging model. A wavelet analysis of phase mixed Alfvén waves indicated that the wavelet transform could provide us with information about the medium the waves are travelling through.

REFERENCES

1. Heyvaerts, J., Priest, E.R., *A&A* **117**, 220 (1983)
2. Cally, P.S., *J.Plasma Phys.* **45**, 453 (1991)
3. Browning, P.K., and Priest, E.R., *A&A* **131**, 283 (1984)
4. Nocera, L., Leroy, B., Priest, E.R., *A&A* **133**, 387 (1984)
5. Hood, A.W., Ireland, J., Priest, E.R., *A&A* **318**, 957 (1997)
6. Hood, A.W., Gonzalés-Delgado, D., Ireland, J., *A&A* **324**, 11 (1997)
7. Ireland, J., *Ann. Geophys.* **14**, 485 (1996)
8. Ruderman, M.S., Nakariakov, V.M., Roberts, B., *A&A* **338**, 1118 (1998)
9. De Moortel, I., Hood, A.W., Ireland, J., Arber, T.D.,*A&A* **346**, 641 (1999)
10. De Moortel, I., Hood, A.W., Arber, T.D., *A&A* **354**, 334 (2000)
11. Torrence, C., Compo, G.P., *Bull. Amer. Meteor. Soc.* **79**, 61 (1998)

Disintegration and reformation of intermediate shock segments in 3D MHD bow shock flows

H. De Sterck[*][†] and S. Poedts[*][1]

[*]Centre for Plasma Astrophysics, K.U. Leuven, Celestijnenlaan 200B, 3001 Leuven, Belgium
[†]von Karman Institute for Fluid Dynamics, Waterloosesteenweg 72, 1640 Sint-Genesius-Rode, Belgium

Abstract. Recently it has been shown that for strong upstream magnetic field stationary three-dimensional (3D) magnetohydrodynamic (MHD) bow shock flows exhibit a complex double-front shock topology with particular segments of the shock fronts being of the intermediate MHD shock type. The large-scale stability of this new bow shock topology is investigated. It is found that large-amplitude perturbations may cause the disintegration of the intermediate shocks — which are indeed known to be unstable against perturbations with integrated amplitudes above critical values—, but that in the driven bow shock problem there are always shock front segments where intermediate shocks are reformed dynamically, resulting in the reappearance of the new double-front topology. This shows that the new bow shock topology, and shock segments of intermediate type in general, may be found in MHD plasma flows even when there are large-amplitude perturbations.

INTRODUCTION

Many phenomena in astrophysical and laboratory plasmas may be described by the equations of magnetohydrodynamics (MHD) [1]. MHD allows for three different anisotropic wave modes, the fast, the Alfvén and the slow wave, with phase speeds in arbitrary direction x denoted by c_{fx}, c_{Ax} and c_{sx}, respectively. Corresponding to the three types of waves, the nonlinear MHD equations allow for three different types of shocks, namely the fast, intermediate and slow shocks. Important examples of shock phenomena in solar and space plasmas are the bow shocks induced by obstacles in fast plasma streams, e.g. the Earth's bow shock in the solar wind [2,3], or the leading shocks induced by fast solar Coronal Mass Ejections (CMEs) [4,3].

Recent simulations of stationary three-dimensional (3D) bow shock flows in MHD plasmas with small dissipation [2,5,4,3] have shown that a new complex double-

---

[1] Research Associate of the Fund for Scientific Research – Flanders (Belgium)

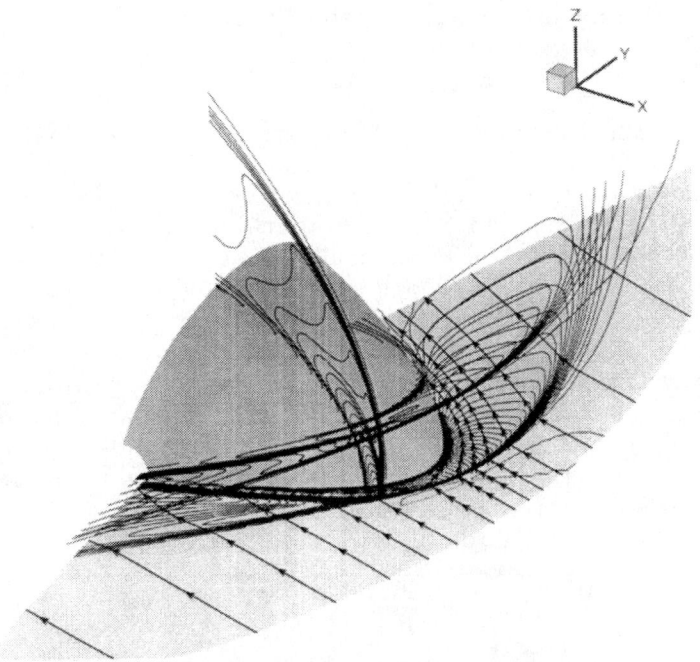

FIGURE 1. 3D MHD bow shock flow around a paraboloid surface with strong upstream magnetic field. The flow comes in from the right. Density contours in three planes and magnetic field lines (with arrows) are shown. A complex double-front topology is obtained. The upstream parameters are $\beta = 0.4$, Mach number $M = v/c = 2.6$ (with c the sound speed), and angle between the velocity and magnetic field $\theta_{vB} = 15°$. The upstream velocity is parallel to the x axis, and the magnetic field is parallel to the xy plane, which is a plane of symmetry.

front bow shock topology (Fig. 1) arises when the flow upstream from the obstacle satisfies the following conditions of strong magnetic field B:

$$B^2 > \gamma p \qquad \text{and} \qquad \rho v_x^2 > B^2 > \rho v_x^2 \frac{\gamma - 1}{\gamma(1 - \beta) + 1}, \tag{1}$$

with p the upstream pressure, v_x the velocity along the magnetic field, ρ the plasma density, γ the adiabatic index (which we take $\gamma = 5/3$), and $\beta = 2p/B^2$ the plasma β. The magnetic permeability $\mu = 1$ in our units. We call a state satisfying these conditions 'magnetically dominated', as opposed to 'pressure-dominated'.

It has been shown that in this new shock topology particular segments of the shock fronts are of the intermediate MHD shock type [5,3]. Intermediate shocks are known to be completely unstable in ideal MHD [1], and to be stable in MHD with small dissipation only when perturbations are small enough [6–10]. This calls for an investigation of the large-scale stability of the new bow shock topology with intermediate shocks against perturbations, which is done in the present paper.

MHD SHOCKS AND SHOCK STABILITY

There are three types of MHD shocks, connecting plasma states which are traditionally labeled from 1 to 4, with state 1 super-fast ($v_n > c_{fn}$ in the shockframe, with n the direction of the shock normal), state 2 sub-fast but super-Alfvénic, state 3 sub-Alfvénic but super-slow, and state 4 sub-slow. The 1–2 fast shock refracts the magnetic field away from the shock normal, while the 3–4 slow shock refracts the field towards the normal. Intermediate shocks (1–3, 1–4, 2–3 and 2–4) bring a super-Alfvénic upstream plasma to a sub-Alfvénic downstream state, while the magnetic field is flipped over the shock normal — the tangential component of the magnetic field changes sign. All MHD shocks have the property of coplanarity, which means that the downstream magnetic field lies in the plane defined by the upstream magnetic field and the shock normal.

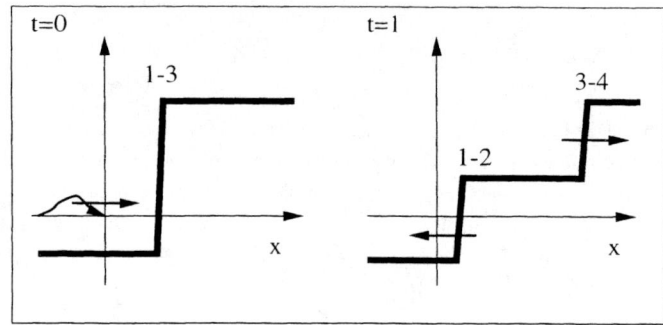

FIGURE 2. A 1–3 intermediate shock splits up into a 1–2 fast shock and a 3–4 slow shock when perturbed by an Alfvén wave.

While fast and slow MHD shocks are known to occur in plasma flows, it has been believed for a long time that intermediate MHD shocks are unphysical [1], a view that is still expressed in most present-day textbooks on MHD. In the dissipationless — or ideal — MHD system intermediate shocks are indeed unstable as they disintegrate instantaneously and split up into fast and slow shocks upon arbitrary small perturbation of the magnetic field component out of the plane of coplanarity (by Alfvén waves) [1,6] (Fig. 2). However, recently it has been shown that intermediate shocks can be stable when dissipation is taken into account [6–10]. The precise influence of dissipation mechanisms and magnitudes on the stability of intermediate MHD shocks is complicated and the analysis remains incomplete. Nevertheless the following general statements can be made. Intermediate shocks are stable in the dissipative MHD system for wide ranges of the dissipative coefficients [6–10].

They can be destabilized by Alfvén waves (Fig. 2), but only when the integrated amplitude $I_z = \int B_z \, dx$ of the non-coplanar magnetic field component B_z of the perturbation is larger than a critical value [6]. This critical value depends on the left and the right states and on the magnitudes of the dissipation coefficients, and vanishes with vanishing dissipation [9]. The stability issues involving intermediate shocks are due to mathematical properties peculiar to MHD, namely non-strict hyperbolicity [7,9], non-convexity [7,10,11], and rotational invariance [9].

These theoretical results on intermediate shock stability were initially confirmed in one-dimensional (1D) simulations, but 1D simulations are of limited generality because coplanarity of left and right states has to be imposed explicitly in order to obtain persistent intermediate shocks. The first clear confirmation of the natural occurrence of intermediate shocks in general 3D MHD flows — where coplanarity is not explicitly imposed — was provided by the simulations of the new complex stationary bow shock topology (Fig. 1) [5,3]. Some concerns regarding the stability and occurrence of intermediate shocks still remain unaddressed, however. Indeed, it has been argued that intermediate shocks cannot be observed 'at large times' in real plasma flows [12,10], because initially present intermediate shocks would disintegrate after short times due to Alfvénic perturbations with super-critical I_z (Fig. 2), which are believed to occur in most real plasma flows with small dissipation. The question whether intermediate shocks can be present 'at large times' in plasma flows with perturbations is addressed in the present paper.

STATIONARY 3D BOW SHOCK FLOWS WITH INTERMEDIATE SHOCK SEGMENTS

In the present Section we briefly explain the topology of the magnetically dominated double-front bow shock flow of Fig. 1 [5,3]. Fig. 3a shows that for a pressure-dominated upstream flow (with a *weak* upstream magnetic field) the classical single-front bow shock topology (Fig. 3c) is obtained that is well-known from hydrodynamic bow shocks, while for a magnetically dominated flow (Fig. 3b) the leading bow shock front is followed by a secondary shock front. In this complex topology (Fig. 3d) shock fronts AB and DE are 1–2 fast, BD is 1–3 intermediate, and DG is 2–4 intermediate close to point D, evolving into 3–4 slow along the front [2,3]. The need for this complex topology in the case of magnetically dominated upstream parameters can be explained in terms of the geometrical properties of MHD shocks [5,3,13].

In the simulations the ideal MHD equations are solved using a conservative finite volume shock capturing scheme which is second order accurate in space and time, employing a slope-limiter approach [13,3]. The stationary 3D bow shock flows are obtained starting from a uniform initial condition and by advancing the time-dependent MHD equations until a steady state solution is reached. These steady solutions are then perturbed in the simulations to be described in the next Section, by varying the upstream conditions at the boundary in a time-dependent

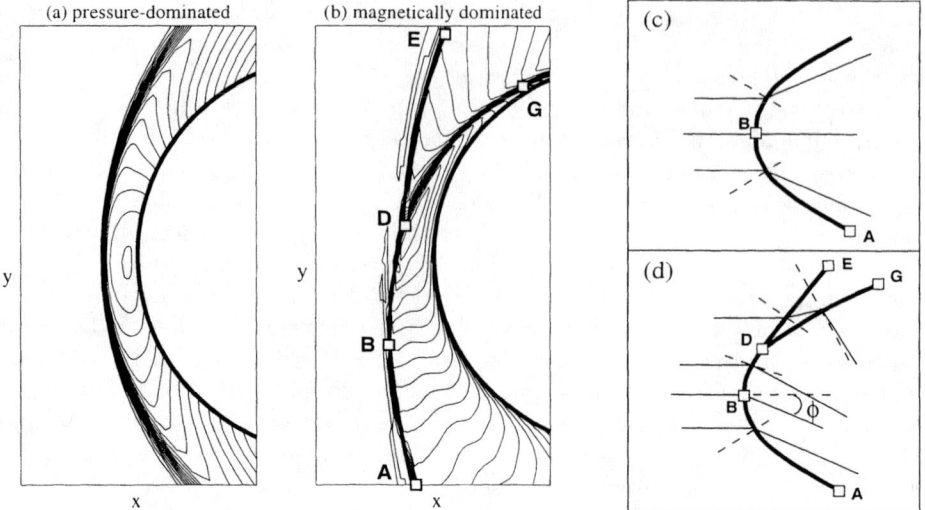

FIGURE 3. (a–b) Bow shock flows over a sphere (thick solid). The flow comes in from the left. Density contours (thin solid) in a plane through the sphere center are shown. The incoming magnetic field is aligned with the x-axis. (a) Pressure-dominated flow ($M_{Ax} = v_x/c_{Ax} = 3.985$, $\beta = 0.4$, $\theta_{vB} = 5°$). (b) Magnetically dominated flow ($M_{Ax} = 1.5$, $\beta = 0.4$, $\theta_{vB} = 3.8°$). (c–d) The two bow shock topologies in the xy symmetry plane. Thick lines are shock fronts, thin lines are magnetic field lines, and shock normals are dashed. (c) Pressure-dominated flow topology. (d) Magnetically dominated flow topology.

manner. Numerical dissipation plays a role analogous to a small physical dissipation in the simulations. As the stability of intermediate shocks depends on the magnitudes of the dissipation coefficients, it would be preferable to perform simulations with explicit discretization of the dissipative terms of the MHD equations. However, realistically small dissipation levels would require a very fine grid resolution, which cannot be achieved because of limited computing resources. Also, our results on steady and perturbed bow shock flows have a general character because qualitatively we obtain the same physical effects using various grid sizes and various numerical schemes — i.e. various effective dissipation. In anticipation of future simulations with explicit discretization of the dissipative terms, the numerical method employed in this paper is suitable to give a qualitative picture of the general large-scale stability of MHD bow shock flows with intermediate shock segments.

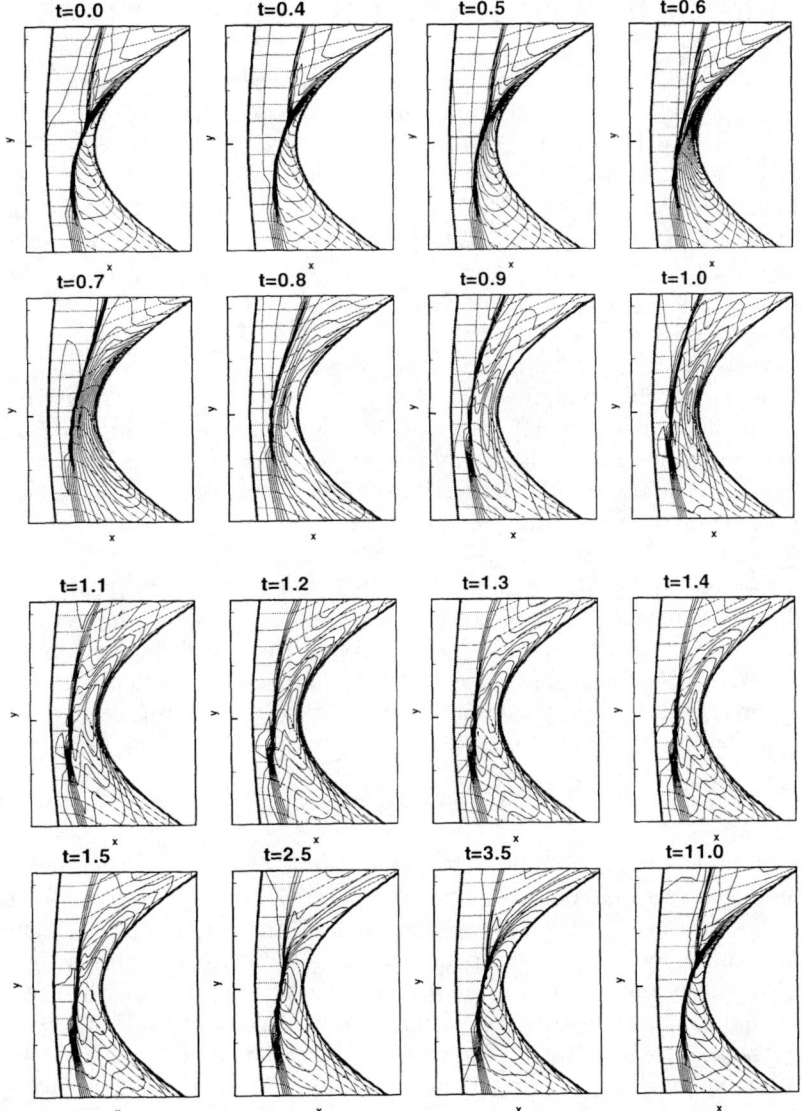

FIGURE 4. Temporal evolution of the bow shock flow in the xy symmetry plane, during and after perturbation of B_z with a Gaussian profile, centered around $t = 0.5$ and with half-width 0.2. Density contours and integral curves of the magnetic field in the xy plane are shown ($40 \times 60 \times 60$ grid).

PERTURBATION OF BOW SHOCK FLOWS

In the present Section we perform a numerical experiment in which we perturb an initial stationary magnetically dominated bow shock flow around a conducting paraboloid with upstream parameters $\rho = 1, p = 0.2, B_x = 1, B_y = 0, B_z = 0, v_x = 1.3 \cos(5°), v_y = 1.3 \sin(5°), v_z = 0$, or equivalently, $\beta = 0.4, M_{Ax} = 1.295, \theta_{vB} = 5°$. The x axis is an axis of rotational symmetry for the paraboloid. The initial flow has the topology of Fig. 1. The (total) Alfvén speed $c_A = 1$. The perturbations are chosen large (of the order of the background field), such that we can be sure that the intermediate shocks become unstable, regardless of the precise values of the dissipation coefficients.

We perturb the non-coplanar magnetic field component at the inflow boundary with a Gaussian profile in time — $B_z = \exp(-((t - 0.5)/0.2)^2)$ —, centered around $t = 0.5$ and with half-width 0.2 (Fig. 4). During the initial evolution (from $t = 0.5$ to $t = 1$) the topology changes substantially as the secondary shock front DG disappears. The intermediate shock segment BD disappears as well, as the leading shock front seems to be entirely of the fast type — magnetic field integral curves are refracted away from the normal — between approximately $t = 0.7$ and $t = 1.4$. Between $t = 0.8$ and $t = 1.0$ there are traces of a second discontinuity following the leading shock front, which may indicate that the 1–3 intermediate shock BD has split up into two shocks as in Fig. 2. When the perturbation has passed the 1–3 and 2–4 intermediate shock segments BD and DG are dynamically reformed starting from $t = 1.4$, and at $t = 11$ the initial steady state topology is recovered.

CONCLUSIONS

We have shown that the new bow shock topology with intermediate shocks is stable, in the sense that perturbations may cause the disintegration of the intermediate shocks, but that the intermediate shocks are dynamically reformed in the driven bow shock flow such that the new topology is regained. This means that the new topology may be observed in solar and space plasma flows with perturbations, provided that super-critical perturbations are not so frequent that the intermediate shock segments do not get time to form. More generally, we have shown that intermediate shocks can be present 'at large times' in plasma flows with perturbations. It is true that initially present intermediate shocks may disintegrate after short times due to Alfvénic perturbations with super-critical I_z, but in driven flows intermediate shock segments may be reformed such that they may be present 'at large times' as well.

Qualitatively we can say that in the context of a given physical plasma with small dissipation the distribution of perturbation amplitudes and frequencies, as related to critical I_z values and time of intermediate shock formation, will ultimately determine if intermediate shocks can occur for long enough times to be observed. It is a complicated task to quantify these requirements, e.g., for the Earth's magnetosphere plasma. Moreover, in the case of the Earth's bow shock, kinetic effects and the collisionless nature of the plasma complicate the stability of shocks [14,15]. There is evidence from observations and simulations that intermediate shocks may form in collisionless plasmas [15,14]. The ultimate test for the applicability of our predictive theoretical results is confrontation with observations. Satellites to be launched in the near future (CLUSTER II and STEREO) may provide observations of the new bow shock topology with intermediate shock segments.

Acknowledgments

The simulations presented in this paper were performed by HDS during a short research visit at the High Altitude Observatory (NCAR), which is gratefully acknowledged for its hospitality.

REFERENCES

1. L. D. Landau and E. M. Lifshitz, *Electrodynamics of continuous media* (Pergamon Press, Oxford, 1984).
2. H. De Sterck and S. Poedts, J. Geophys. Res. **104**, 22,401 (1999).
3. H. De Sterck, Ph.D. thesis, Katholieke Universiteit Leuven, Belgium, and National Center for Atmospheric Research, Boulder, Colorado, USA, 1999.
4. H. De Sterck and S. Poedts, in *Proceedings of the 9th European Meeting on Solar Physics*, Florence, **ESA-SP-448**, 935 (1999).
5. H. De Sterck and S. Poedts, *Phys. Rev. Lett.* **84 (24)** (2000).
6. C. C. Wu, J. Geophys. Res. **93**, 987 (1988).
7. C. C. Wu, in *Viscous profiles and numerical methods for shock waves, Siam Proceedings Series*, edited by M. Shearer (SIAM, Philadelphia, 1991), pp. 209–236.
8. H. Freistuehler, J. Geophys. Res. **96**, 3825 (1991).
9. H. Freistuehler, Phys. Scripta **T74**, 26 (1998).
10. R. S. Myong and P. L. Roe, J. Plasma Physics **58**, 521 (1997).
11. H. De Sterck, B. C. Low, and S. Poedts, Phys. Plasmas **6**, 954 (1999).
12. S. A. E. G. Falle and S. S. Komissarov, On the inadmissibility of non-evolutionary shocks, submitted to *J. Plasma Phys.*, (1999).
13. H. De Sterck, B. C. Low, and S. Poedts, Phys. Plasmas **5**, 4015 (1998).
14. C. C. Wu and T. Hada, J. Geophys. Res. **96**, 3769 (1991).
15. M. Kivelson, C. Kennel, R. McPherron, C. Russell *et al.*, Science **253**, 1518 (1991).

The influence of a buried magnetic field on solar p-modes

C. Foullon[1] and B. Roberts

School of Mathematics and Statistics
University of St Andrews
St Andrews KY16 9SS, Scotland

Abstract. The magnetic field considered to reside at the base of the convection zone is presumed to vary over the solar activity cycle. We examine the effect of such a buried magnetic layer on the properties of solar p-modes. Analytical and numerical solutions to the dispersion relation for these modes are presented. Frequency changes due to the stored magnetic field are found to be negligible in comparison with the low- and intermediate-degree frequency shifts reported over the solar activity cycle. Nonetheless, there are grounds for inferring the signature of such a buried field through examining shifts of various degree. The p-mode frequencies are increased proportionally to the square of the field strength at the base of the convection zone and depend upon the thickness of the magnetic layer.

INTRODUCTION

Changes in solar p-mode oscillations have been observed (e.g. [1]) both in the multiplet frequencies ν_{nl} and the even-index splitting coefficients[2] $\alpha(n, l)$ of modes of radial order n and angular degree l. Early considerations of the shifts in ν_{nl} on the basis of an evolving magnetic field considered to reside at the base of the convection zone [2] suggested a field strength larger than 1 MG [3,4]. Through an analysis of the even splitting coefficients, Basu [5] deduced that a toroidal field modelled by Gough and Thompson [6] would be of strength not larger than 0.3 MG, slightly above but in broad agreement with the range $3 \times 10^4 - 10^5$ G suggested by numerical simulations of active region emergence and evolution at the solar surface (see [7,8], [9] and references therein).

The central frequencies ν_{nl} are determined by the spherically symmetric component of solar structure. The shifts of these frequencies observed over the solar cycle have the same increasing trends with frequency for both intermediate and

---

[1] Email: clairef@mcs.st-and.ac.uk
[2] The measured frequencies are fitted to a sum of Legendre polynomials P_i, each weighted by the coefficients α_i. The odd-α coefficients arise from solar rotation, while the even-α's arise from asphericities, such as caused by magnetic and centrifugal forces.

CP537, *Waves in Dusty, Solar, and Space Plasmas*, edited by F. Verheest, et al.
© 2000 American Institute of Physics 1-56396-962-9/00/$17.00

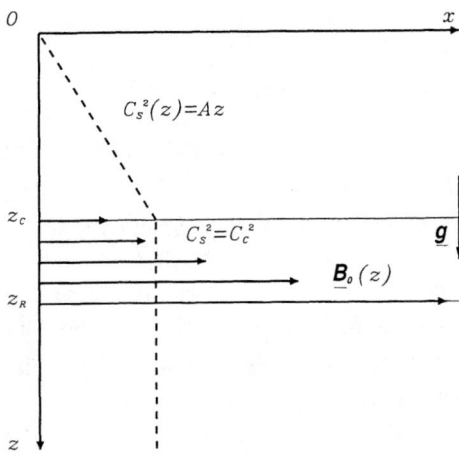

FIGURE 1. A buried magnetic field $\mathbf{B}_o(z)$ lying below a polytrope (with sound speed squared $c_s^2 = Az = \frac{g}{m}z$) below which is an isothermal field-free zone.

low-degree modes, but at high frequencies, low-degree modes are less sensitive to the solar cycle than high-degree modes. Over a solar cycle, the peak shifts at a frequency of 3900 μHz are of order 900 nHz for intermediate-degree modes [1,10] and 750 nHz for low-degree modes [11]. The intermediate frequency shifts are explained by time-dependent variations in the near surface magnetism and temperature (see e.g. [12–16]). Furthermore, it has recently been suggested that the low-degree shifts are consistent with near surface magnetic flux distributions [17].

Aside from the influence of a buried magnetic field on p-mode frequency shifts, which, for a realistic value of the field, might prove much smaller than the observed frequency shifts over a solar cycle [4], it remains important to ascertain the role of such field structure on the physics of the base of the convection zone and to understand in greater detail how the p-modes are affected (see also [18–20]). We consider a semi-infinite region of fluid in a Cartesian system (x, z), with z measured downwards from the solar surface (at $z = 0$). The model consists of a hydrostatically stratified polytrope, representing the convection zone, below which is an isothermal magnetic slab of constant Alfvén speed, underlaid by an isothermal field-free region, standing for the radiative interior; see Figure 1. The polytropic index of the adiabatically stratified medium is $m = 1/(\gamma - 1) = 3/2$, where $\gamma = 5/3$ is the ratio of specific heats. The squared normalised frequency is $\Omega^2 = \omega^2/gk_x$, where ω is the angular frequency of a mode of horizontal wavenumber k_x (related to the degree l through $k_x = L/R_\odot$, for $L = \sqrt{l(l+1)}$ and solar radius $R_\odot$) and g is the constant gravitational acceleration (taken to be the solar surface value $g = 274$ m s^{-2}). The magnetic slab lies between depths $z = z_c$ and $z = z_R$. The square of the sound speed has linear profile and is zero at the solar surface: $c_s^2(z) = \frac{g}{m}z$ for $0 \le z \le z_c$. The sound speed is assumed to be continuous across each magnetic interface, giving

$c_s^2(z) = c_c^2 = \frac{g}{m}z_c$ for $z \geq z_c$. Note that the ratio $\beta = c_c^2/v_A^2$ between the squares of the sound speed c_c and the Alfvén speed v_A in the magnetic slab is very large. In our model, for $c_c = 223$ km s^{-1}, pressure $p(z_c) = 5.26 \times 10^{12}$ Pa at the field-free base of the convection zone [21], and field strength $B(z_c) = 0.3$ MG, the requirement of continuous total pressure gives $v_A = 2.0$ km s^{-1} and $\beta = 12237$.

WAVE EQUATIONS

To derive the dispersion relation, we use the equations of ideal linear MHD (see e.g. [22], and for this particular model see [19]) and Fourier analyse perturbations proportional to $\exp i(\omega t - k_x x)$. We consider the wave equations associated with each of the three layers.

The polytropic layer

Lamb's equation for a field-free medium [23] can be written in the standard form

$$\frac{d^2Q}{dz^2} + K^2(z)Q = 0,\tag{1}$$

where $Q = \rho^{1/2}(z)c_s^2(z)\Delta$, $\Delta = \nabla \cdot \mathbf{V}$ is the compressibility for velocity $\mathbf{V}$ and, in the case of an adiabatic polytrope (see [19,20]),

$$K^2(z) = \frac{k_x}{z}\left[m\Omega^2 - k_x z - \frac{m(m+2)}{4k_x z}\right].\tag{2}$$

With vanishing sound speed at the surface, the solution to Lamb's equation for a polytrope is

$$\Delta = \alpha e^{-k_x z}M(-a, m+2, 2k_x z)\tag{3}$$

where α is an arbitrary constant, M is the confluent hypergeometric Kummer function, and $2a = m\Omega^2 - (m+2)$. In a simple polytrope model [24], which does not contain a magnetic slab, the boundary condition as $z \to \infty$ requires the M functions to terminate in the form of generalised Laguerre polynomials, and the parameter a has to be a non-negative integer. The solutions are the p-modes, given by

$$\Omega^2 = \Omega_n^2 \equiv 1 + \frac{2n}{m}\tag{4}$$

for integer $n \geq 0$.

The solutions $Q(z)$ are oscillatory when $K^2(z) > 0$ and evanescent when $K^2(z) < 0$, with the division between propagation and evanescence, occurring at a depth $z = z_t$ where

$$m\Omega^2 = k_x z_t + \frac{m(m+2)}{4k_x z_t} . \tag{5}$$

For small wavelengths ($k_x \to \infty$), the depth z_t is determined by the Lamb frequency $\omega = k_x c_s(z_t)$.

Isothermal layers

In the magnetic layer, with the assumption that the sound and Alfvén speeds are constants, the vertical component V_z of the velocity perturbation satisfies the second order differential equation [12]

$$\frac{d^2 V_z}{dz^2} + \frac{1}{H_o} \frac{dV_z}{dz} + A_o V_z = 0 , \tag{6}$$

where

$$A_o = \frac{k_x m}{z_c} \cdot \frac{(\Omega^2 - \frac{k_x z_c}{m})(\Omega^2 - \frac{k_x z_c}{m\beta}) + (\Gamma_o - 1)}{(1 + \frac{1}{\beta})(\Omega^2 - \frac{k_x z_c}{m(\beta+1)})} ; \tag{7}$$

$\Gamma_o = \gamma/(1 + \gamma/2\beta)$ is the magnetically-modified adiabatic exponent, and $H_o = \rho_o/\rho_o' = z_c/m\Gamma_o$ is the density scale height.

The determinant $1 - 4A_o H_o{}^2$ indicates the behaviour of the modes. Evanescent surface modes ($4A_o H_o^2 < 1$) are of the form

$$V_z = D_1 e^{\Lambda^- z} + D_2 e^{\Lambda^+ z} , \tag{8}$$

with

$$\Lambda^\pm = \frac{-1 \pm \sqrt{1 - 4A_o H_o^2}}{2H_o} = \frac{-1}{2H_o} \pm \kappa , \tag{9}$$

where

$$\kappa = \frac{\sqrt{1 - 4A_o H_o^2}}{2H_o} . \tag{10}$$

The constants D_1 and D_2 are arbitrary. Oscillatory, body modes ($4A_o H_o^2 > 1$) are not treated here.

In the field-free limit ($\beta \to \infty$), $\Gamma_o \to \gamma$ and $H_o \to \frac{z_c}{m+1}$, so

$$A_o \to \frac{k_x}{z_c} \left[m\Omega^2 - k_x z_c + \frac{1}{\Omega^2} \right] . \tag{11}$$

Hence, the above analysis for an isothermal magnetic layer applies to the isothermal field-free region by taking the limit $\beta \to \infty$.

Evanescent modes exist in between two acoustic cut-offs given by the equation $4A_o H_o{}^2 = 1$ (see the diagnostic diagrams in [19,20]). When Ω^2 and k_x are large, both high frequency separations in the magnetic or field-free case reduce to the Lamb mode at depth $z = z_c$ (i.e. $\omega = k_x c_c$).

THE DISPERSION RELATION

Our three layer model of the solar interior is subject to six boundary conditions: two at each interface between a magnetic and a field-free region at $z = z_i$, one at the surface $z = 0$, and one at $z \to \infty$. Across an interface at $z = z_i$, the continuity of the vertical velocity perturbation V_z and the continuity of the total Lagrangian pressure perturbation

$$P = -\rho_o c_s^2 \left(\Delta + \frac{1}{\beta} \frac{dV_z}{dz} \right) \tag{12}$$

combine in the form $\mathcal{D} \mid_{z=z_i^{nonmag}} = \mathcal{D}_o \mid_{z=z_i^{mag}}$, where $\mathcal{D}(\Omega^2, k_x, c_s^2(z))$ is the ratio V_z / Δ for a field-free medium,

$$\mathcal{D} = \frac{1}{\Omega^4 - 1} \left[\frac{c_s^2}{g} - \frac{\gamma \Omega^2}{k_x} - \frac{c_s^2 \Omega^2}{g k_x} \frac{1}{\Delta} \frac{d\Delta}{dz} \right] , \tag{13}$$

and $\mathcal{D}_o(\Omega^2, k_x, c_c^2, \beta)$ is a magnetic part that depends upon the ratio $\beta = c_c^2 / v_A^2$,

$$\mathcal{D}_o = \frac{Q}{P \frac{1}{V_z} \frac{dV_z}{dz} + k_x} ; \tag{14}$$

$$Q = \frac{\gamma}{\Gamma_o} \left(\Omega^2 - \frac{k_x c_c^2}{g} \right) , \qquad P = \left(1 + \frac{1}{\beta} \right) \left(\Omega^2 - \frac{k_x c_c^2}{g(\beta + 1)} \right) . \tag{15}$$

The expression $\frac{dV_z}{dz}$ in (14) at the magnetic interface z_i can be derived from (8) in the case of surface modes. The condition of continuity at one interface is then

$$\mathcal{D} \left(k_x + P \frac{D_1 \Lambda^+ e^{\Lambda^+ z_i} + D_2 \Lambda^- e^{\Lambda^- z_i}}{D_1 e^{\Lambda^+ z_i} + D_2 e^{\Lambda^- z_i}} \right) - Q = 0 ,$$

which can be developed to yield

$$D_1 [(k_x + P\Lambda^+)\mathcal{D} - Q] e^{\Lambda^+ z_i} + D_2 [(k_x + P\Lambda^-)\mathcal{D} - Q] e^{\Lambda^- z_i} = 0 . \tag{16}$$

Taking $z_i = z_c$ and $z_i = z_R$ in turn, we obtain the dispersion relation in the form

$$\mathcal{E}_p^- \mathcal{E}_i^+ e^{2\kappa(z_R - z_c)} - \mathcal{E}_p^+ \mathcal{E}_i^- = 0 , \tag{17}$$

where

$$\mathcal{E}_p^\pm = (k_x + P\Lambda^\pm)\mathcal{D}_p - Q , \qquad \mathcal{E}_i^\pm = (k_x + P\Lambda^\pm)\mathcal{D}_i - Q . \tag{18}$$

$\mathcal{D}_p$ and $\mathcal{D}_i$ are the field-free parts given by Equation (13), applied respectively for a polytropic medium at $z_i = z_c$ and for an isothermal field-free region at $z_i = z_R$.

The requirement of vanishing Lagrangian pressure perturbation at the surface secures that we study the isolated effect of a buried magnetic field on p-modes. This is readily obtained for vanishing sound speed at the surface. Thus

$$\mathcal{D}_p = \frac{\gamma z_c}{(m+1)(\Omega^4 - 1)} \left[1 - \frac{(m+1)}{k_x z_c}\Omega^2 + \Omega^2 - 2\Omega^2 \frac{M'(-a, m+2, 2k_x z_c)}{M(-a, m+2, 2k_x z_c)} \right],$$

where the prime denotes the derivative of $M(-a, m+2, Z)$ with respect to $Z = 2k_x z$.

The requirement of vanishing kinetic energy density at $z \to \infty$ is satisfied by selecting the surface mode solution Λ_i^- (see Equation (9)), so that

$$\mathcal{D}_i = \frac{\gamma}{k_x(\Omega^4 - 1)} \left[\frac{k_x z_c}{m+1} - \Omega^2 - \frac{z_c}{m+1}\Omega^2 \Lambda_i^- \right].$$

Note that the case of a semi-infinite magnetic region ($z_R \to \infty$) is recovered here as $\mathcal{E}_p^- = 0$, which corresponds to the dispersion relation

$$\mathcal{D}_p = \frac{Q}{P\Lambda^- + k_x} \tag{19}$$

derived in [19]. In the limit of a field-free region underlying the polytrope ($\beta \to \infty$ or $z_R \to z_c$), the right handside of (19) remains non-zero and the dispersion relation is equivalent to $\mathcal{D}_p = \mathcal{D}_i$. All dispersion relations are functions of the combination $k_x z_c$. Thus when $k_x \to \infty$, we recover the results (4) on the basis of a simple polytrope model where $z_c \to \infty$.

FREQUENCY SHIFTS

Frequency shifts $\Delta\nu$ between a maximum and a minimum of activity,

$$\Delta\nu = \nu(max) - \nu(min), \tag{20}$$

can be plotted as a function of base cyclic frequency $\nu(min)$ ($= \omega/2\pi$) or angular degree l. We present in Figure 2 the frequency shifts as a function of frequency, with $\nu(max)$ obtained for a magnetic slab of layer thickness $z_R - z_c = 0.05R_\odot$ and field strength $B_o(z_c) = 0.3$ MG; the base frequency $\nu(min)$ is determined for a semi-infinite isothermal field-free region located at the base of the convection zone. No averaging of ν is carried out; therefore we do not lose information about the radial order n and degree l. We have drawn lines joining frequency shifts with the same radial order n.

We see that an increase of the layer thickness from 0 to $0.05R_\odot$ produces very small frequency shifts, of order not larger than 1 nHz at base frequency 3900 μHz; such shifts are much smaller than actually observed. In a diagnostic diagram, where ν is plotted against degree l, p-mode frequencies form parabolic ridges of order n. In our model, the closer to the cut-off the p-modes get, the more they behave as

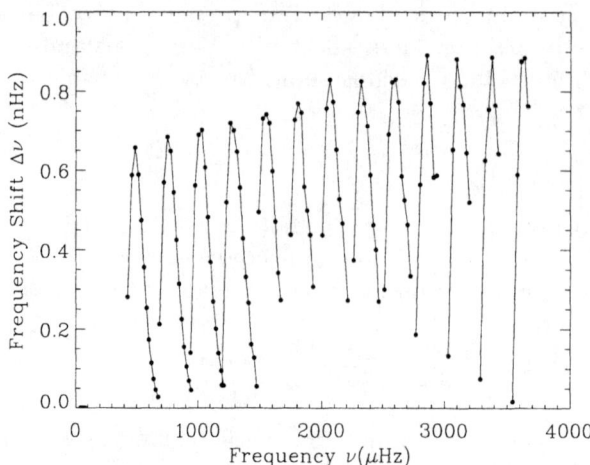

FIGURE 2. The difference $\Delta\nu$ in cyclic frequency between $\nu_{nl}(max)$ obtained with a magnetic slab and $\nu_{nl}(min)$ in the absence of a magnetic field, as a function of base frequency $\nu = \nu_{nl}(min)$. The slab is of thickness $0.05R_\odot$ and field strength $B_o(z_c) = 0.3$ MG. Each p-mode of radial order n (here for $n = 1$ to $n = 13$) shows a peak in frequency shift.

surface waves, and the more the frequency of a p-mode is modified [20]. Hence, shifts from any given p-mode ridge of order n occur close to the Lamb cut-off; the shifts are larger where $\omega_n \approx Lc_c/R_\odot$ or $L \approx gR_\odot\Omega_n^2/c_c^2$, i.e. for l close to

$$\frac{gR_\odot}{c_c^2}\left(1 + \frac{2n}{m}\right) \approx 3.8\left(1 + \frac{2n}{m}\right).$$

Shifts in the ridge of a given p-mode become negligible for degrees higher than this value. However, surface effects have a greater influence on high-degree p-modes (e.g. [13–16]). In general, the frequency ν observed at some period of the solar cycle is an average of the central multiplets ν_{nl} over all degrees and frequency bins. Frequency bins used by observers contain branches of p-mode ridges increasing in number and of radial orders decreasing in n as the range of degrees considered increases. Therefore, frequency shifts obtained with a set of intermediate-degree observations would be averaged over a number of points which do not contribute to the shifts we consider, but to shifts due to variations in the surface layers.

The observed frequency shifts are more likely a consequence of surface magnetism than of a buried magnetic field. Nevertheless, the signals to be expected if one could remove contributions due to the near surface perturbations [25] would be of the type presented in Figure 2. An increase in magnetic field raises the shifts proportionally to the square of the field strength [3]. For instance, a magnetic slab of field strength ten times larger than the original yields shifts of amplitude scaled hundred times bigger. Note also that the degrees of the modes affected increase linearly with radial order n. Thus the graph of $\Delta\nu$ plotted as a function of l will be similar to

Figure 2. The role of the thickness of the layer has yet to be clarified. We expect that modes are affected not only as a function of how deep they propagate, but also on the mode wavelength relative to the layer thickness. It follows that, as shown in Figure 2, a thin buried magnetic slab affects high frequency (or high degree) modes rather more than low frequency (or low degree) modes.

A refinement for future work is the inclusion of an increasing temperature profile in the radiative zone, so that p-modes may undergo refraction there [18]. Of possible interest is the application of our model to magnetic stars.

ACKNOWLEDGEMENTS

C. Foullon would like to express her gratitude to V.M. Čadež for enlightening discussions, P. Vanlommel for her tips in programming and W.J. Chaplin for shedding light on the frequency shift observations. Her thanks also to D. Boddie and E. Verwichte. Financial support is given by PPARC and the University of St Andrews.

REFERENCES

1. Libbrecht, K. G. and Woodard, M. F. (1991) *Science* **253**, 152.
2. Spiegel, E.A. and Weiss, N.O. (1980) *Nature* **287**, 616.
3. Roberts, B. and Campbell, W.R. (1986) *Nature* **323**, 603.
4. Vorontsov (1988) *IAU Symp.* **123**, 151.
5. Basu, S. (1997) *Mon. Not. R. Astron. Soc.* **288**, 572.
6. Gough, D.O. and Thompson, M.J. (1990) *Mon. Not. R. Astron. Soc.* **242**, 25.
7. Choudhuri, A.R. and Gilman, P.A. (1987) *Astrophys. J.* **316**, 788.
8. Schüssler, M. *et al.* (1994) *Astron. Astrophys.* **281**, L69.
9. Fisher, G.H. *et al.* (2000) *Phys. Plasmas* **7(5)**, 2173.
10. Howe, R., Komm, R. and Hill, F. (1999) *Astrophys. J.* **524**, 1084.
11. Chaplin, W.J. *et al.* (1998) *Mon. Not. R. Astron. Soc.* **300**, 1077.
12. Campbell, W.R. and Roberts, B. (1989) *Astrophys. J.* **338**, 538.
13. Evans, D.J. and Roberts, B. (1990) *Astrophys. J.* **356**, 704.
14. Goldreich, P. *et al.* (1991) *Astrophys. J.* **370**, 752.
15. Evans, D.J. and Roberts, B. (1992) *Nature* **355**, 230.
16. Jain, R. and Roberts, B. (1994) *Sol. Phys.* **152**, 261.
17. Moreno-Insertis, F. and Solanki, S. K. (2000) *Mon. Not. R. Astron. Soc.* **313**, 411.
18. Daniell, M. (1998) *Ph.D. thesis*, University of St Andrews, Scotland.
19. Foullon, C. (1999) *Ninth European Meeting on Solar Physics*. ESA SP-448, 87.
20. Foullon, C. and Roberts, B. (2000) *The Dynamic Sun*, ASSL, in press.
21. Guenther, D.B. *et al.* (1992) *Astrophys. J.* **387**, 372.
22. Priest, E.R. (1982) *Solar Magnetohydrodynamics*, 73. D. Reidel Publ. Co.
23. Lamb, H. (1932) *Hydrodynamics*, 549. Cambridge University Press.
24. Christensen-Dalsgaard, J. (1980) *Mon. Not. R. Astron. Soc.* **190**, 765.
25. Dziembowski, W. A. and Goode, P. R. (1997) *Astron. Astrophys.* **317**, 919.

Non-Linear Kink Instabilities in Line-Tied Coronal Loops

C.L.Gerrard, T.D.Arber, A.W.Hood, R.A.M.Van Der Linden*

School of Mathematical and Computational Sciences, University of St Andrews, St Andrews, Fife, KY16 9SS, Scotland
* Koninklijke Sterrenwacht van Belgie, Ringlaan 3, 1180 Ukkel, Belgium

Abstract. Photospheric line-tying has a stabilising effect that allows magnetic energy to build up in coronal loops until critical conditions are reached and the loop becomes unstable to the m=1 kink instability [1].

Recent research has concentrated on the non-linear evolution of instabilities in line-tied coronal loops. There are suggestions [2–5] that current sheets form during the non-linear evolution of the kink instability. We present 3D MHD simulations of the non-linear evolution of MHD instabilities in line-tied coronal loops. These simulations are carried out on a multi-processor cluster at St Andrews using a new 3D MHD Lagrangian remap code (Lare3d) which we shall discuss briefly. Results are presented for loops with different shear profiles to test the conditions for current sheet formation.

We begin by presenting the test case of the Gold-Hoyle field, and compare our results with previous results [1,6]. New results for equilibria with no net axial current are presented and the formation of current sheets discussed.

INTRODUCTION

It has been suggested that m=1 kink instabilities in line-tied coronal loops may provide the mechanism for releasing energy in compact loop flares. The footpoints of coronal loops are anchored in the photosphere and since it is much denser than the corona ($\rho = 8 \times 10^{-5}$ kg m^{-3} in the photosphere compared to 1×10^{-11} kg m^{-3} in the corona), perturbations from the corona do not propagate into the photosphere. Raadu [7] and Hood and Priest [1,8] investigated the stability of coronal loops including this line-tying effect. They found that the loop would become unstable to the m=1 kink mode once the twist ($\phi = \frac{LB_\theta}{rB_z}$) has exceeded some critical value ϕ_{crit}. This allows the loop to be twisted by photospheric motions, building up energy until the twist reaches ϕ_{crit} and it becomes unstable, releasing magnetic energy.

Recent research on kink instabilities in coronal loops has been carried out using numerical simulations [2–6,9–12]. It has been suggested that current sheets can

CP537, *Waves in Dusty, Solar, and Space Plasmas,* edited by F. Verheest, et al.
© 2000 American Institute of Physics 1-56396-962-9/00/$17.00

form during the non-linear evolution of the kink instability [2–5] and that reconnection occurs if the plasma is resistive [3,5,11]. However, there has been some discussion on the question of current sheet formation. While some simulations indicate current sheet formation, others [9,10] do not. This may be due to the way in which the different codes used treat small scales [12] or that current sheets do not form for loops of small shear (shear < 6), where shear is defined as $= \frac{r\phi'}{\phi}$ [11].

In this paper we present results from numerical simulations carried out using a 3D MHD Lagrangian remap code described in Section II. We consider the Gold-Hoyle equilibrium and two new equilibria which we describe in Section III. Next we present the results for the non-linear evolution of the kink instability concentrating on the way the shear of the equilibrium configuration influences the formation of current sheets. Finally, we discuss the results in Section V.

NUMERICAL DETAILS

The simulations are carried out using a 3D MHD shock capturing code (Lare3d). At present, we consider only the non-linear (ideal) evolution and, therefore, the code solves the ideal MHD equations,

$$\frac{\partial \rho}{\partial t} = -\nabla.(\rho \underline{v}), \tag{1}$$

$$\frac{\partial}{\partial t}(\rho \underline{v}) = -\nabla.(\rho \underline{v}\underline{v}) + (\nabla \times \underline{B}) \times \underline{B} - \nabla P, \tag{2}$$

$$\frac{\partial \underline{B}}{\partial t} = \nabla \times (\underline{v} \times \underline{B}), \tag{3}$$

$$\frac{\partial}{\partial t}(\rho \epsilon) = -\nabla.(\rho \epsilon \underline{v}) - P\nabla.\underline{v}, \tag{4}$$

with specific energy density,

$$\epsilon = \frac{P}{(\gamma - 1)\rho}. \tag{5}$$

Lare3d is a Lagrangian remap code. The Lagrangian step uses the fact that mass is a conserved quantity to conserve kinetic energy to machine precision and the remap step uses Van Leer gradient limiters [13] to ensure that it is monotonicity preserving. Furthermore, Lare3d uses Evans and Hawley constrained transport [14] to guarantee that if $\nabla.\underline{B}$ is initially zero then it will remain zero to machine precision for all time. The numerical grid is staggered so that the density, energy (and pressure) are defined at the cell centres, the magnetic field components are

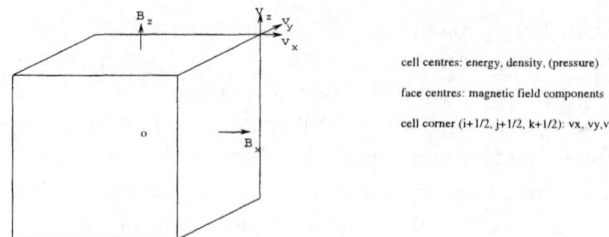

FIGURE 1. A typical cell for Lare3D

defined at the cell faces and the velocity components are defined at the vertex as shown in Figure 1. The code is described in Arber *et al* [15].

As in previous simulations, the coronal loop is modelled as an initially straight cylinder since coronal loops have a large aspect ratio (ratio of length to width) the order of 10. The loop has length L_z with line-tied boundary conditions at $z = -L_z/2$ and $z = L_z/2$. The numerical box has width L_x in the x-direction and L_y in the y-direction with L_x and L_y chosen to be large enough that the boundary conditions in x and y do not effect the evolution of the instability.

EQUILIBRIA

Consider three initial equilibrium configurations. We begin with the Gold-Hoyle equilibrium [16] as a test case. We then consider an equilibrium with shear of 6.32 and a second equilibrium for which we can vary the shear.

Gold-Hoyle

In dimensionless variables, we have,

$$B_\theta = \frac{r}{1 + r^2}, \tag{6}$$

$$B_z = \frac{-\lambda}{1 + r^2}, \tag{7}$$

$$p = p_0 + \frac{1 - \lambda^2}{2(1 + r^2)^2}, \tag{8}$$

and take the force-free case, $\lambda = 1$, $p_0 = 0$. We assume,

$$\rho = \frac{1}{1 + r^2}, \tag{9}$$

giving constant Alfvén speed, and this is one of the configurations considered by Mikic *et al* [6]. To save computational time we use a linear m=1 mode velocity perturbation.

Equilibrium 1

We define this equilibrium by specifying B_θ,

$$B_\theta = \begin{cases} r(1 - r^2)^2 & r < 1.0, \\ 0.0 & r \geq 1.0, \end{cases} \qquad (10)$$

This gives $B_\theta = 0.0$ at $r = 1.0$ and $j_z = 0.0$ at $r = 1.0$. In addition,

$$B_z = \begin{cases} \sqrt{(B_0^2 - r^2(1 - r^2)^4 - r^2 + 2r^4 - 2r^6 + r^8 - \frac{1}{5}r^{10})}, & r < 1.0, \\ \sqrt{B_0^2 - \frac{1}{5}}, & r \geq 1.0, \end{cases} \qquad (11)$$

where B_0^2 is chosen so that $B_z \geq 0$ for $r < 1.0$. We choose $B_0 = 0.5$, and run the equilibrium for a density of 1.0 and a pressure of 0.0. This equilibrium has a shear of 6.32 at the mode rational surface, $r = 0.69$. We take the length of the loop to be 3π as this exceeds the critical length suggested by the linear theory and we use an m=1 mode velocity perturbation to speed up the initial development of the instability.

Equilibrium 2

We define the final equilibrium by specifying the twist of the loop,

$$\phi = \begin{cases} \phi_0 & r < a, \\ \frac{\phi_0}{2}\left(1 + \cos\pi\left(\frac{r-a}{b-a}\right)\right) & a < r < b, \\ 0.0 & r > b \end{cases} \qquad (12)$$

where we can vary ϕ_0, a and b to modify the shear. We use a numerical routine to calculate B_z and hence B_θ. We consider $\rho = 1.0$ and pressure, $P = 0.0$. The shear and position of the mode rational surface vary with ϕ. We use a Hain-Lust equation solver for the infinite non-line-tied case to find the value of k giving the largest growth rate and the eigenmodes of the equilibrium. We take L_z to be larger than L_{crit} and, as above, use a linear velocity perturbation to speed up the instability.

RESULTS

Gold-Hoyle

Due to the fact that the Gold-Hoyle equilibrium does not have a mode rational surface current sheets do not form during the non-linear evolution of the instability [6,10]. However we find that the equilibrium does becomes unstable for a length between 2.4π and 2.6π agreeing with previous results which give a value of 2.49π [8]. For a length of 3π, the loop has a growth rate of 0.11, which agrees with the results of Mikic *et al* [6]. We conclude that our results show good correlation with previous results.

Equilibrium 1

We run the code for 40 Alfvén times on both an 81^3 and 161^3 grid. This allows us to observe how the maximum value of the current scales with higher resolution. We find that the instability has a growth rate of 0.09 which agrees with the linear results. A current concentration is observed at $t = 5$ at $r = 0.75$ as shown in Figure 2. This is as expected since the current sheet should form near the mode rational surface but will be shifted away from it slightly due to non-linear effects. On the 81^3 grid the maximum of the current is 8.0 whereas on the 161^3 grid the value is 18.0. This demonstrates that the current does scale with increased resolution suggestive of the formation of a current sheet. The sheet can be observed as a helical ribbon of current wrapped around the kinked central column of initial current as shown in Figure 3.

Equilibrium 2

We run the code an both an 81^3 grid and 161^3 grid varying a, b and ϕ_0 to adjust the shear. We consider loops with shear between 1.57 and 9.42. As for eq1 we observe the formation of a current concentration near the mode-rational surface of the equilibrium. For example, if we take $a = 0.5$, $\phi_0 = 1.0$, and $b = 2.0$ we find that the mode-rational surface is at $r = 1.1$ and the current concentration forms at $r = 1.3$. We also observe that the maximum value of the current scales with higher resolution. For $a = 0.5$, $\phi_0 = 1.0$, and $b = 2.0$ we find that at $t = 60$ Alfvén times $j_{max} = 8.0$ on an 81^3 grid and 14.0 on a 161^3 grid. For this configuration the shear is 1.67 and the current sheet appears as a very thin, helical, thread-like structure wrapped around the central column of kinked initial current as shown in Figure 4. We can then take $a = 1.5$, $\phi_0 = 1.5$ and $b = 1.1$ to give a shear of 3.01. The current sheet is observed as an axially thicker ribbon as shown in Figure 5. We increase the shear to 9.42 by choosing $a = 0.5$, $\phi_0 = 1.5$, and $b = 0.7$ and find that the current sheet becomes thicker again as shown in Figure 6.

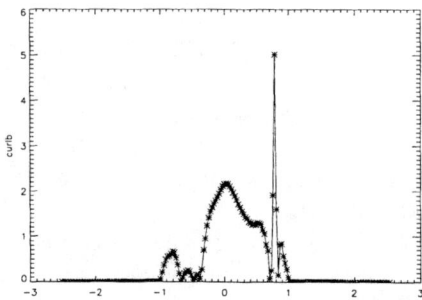

FIGURE 2. current sheet formation - plot of current showing background current and a spike of current at $r = 0.75$.

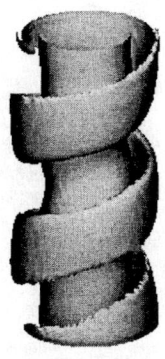

FIGURE 3. isosurface of $|j|$

FIGURE 4. The isosurface of the current with the current sheet seen as a thin ribbon wrapped around the central current

DISCUSSION AND CONCLUSIONS

We initially consider the Gold-Hoyle equilibrium and note that our results show good agreement with previous research [8,6]. Then we investigate the effect of shear on current sheet formation. For Equilibrium 1 we have a shear of 6.32 and find that a current sheet forms near the mode-rational surface. The current sheet can be observed as a helical sheet wrapped around the kinked initial current. Next we consider Equilibrium 2 for which we can vary the shear. We observe current sheet formation for configurations with shear as small as 1.57 but these sheets are extremely thin and thread-like compared to that observed for Equilibrium 1. Increasing the value of the shear up to 9.42, we find that the current sheet becomes axially thicker. It is, therefore, possible that for loops of small shear current sheet formation cannot be observed using certain numerical codes because of the way

FIGURE 5. The isosurface of the current for the *shear* $= 3.01$ configuration showing the current sheet as a helical ribbon wrapped around the central equilibrium current.

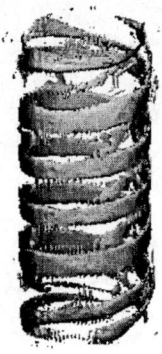

FIGURE 6. The isosurface of the current for the large shear configuration showing the current sheet as a helical ribbon wrapped around the central equilibrium current.

they resolve the current. We conclude that current sheet formation occurs for equilibrium configurations with small shear and that the effect of increasing the shear is to increase the axial thickness of the current sheet. This suggests that reconnection will be less efficient for loops with small shear as they have smaller current sheets.

Acknowledgments

The authors would like to thank Aaron Longbottom for many helpful discussions. The simulations were carried out using the JREI SHEFC funded Compaq MHD cluster in St Andrews.

REFERENCES

1. Hood, A.W., Priest, E.R., *Sol. Phys* **64**, 303 (1979)
2. Velli, M., Lionello, R., Einaudi, G., *Sol. Phys* **172**, 257 (1997)
3. Lionello, R., Velli, M., Einaudi, G., Mikic, Z., *Ap. J* **494**, 840 (1998a)
4. Lionello,R., Schnack, D.D., Einaudi, G., Velli, M., *Phys. of plasmas* **5**, 3722 (1998b)
5. Arber, T.D., Longbottom, A.W., Van Der Linden, R.A.M., *Ap. J* **1516**, 990 (1999)
6. Mikic, Z., Schnack, D.D., Van Hoven, G., *Ap. J* **361**, 690 (1990)
7. Raadu, M.A., *Sol. Phys* **22**, 425 (1972)
8. Hood, A.W., Priest, E.R., *Geophys Astrophys Fluid Dynamics* **17**, 297 (1981)
9. Baty, H., Heyvaerts, J., *Astron & Astrophys* **308**, 935 (1996)
10. Baty, H., *Astron & Astrophys* **318**, 621 (1997)
11. Baty, H., *Astron & Astrophys* **353**, 1074 (2000)
12. Baty, H., Einaudi, G., Lionello, R., Velli, M., *Astron & Astrophys* **333**, 305 (1998)
13. Van Leer, B., *J Comput. Phys.* **135**, 229 (1997)
14. Evans, C.R., Hawley, J.F., *Ap. J* **332**, 659 (1988)
15. Arber, T.D.,*et al, Journal comp. phys.* In Preparation
16. Gold, T, Hoyle, F, *Monthly Notices Roy. Astron. Soc.* **120**, 89 (1960)

Angular Momentum Transport by Internal Gravity Waves

Eun-jin Kim and Keith B. MacGregor

HAO, National Center for Atmospheric Research[1], PO Box 3000, Boulder, CO80307, USA

Abstract. We discuss the effects of internal gravity waves on the mean radial differential rotation profile in the solar tachocline. Vertical transport of horizontal momentum arises from the radiative damping of inwardly traveling waves that are generated by low-frequency, convective fluid motions. For a viscosity typical of the radiative layers below the convection zone, the equilibrium profile of radial differential rotation is demonstrated to be unstable, possibly leading to turbulent mixing in the tachocline. The effect of a uniform, toroidal magnetic field is discussed.

INTRODUCTION

In view of the near-uniform rotation of the solar radiative interior, as revealed by helioseismology, a possible mechanism for angular momentum transport in the solar interior has received much attention. The basic problem is that the angular momentum transport by molecular viscosity (on time scale of order 10^{12} yr) is not fast enough to account for the loss of angular momentum by solar wind from the surface layers of the Sun, which gives rise to a spin-down on time scale of order 10^9 yr at the present time. In order to maintain the apparent uniform rotation of the solar radiative core, there must be a mechanism (besides molecular viscosity) by which angular momentum is transferred radially on a time scale that is, at least, shorter than evolutionary time scales ($\sim 4.5 \times 10^9$ yr). One possibility is that the transport takes place via magnetic fields, as discussed by [1]. Another proposed mechanism, which we shall consider in this paper, is that angular momentum is redistributed inside the Sun through the action of gravity waves [2,3].

Gravity waves are thought to be generated at the bottom of the convection zone by turbulent Reynolds stresses or plume penetration [4,5]. The momentum associated with these waves can be deposited in the mean background flow if the waves encounter critical layers (e.g., [6]), or if dissipative processes such as viscous, thermal (either conductive or radiative), or Ohmic diffusion are operative in the

[1] The National Center for Atmospheric Research is sponsored by the National Science Foundation.

fluid. As a result, internal gravity waves can exert a force on the material through which they propagate, and thereby modify the dynamics of an ambient shear flow.

In this paper, we focus on such momentum transport and deposition due to radiatively damped gravity waves in the radiative interior of the Sun. If these waves are generated by turbulent Reynolds stresses, they should have frequencies $\omega \sim 10^{-6}$ s^{-1} (period ~ 30 days), comparable to the eddy turnover frequency in the lower portion of the solar convection zone [4]. As will be discussed in subsequent sections, waves of this kind are subject to vigorous radiative damping, leading to significant diminution of the wave amplitude over a distance that is much smaller than the pressure scale height in the outer layers of the Sun's core. Hence, the waves are most likely to affect the dynamical state of material just below the base of the convection zone, not in the innermost regions of the radiative interior.

With these points in mind, the investigation described herein is aimed at understanding the ways in which gravity wave forcing might contribute to the formation and structure of the solar tachocline. Toward this end, we utilize a simplified physical and computational model that explicitly treats the interaction between a thin shear layer and two waves. The waves are identical in all respects except their propagation directions: one (the 'prograde' wave) has the horizontal component of its propagation vector parallel to the direction of the background shear flow, while the other (the 'retrograde' wave) travels in the opposite sense. We calculate the equilibrium profile of radial differential rotation that obtains within the shear layer when the net gravity wave force is balanced by the force arising from a prescribed viscosity. Furthermore, because the tachocline region probably contains a substantial, toroidal magnetic field, we also study how gravity wave forcing of the rotational shear is modified when a uniform, flow-aligned field is present in the layer.

FORMULATION OF THE PROBLEM

We treat the solar tachocline as a two dimensional Cartesian domain (x,z), with x and z in the azimuthal and (local) radial directions, respectively. To mimic the radial differential rotation and toroidal magnetic field, a mean shear flow $\mathbf{u}_0$ and uniform magnetic field $\mathbf{b}_0$ are assumed to be aligned in x direction, i.e., $\mathbf{u}_0 = u_0(z)\hat{x}$ and $\mathbf{b}_0 = b_0\hat{x}$. Since the thickness of the solar tachocline is less than the pressure scale height (1/10 of solar radius), we adopt the Boussinesq approximation by incorporating the background density variation only in the buoyancy term in equations for fluctuations [7]. Under these assumptions, the linearized equations for fluctuations (magneto-gravity waves) are

$$\bar{\rho}\left[(\partial_t + u_0\partial_x)\mathbf{u}_1 + u_{1z}\partial_z u_0\hat{x}\right] = -\nabla\pi_1 - g\rho_1\hat{z} + \frac{1}{4\pi}b_0\partial_x\mathbf{b}_1 , \tag{1}$$

$$(\partial_t + u_0\partial_x)\mathbf{b}_1 = b_{1z}\partial_z u_0\hat{x} + b_0\partial_x\mathbf{u}_1 , \tag{2}$$

$$(\partial_t + u_0\partial_x)\rho_1 = \frac{\bar{\rho}N^2}{g}u_{1z} + \mu\nabla^2\rho_1 , \tag{3}$$

$$\nabla \cdot \mathbf{u}_1 = \nabla \cdot \mathbf{b}_1 = 0 \,, \tag{4}$$

where suffixes "0" and "1" denote the background and fluctuating quantities, respectively. In equations (1)-(4), $\pi_1 \equiv p_1 + b_0 b_{1x}/4\pi$ is the total pressure, $\bar{\rho}$ is the mean constant background density, and $N^2 \equiv -g(\partial_z \rho_0 + \bar{\rho}g/c_s^2)/\bar{\rho}$ is the Brunt-Väisälä frequency, where c_s is the sound speed and $\rho_0 = \rho_0(z)$ is the background density. The last term on the right-hand side of equation (3) describes the radiative interaction between the wave and the background medium by thermal diffusivity μ arising from radiative transfer. The viscosity ν ($\sim 10^2$ cm^2 s^{-1}) and Ohmic diffusivity ($\sim 10^4$ cm^2 s^{-1}) are ignored in comparison to the thermal diffusivity (i.e., the radiative damping coefficient) μ ($\sim 10^7$ cm^2 s^{-1}) in the above equations.

Under the WKB approximation, that is, by assuming that the waves have vertical wavelengths that are small in comparison to any of the relevant scale heights in the background medium, we assume $q_1 = \tilde{q}_1 \exp\{i(k_x x + k_z z - \omega t)\}$, where q_1 represents any of the fluctuating quantities ρ_1, p_1, $\mathbf{u}_1$, or $\mathbf{b}_1$. Then, by ignoring the effects of the gradient of u_0 and damping term, the above equations yield the following local dispersion relation:

$$\omega_* \equiv \omega - u_0 k_x = \pm \left[\left(\frac{k_x^2}{k^2} \right) N^2 + c_a^2 k_x^2 \right]^{\frac{1}{2}} , \tag{5}$$

where $k^2 = k_x^2 + k_z^2$ and $c_a = b_0/\sqrt{4\pi\bar{\rho}}$ is the Alfvén speed (see also [8]). Thus, the vertical group velocity $v_{gz} \equiv \partial\omega/\partial k_z$ is

$$v_{gz} = -\frac{k_x^2 k_z}{\omega_* k^4} N^2 \,. \tag{6}$$

Stationary Mean Shear Flow

The evolution equation for a mean shear flow is obtained by averaging the full, nonlinear momentum equation, including terms to second order in the amplitudes of fluctuating quantities; in the stationary case, this procedure yields

$$0 = -k_x \partial_z \mathcal{F} + \partial_z(\rho_0 \nu \partial_z u_0) \,. \tag{7}$$

Here $\mathcal{F} \equiv \langle E \rangle v_{gz}/\omega_*$ is the flux of wave action, representing the nonlinear effect of gravity waves on the shear flow, and $\langle E \rangle \equiv \langle (\bar{\rho}u_1^2/2) + (b_1^2/8\pi) + (g^2\rho_1^2/2\bar{\rho}N^2) \rangle$ is the average total energy; the viscosity ν is retained to compute a stationary solution for u_0. The above equation needs be solved in concert with the equation for the wave action (see [9] for more details)

$$\partial_z \mathcal{F} = \frac{-\mathcal{F}}{L} \,. \tag{8}$$

Here L is a damping length defined by

$$L \equiv \frac{v_{gz}\omega_*^2}{\mu k_x^2 N^2} = \text{sign}(v_{gz})\frac{1}{\mu}\left|\frac{\omega_*\omega_c^3}{k_x^3 N^3}\right|\sqrt{1 - \frac{\omega_c^2}{N^2}}, \qquad (9)$$

where $\omega_c \equiv \sqrt{\omega_*^2 - k_x^2 c_a^2}$. As can be seen from the above equations, the sign of the force $-k_x \partial_z \mathcal{F}$ due to waves depends on the sign of vertical group velocity and horizontal wavenumber. For a downward propagating wave with $v_{gz} < 0$ and $\omega_* > 0$ ($\mathcal{F} < 0$, $L < 0$), the force due to a wave with $k_x > 0$ is positive whereas that due to a wave with $k_x < 0$ is negative. The magnitude of the force exerted on the flow depends on the damping length and wave action density flux, with the force being large for small damping length and large wave flux. Since the damping length is proportional to $\omega_*\omega_c^3$, a prograde wave has a shorter damping length, exerting a stronger force, and also damps more quickly than a retrograde wave does.

NUMERICAL RESULTS

We consider two waves that are propagating downward $v_{gz} < 0$ with the same (local) vertical wavenumber $k_z = m > 0$, but with the opposite horizontal wavenumber $k_x = \pm l$. In the following, waves with $k_x = \pm l$ will be called $\pm l$ waves, and l will be taken to be $1/H_0$. Here $H_0 \equiv H(0) = 5.7334 \times 10^9$ cm is the pressure scale height at the bottom of convection zone, calculated by using the parameter values in [10]. We take the bottom of the convection zone to be $z = 0$ and consider one pressure scale height $H(0) \equiv H_0$ downward (i.e., $z \in [-H_0, 0]$). The effect of stratification is incorporated in the equation for the mean shear by using an exponential fit for each mean variable that was obtained from the values of mean pressure, density, temperature, opacity and Brunt-Väisälä frequency given as a function of depth in [10].

A stationary profile of a mean shear flow is sought by numerically solving the coupled equations (7) and (8) with following boundary conditions. First, for the mean shear flow, we impose constant velocity $u_0 = 10^4$ cm s^{-1} at $z = 0$ and $u_0 = 0$ at $z = -H_0$ where $u_0(z = 0) = 10^4$ cm s^{-1} is of order of the average difference in the radial differential rotation across the tachocline. Next, for the wave action flux, we use $u_{1z}(z = 0) = \omega/\overline{m}(0)$ [4], where $\overline{m}(0) = l\sqrt{N(0)^2/\omega^2 - 1}$ with $N(0) = 2.6280 \times 10^{-3}$ s^{-1}.

In the Absence of Magnetic Field

In order to elucidate the effect of gravity waves on a shear flow, we first consider the case with a rather high frequency $\omega/N(0) = 5 \times 10^{-3}$ (i.e., period ~ 5.5 days). Figure 1 shows the profile of the mean shear flow (panels [a]), and the force due to $\pm l$ waves (panel [b]) in the case $\nu = 10^7$ cm^2 s^{-1}. Here solid and dotted lines represent $\pm$force exerted by $+l$ and $-l$ waves, respectively. Note that Figure 1a indicates $\omega_* > 0$ and subsequently $\mathcal{F} < 0$ for both $\pm l$ waves ($v_{gz} < 0$). Thus, the force $k_x \mathcal{F}/L \propto k_x/\omega_*$ due to $+l$ and $-l$ waves is always positive and negative, respectively (see equations (7)–(9)). For these waves, the damping length is of order $10H_0$, with the amplitude of u_0 hardly changing over H_0. Nevertheless, due to the net force exerted by the $\pm l$ waves, u_0 develops a fine structure (see panel [a]) as follows.

At the start, the force due to $+l$ wave — being a prograde wave — is larger than that due to $-l$ wave. However, as the $+l$ wave damps faster than the $-l$ wave, the force due to $-l$ waves soon dominates over that due to $+l$ wave, changing the sign of u_0 at $z/H_0 \sim -5 \times 10^{-4}$ (see panel [a]). Since the $-l$ wave is now a prograde wave aligned in the same direction as u_0, it exerts more force and damps faster than the $+l$ wave until it again leads to the reversal of the sign of u_0 at $z/H_0 \sim -3 \times 10^{-3}$. The turning point of the shear flow occurs at $z/H_0 \sim -2 \times 10^{-3}$ where the net force is locally minimum — the magnitudes of the net force and the curvature of u_0 are locally maximum at that point. This whole process will repeat over and over again until both $\pm l$ waves are completely damped out at $z/H_0 \sim -10$. It is interesting to note that the upper bound on $|u_0|$ is placed by $w_c = 0$ since the force $\propto \mathcal{F}(z)/|L(z)| \propto \exp\left[\int_{z_0}^{z} dz' 1/|L(z')|\right]/|L(z)| \to 0$ $(z < z_0)$ in the limit as $w_c \to 0$. That is, a prograde wave loses too much of its wave flux before a critical critical layer is ever reached.

The minimum Richardson number for the equilibrium profile in this case is 0.15, smaller than the critical value $Ri_c = 1/4$ for stability [11,12]. Therefore, the mean flow is likely unstable. In fact, the critical Richardson number Ri_c for the stability is expected to be larger than $1/4$ in the presence of radiative damping as the

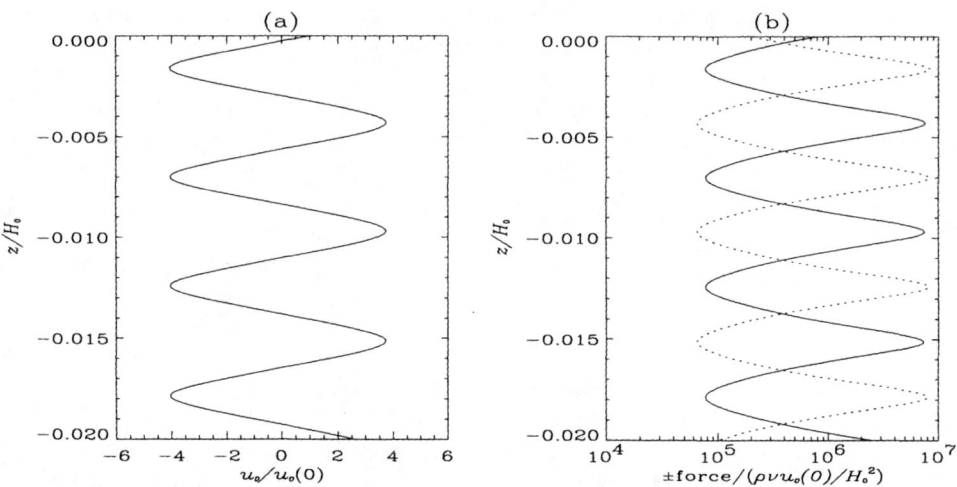

FIGURE 1. Mean shear profiles (panel [a]) and force exerted on the mean flow (panel [b]) for $\omega/N(0) = 5 \times 10^{-3}$ and $b_0 = 0$. $u_0(0) = 10^4$ cm s^{-1}. Solid (dotted) line in panel [b] represents $\pm$force due to $\pm l$) wave.

latter weakens the stabilizing mechanism of buoyancy. Now, as the viscosity is decreased below $\nu = 10^7$ cm^2 s^{-1}, the equilibrium profile of the shear flow will develop even sharper structure in order to balance the force from the gravity waves by the viscous force (see equation (7)). In other words, the spacing between peaks become narrower with their amplitude approaching the maximum set by the critical layer. Therefore, for a more realistic value of the viscosity ($\sim 10^2$ cm^2 s^{-1}) at the bottom of convection zone, the equilibrium shear flow will most likely be unstable.

Figures 2 shows the mean shear profiles for lower frequencies $\omega/N(0) = 3 \times 10^{-3}$ (panel [a]) and 10^{-3} (panel [b]), with all other parameters fixed. As the frequency of the wave is decreased, the wave damps quickly with a short damping length $\sim 3H_0$ and $0.02H_0$ for $\omega/N(0) = 3 \times 10^{-3}$ and 10^{-3}, respectively. In particular, in the case $\omega/N(0) = 10^{-3}$ (period ~ 30 days), corresponding to the most plausible frequency generated by turbulent Reynolds stress, the wave damps so quickly over $z/H_0 \sim 0.02$ as to induce a strong localized shear only near $z = 0$ (see panel [b]). Figures 1 and 2 also reveal that the overall amplitude of the shear flow becomes smaller for lower frequency. It is because the maximum amplitude of u_0, set by $\omega_c = 0$, becomes smaller for lower frequency. Finally, the equilibrium profiles in Figure 2, with the minimum Richardson number 3.2×10^{-4} (panel [a]) and 8.4×10^{-3} (panel [b]), again seem to be unstable.

In the Presence of Magnetic Field

The gravity waves can propagate downward (m real) only when the Alfvén frequency $\omega_B = lc_a = b_0 l/\sqrt{4\pi\bar{\rho}}$ is less than ω_* (see equation (5)). For the adopted

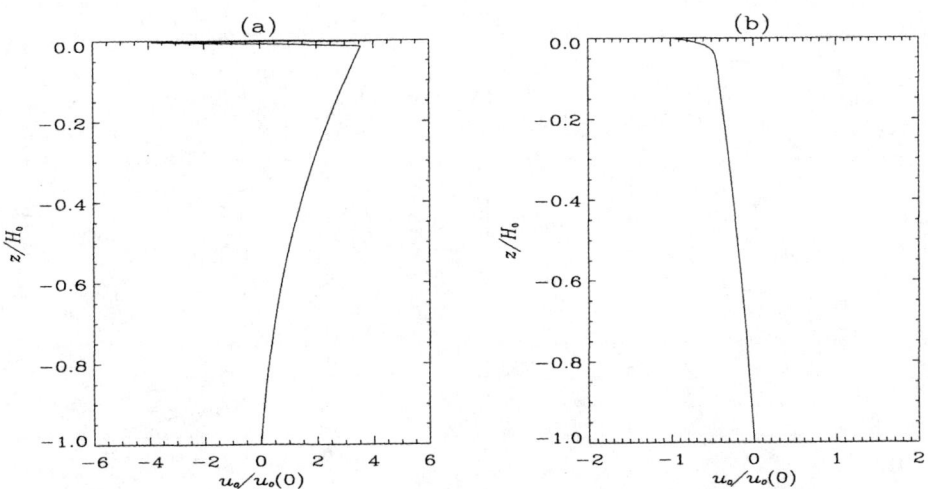

FIGURE 2. Mean shear profiles for $\omega/N(0) = 3 \times 10^{-3}$ (panel [a]) and 10^{-3} (panel [b]).

parameter values, the minimum frequency of the gravity waves that can propagate is $\omega/N(0) = 4.3 \times 10^{-3}$ (period of 6.4 days) for $b_0 = 10^5$ G, for instance. Alternatively, gravity waves with frequency $9.2 \times 10^{-4} N(0)$ [$2.8 \times 10^{-2} N(0)$] (corresponding to period of 30 days [1 day]) will propagate downward for $b_0 < 2.1 \times 10^4$ G [$b_0 < 6.5 \times 10^5$ G]. The upper bound on the magnetic field strength that permits the propagation of gravity waves becomes small for large horizontal wave number l. In other words, for a fixed magnetic field, it is harder for waves with larger l to propagate. The foregoing analysis implies a significant role that a strong magnetic field may play in angular momentum transport by gravity waves.

In the case where a magnetic field is weak enough to permit wave propagation, the equilibrium profile of the shear flow tends to be somewhat smoother compared to the case without a magnetic field, as can be seen from Figure 3. The latter depicts the profile of the mean shear flow in the case $\omega/N(0) = 5 \times 10^{-3}$, $\nu = 10^7$ cm^2 s^{-1}, and $b_0/\sqrt{4\pi} = 2 \times 10^4$ G. In comparison with Figure 1, the profile is smoother (minimum Richardson number 1.8 in contrast to 0.15) with a smaller amplitude u_0. This is because in the presence of a magnetic field, the critical layer where $\omega_c = 0$ (which sets the maximum amplitude of u_0) occurs for a smaller value of u_0. Moreover, despite the destabilizing effect of radiative damping, the equilibrium profile in Figure 3a may be stable due to the opposite, stabilizing effect of magnetic field.

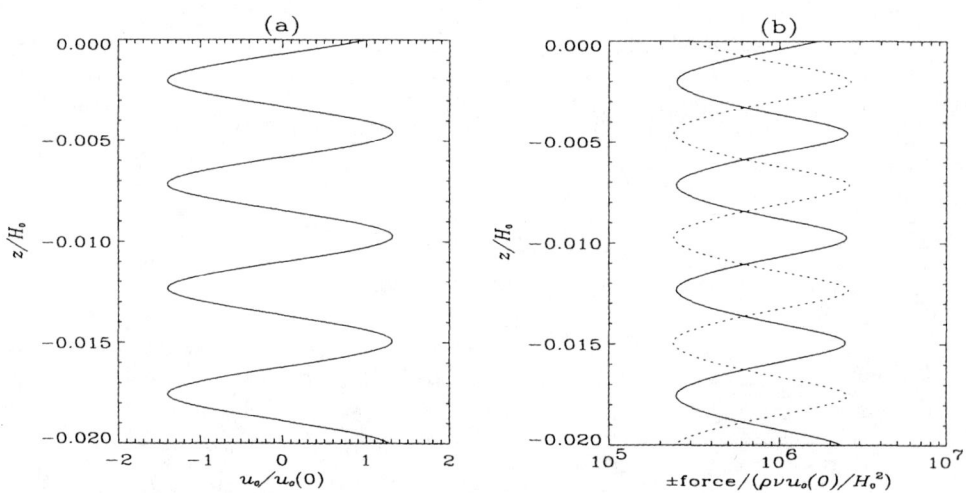

FIGURE 3. Same as Fig. 1 but for $b_0/\sqrt{4\pi} = 2 \times 10^4$ G.

CONCLUSIONS

We have demonstrated that in both hydrodynamic and hydromagnetic cases, radiatively damped gravity waves tend to create and accentuate flow structures with strong radial gradients (see also [13–15]), with the region over which a strong shear resides depending on the frequency of the waves. Firstly, waves with period larger than $5 \sim 6$ days ($\omega/N(0) < 5 \times 10^{-3}$) and $l = 1/H_0$ are likely to be all damped out before entering the solar interior, leading to the localized strong shear near the top portion of the solar tachocline (see Figs. 2–3). Secondly, high frequency waves with period shorter than $5 \sim 6$ days and $l \lesssim 1/H_0$ seem to be able to propagate through the tachocline, thereby entering the solar interior, even in the presence of a toroidal magnetic field of order 10^5 G in the solar tachocline. If there is no strong toroidal magnetic field in the solar interior, these gravity waves are mostly likely to enhance any radial differential rotation that may be present in the interior as they propagate through that region. Nevertheless, if we take into account a plausible power spectrum $P(\omega)$ of gravity waves generated by Reynolds stress, which rapidly decreases for high ω as $P(\omega) \sim \omega^{-4.5}$ (see, for instance, [2]), there might be only little power of gravity waves that can actually propagate into the solar interior. In either cases, gravity waves cannot bring the solar interior to the uniform rotation.

Acknowledgments

E.K. acknowledges the International Travel Grant of American Astronomical Society to attend the workshop on Waves in Dusty, Solar and Space Plasmas.

REFERENCES

1. MacGregor, K., and Charbonneau, P., *ApJ*, **519**, 911–917 (1999).
2. Kumar, P., and Quataert, E. J., 1997, *ApJ*, **475**, L143–146 (1997).
3. Zahn, J.-P, Talon, S., and Matias, J, *A&A*, **322**, 320–328 (1997).
4. Press, W. H. 1981, *ApJ*, **245**, 286–303 (1981).
5. Fritts, D. C, Vadas, S. L., and Andreassen, Ø., A., *A&A*, **333**, 343–361 (1998).
6. Booker, J. R., and Bretherton, F. P., *J. Fluid. Mech.*, **27**, 513–539 (1967).
7. Spiegel, E. A., and Veronis, G., *ApJ*, **131**, 442–447 (1960).
8. Barnes, G., MacGregor, K. B., and Charbonneau, P., *ApJ*, **498**, L169–172 (1998).
9. Kim, E., and MacGregor, K. B., in preparation (2000).
10. Bahcall, J. N., and Pinsonneault, M. H., *Rev. Mod. Phys.*, **67**, 781–808 (1995).
11. Miles, J. W., *J. Fluid Mech.*, **10**, 496–508 (1961).
12. Howard, L. N. 1961, *J. Fluid. Mech.*, **10**, 509–512 (1961).
13. Gough, D. O., and McIntyre, M. E., *Nature*, **394**, 755–757 (1998).
14. Ringot, O., *A&A*, **355**, L89–92 (1998).
15. Kumar, P., Talon, S., and Zahn, J.-P., *ApJ*, **520**, 859–870 (1999).

Flare-Generated Coronal Loop Oscillations: A Tool for MHD Coronal Seismology

Valery M. Nakariakov

Physics Department, University of Warwick, Coventry CV4 7AL, UK

Abstract. Quasi-periodical oscillations of positions of coronal loops were observed in the extreme- ultraviolet band (171A, FeIX) with the imaging telescope onboard the TRACE spacecraft. Oscillating transversal displacements (swinging) of a long ($\sim$130 Mm) thin (diameter $\sim$2 Mm) bright loop, were detected. The oscillations were excited by a flare (14th July 1998 at about 12:55 UT) at distance of about 60 Mm from the loop. The frequency of the oscillations was about 4 mHz (the period about 265 s) and was determined by the wavelet analysis. Neighbouring perpendicular slits show synphase temporal behaviour, suggesting that the oscillations are produced by a *kink global fast magnetoacoustic mode* of the loop. With this interpretation, we determine the kink speed in the loop as 1040 km/s, which gives the Alfvén speed of about 770$\pm$40 km/s.

The detection and analysis of post-flare oscillations of coronal loops provides us with an efficient tool for indirect determination of coronal parameters, e.g. the magnetic field found to be of about 20 G.

INTRODUCTION

Investigation of the structure and dynamics of the corona of the Sun, the upper, hottest and magnetically dominated part of the solar atmosphere, is one of the most important and interesting branches of modern astrophysics. Despite significant progress in solar physics over several decades, a number of fundamental questions, such as what are the physical mechanisms responsible for coronal heating, the solar wind acceleration and solar flares, remain to be answered.

The solution to these problems will bring us important knowledge crucially required for further development not only in solar physics but also in stellar and magnetospheric physics, and laboratory plasma physics; and for better understanding of solar–terrestrial connections. All these questions, however, require detailed knowledge of physical conditions and parameters in the corona, which cannot yet be measured accurately. In particular, the exact value of the coronal magnetic field remains unknown, because radio methods and the extrapolation of chromospheric magnetic fields do not allow us to reach the required precision. Also, the coronal

CP537, *Waves in Dusty, Solar, and Space Plasmas*, edited by F. Verheest, et al.

transport coefficients, such as volume and shear viscosity, resistivity and thermal conduction remain unknown even by the order of magnitude.

Imaging EUV telescopes onboard SOHO and TRACE missions provide us with a new window for observation of dynamical processes in the corona and determination of physical parameters of the corona. Recent discoveries of compressive waves in polar plumes [3] and long loops [2], [4], interpreted as slow magnetoacoustic waves [7]; coronal Moreton waves [10]; and flare-generated kink oscillations of coronal loops [1], [6] provide us with an additional tool for determination of the unknown parameters of the corona - *MHD seismology of the corona* (see, also, [9]). Measuring the properties of MHD waves and oscillations (periods, wavelengths, amplitudes, temporal and spatial signatures, characteristic scenaria of the wave evolution), combined with theoretical modeling of the wave phenomena (dispersion relations, evolutionary equations, etc.), we can determine values of the mean parameters of the corona, such as the magnetic field strength and transport coefficients.

OBSERVATIONAL FINDINGS

On 14th July 1998, the imaging telescope onboard the Transition Region and Coronal Explorer (TRACE) registered, in both 171A and 195A lines, an interesting event in the solar corona: after a flare happening in active region AR8270, coronal loops surrounding the flare epicentre experienced oscillations seen as periodic transversal displacements of the loops [1], [6]. Probably, a similar event has been observed in the green coronal line [5]. Perhaps, the loop oscillations are excited by blast waves propagating out of the flare epicentre, (coronal Moreton waves

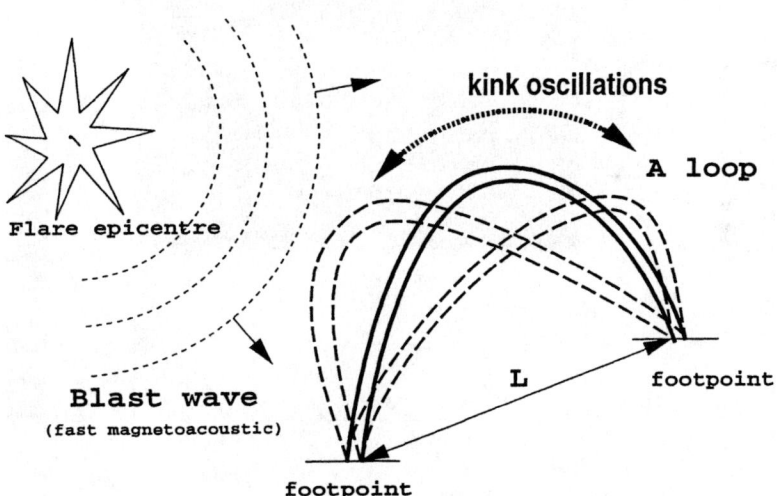

FIGURE 1. A sketch of flare-generated kink oscillations of coronal loops.

[10]). Oscillations of different loops were not synphase. The highest amplitude was seen near loop apices. The sketch of the phenomenon is shown in Figure 1.

The periodogram analysis of the loop displacement, shows that the oscillations are almost harmonic with the periods of about 256 s (the frequency about 4 mHz). According to the wavelet analysis of the loop displacement as a function of time (see Figure 2), the period remains almost constant during the oscillations. Displacement amplitudes are several Mm for the distance between the loop footpoints estimated about 83 Mm. The displacement amplitude is several times larger than the loop cross-section radius, which is about 1 Mm. The oscillations show evidence of strong damping (the best-fit exponential function gives the decay time of about 14.5 min, see [6] for the details). The quasi-periodic oscillations have been found for several loops at the distance of several Mm to 60-70 Mm from the flare epicentre [1].

Neighbouring perpendicular slits across the loop show synphase temporal behaviour (there is no evidence that the perturbation is propagating along the loop). This fact, together with the quasi-harmonic temporal behaviour, suggests that the oscillations are produced by a *kink global standing mode* of the loop.

MHD MODES OF MAGNETIC TUBES

The theoretical analysis of MHD modes of a magnetic flux tube allows us to connect the observationally detected parameters of the loop oscillations with the

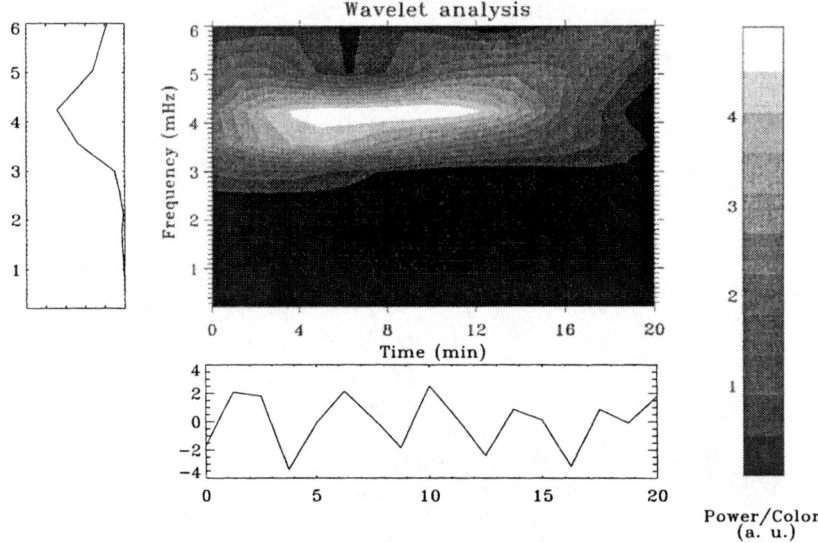

FIGURE 2. The wavelet analysis of the temporal evolution of the loop displacement near the loop apex. The spectral intensity of the oscillations is measured in normalized units.

physical parameters of the coronal plasma. Following [8], we consider a straight untwisted magnetic cylinder of the radius a embedded in a magnetic environment. The magnetic field, density and the kinetic pressure vary in the radial direction:

$$B_0(r) = \begin{cases} B_0, & r < a, \\ B_e, & r > a, \end{cases} \quad \rho_0(r) = \begin{cases} \rho_0, & r < a, \\ \rho_e, & r > a, \end{cases} \quad p_0(r) = \begin{cases} p_0, & r < a, \\ p_e, & r > a \end{cases}, \quad (1)$$

which gives the sound speeds C_{s0} and C_{se}, Alfvén speeds C_{A0} and C_{Ae}, and tube speeds C_{T0} and C_{Te} inside and outside the tube, respectively. The total pressure balance across the tube is to be held.

For the expected plasma parameters in the coronal loop, the magnetic field $B \approx 10 - 20$ G, concentration $n \approx 5 \times 10^{14}$ m^{-3} and temperature $T \approx 1.3$ MK, the ion gyroradius is about 1 m, and the mean free path length is about 0.2 Mm. Consequently, oscillations with typical scales over several Mm can be satisfactory described by MHD.

In the observed range of temporal and spatial parameters, there are two types of wave modes propagating along the tube: torsional modes, which are practically incompressive and do not perturb the tube boundary (and, consequently, they can not be detected with an imaging telescope), and magnetoacoustic modes (see, e.g., [8] for details). The magnetoacoustic modes have the dispersion relation

$$\rho_e(\omega^2 - k^2 C_{Ae}^2) m_0 \frac{I_n'(m_0 a)}{I_n(m_0 a)} + \rho_0(k^2 C_{A0}^2 - \omega^2) m_e \frac{K_n'(m_e a)}{K_n(m_e a)} = 0, \quad (2)$$

where

$$m_\alpha^2 = \frac{(k^2 C_{s\alpha}^2 - \omega^2)(k^2 C_{A\alpha}^2 - \omega^2)}{(C_{s\alpha}^2 + C_{A\alpha}^2)(k^2 C_{T\alpha}^2 - \omega^2)}, \quad (3)$$

where ω and k are the frequency and the longitudinal wave number, respectively; the indices $\alpha = 0, e$ are for internal and external media, respectively; $I_n(x)$ and

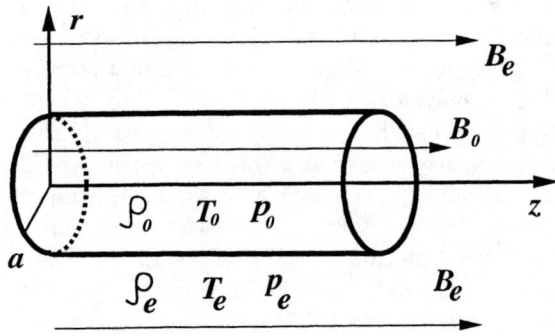

FIGURE 3. A sketch of a magnetic tube modeling a coronal loop.

267

$K_n(x)$ are modified Bessel functions of order n; the prime denotes the derivative of a function with respect to its argument. For the trapped modes, which are evanescent outside the tube, the condition $m_e > 0$ has to be fulfilled. The number n determines the mode structure, for the kink modes considered, $n = 1$.

According to the observational findings, the wavelength of the standing global kink mode is the doubled length of the loop,

$$\lambda = 2\pi L, \tag{4}$$

where L is the measured distance between the loop footpoints, $L \approx 83$ Mm, and a semi-circular shape of the loop is assumed. The comparison of this value with the observed radius of the tube $a \approx 1$ Mm shows that the dispersion relation can be analyzed in the long wave length limit, $ka \ll 1$. There are two kink modes in the limit, slow and fast. The slow kink mode has the phase speed about

$$\frac{\omega}{k} \approx C_{T0} \equiv \frac{C_{s0}C_{A0}}{(C_{s0} + C_{A0})^{1/2}}. \tag{5}$$

The phase speed of the fast kink mode is

$$\frac{\omega}{k} \approx C_k \equiv \left(\frac{2}{1 + \rho_e/\rho_0}\right)^{1/2} C_{A0}, \tag{6}$$

where C_k is a so-called *kink* speed.

DETERMINATION OF THE MAGNETIC FIELD

Taking the observed periods $P \approx 256$ s and the distance between the loop footpoints, we estimate the phase speed required as

$$\frac{\omega}{k} = \frac{\pi L}{P} \approx 1040 \text{km/s}. \tag{7}$$

The speed of slow waves, C_{T0} is *below* than both the Alfvén C_{A0} and the sound C_{s0} speeds. Estimating the sound speed from the loop temperature $T \approx 1.3$ MK as $C_{s0} = 152\,T^{1/2}(\text{MK}) \approx 173$ km/s, we conclude that $C_{T0} < 173$ km/s and, consequently, the slow mode has to be excluded from the consideration. On the other hand, the fast wave can have the phase speed required. The expression for the kink speed contains two unknown parameters, the Alfvén speed C_{A0} and the density ratio ρ_e/ρ_0. Considering the density ratio as a parameter, we can determine the Alfvén speed in the loop. Figure 4 shows the dependence of the Alfvén speed in the loop analyzed as a function of the density ratio. The dashed lines indicate the upper and lower possible values of the kink speed, connected with the errors in the determination of the loop length. Assuming $\rho_e/\rho_0 = 0.1$, we obtain that

$$C_A = 770 \pm 50 \text{ km/s}. \tag{8}$$

The Alfvén speed is defined by the magnetic field strength and the density of the medium. Consequently, applying certain assumption about the density or concentration of the plasma in the loop, we can estimate the value of the magnetic field in the loop:

$$B_0 = (4\pi\rho_0)^{1/2} C_{A0} = \frac{\sqrt{2}\,\pi^{3/2} L}{P}\sqrt{\rho_0(1 + \rho_e/\rho_0)}. \qquad (9)$$

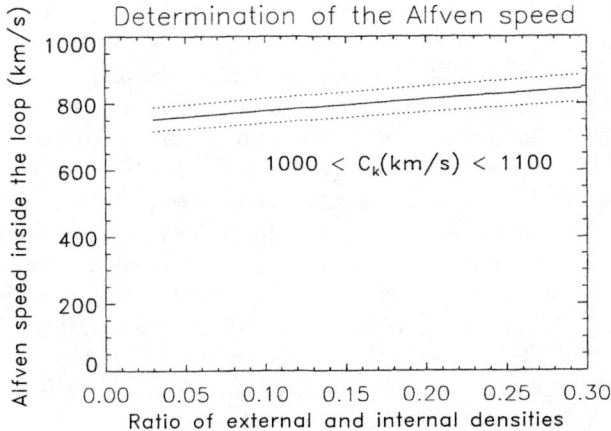

FIGURE 4. The Alfvén speed inside a coronal loop for the observationally determined kink speed $1000 < C_k < 1100$.

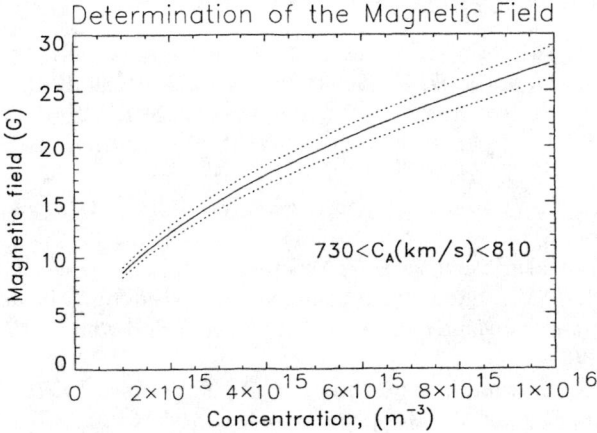

FIGURE 5. The magnetic field inside a coronal loop for the observationally determined Alfvén speed inside the loop as function of ion concentration inside the loop.

269

The density ρ_0 can be determined independently from the intensity of the coronal emission. However, the determination of the magnetic field is weakly sensitive to errors in the determination of the density, because the magnetic field is proportional to the square root from the density. Figure 5 shows that for a quite wide range of plasma concentration, from 1.0×10^{15} to 1.0×10^{16} m^{-3}, the value of the magnetic field is between 10 and 30 G.

CONCLUSIONS

The combination of observations of oscillations and wave motions in the solar corona with the theory of MHD waves trapped in coronal structures provides us with a new tool for determination of the physical parameters of the coronal plasma. In this paper, we have determined the coronal magnetic field to be about 10-30 G in coronal loops by measuring periods of standing MHD waves in a coronal loop and observationally determining the distance between the loop footpoints (formula (9)). Some rough assumptions were used for values of plasma concentration and density ratio inside and outside the loop. An independent measurement of the concentration and density ratio can significantly improve the method. Also, statistical information on the relation between periods of loop oscillations and loop lengths is highly desirable. On the other hand, the method proposed requires to develop theory of MHD oscillations of coronal loops, incorporating effects of stratification and 3D effects (loop curvature, loop twisting).

The method of MHD coronal seismology can become a powerful tool for determination of physical parameters in the solar corona.

REFERENCES

1. Aschwanden, M., et al., *Astrophys. J* **520**, 880-894 (1999).
2. Berghmans, D. & Clette, F., *Solar Phys.* **186**, 207-229 (1999).
3. Deforest, C. E. & Gurman, J. B., *Astrophys. J.* **501**, L217-L220 (1998).
4. De Moortel, I., Ireland, J. and Walsh, R. W., *Astron. Astrophys.* **355**, L23-L26 (2000).
5. Koutchmy, S., Zugzda, Y.D., and Locans, V., *Astron. Astrophys.* **10**, 185-191 (1983).
6. Nakariakov, V.M., et al., *Science* **285**, 862-864 (1999).
7. Ofman, L., Nakariakov, V. M. & Deforest, C. E., *Astrophys. J.* **514**, 441-447 (1999).
8. Roberts, B., "Magnetohydrodynamic Waves in the Sun", in *Advances in Solar System Magnetohydrodynamics* (edited by E. R. Priest and A. W. Hood), CUP, 1991, pp. 105-136.
9. Roberts, B., Edwin, P.M. and Benz, A.O., *Astrophys. J.* **279**, 857-865 (1984).
10. Thompson, B.J., et al. *Astrophys. J* **517**, L151-L154 (1999)

Slow magnetoacoustic waves in coronal loops: EIT vs TRACE

E. Robbrecht°, E. Verwichte*, D. Berghmans*, J.F. Hochedez*,
S. Poedts°¹

°*Centre for Plasma Astrophysics , K.U.Leuven, Celestijnenlaan 200B, B-3001 Heverlee, Belgium*
email: Eva.Robbrecht@wis.kuleuven.ac.be; Stefaan.Poedts@wis.kuleuven.ac.be
**Royal Observatory of Belgium, Ringlaan 3, B-1180 Brussel, Belgium*
email: Erwin.Verwichte@ksb-orb.oma.be; David.Berghmans@ksb-orb.oma.be; hochedez@oma.be

Abstract. On May 13, 1998 the EIT (Extreme-Ultraviolet Imaging Telescope) and TRACE (Transition Region And Coronal Explorer) instruments produced simultaneous high cadence image sequences of the same active region (AR 8218). TRACE achieved a 25 sec cadence in the $Fe\,IX/X$ (171 Å) bandpass while EIT achieved a 15 sec cadence (operating in 'shutterless mode', SOHO JOP 80) in the $Fe\,XII$ (195 Å) bandpass. These high cadence observations in two complementary wavelengths have revealed the existence of weak transient disturbances in an extended coronal loop system. These propagating disturbances (PDs) seem to be a common phenomenon in this part of the active region. The disturbances originate from small scale brightenings at the footpoints of the loops and propagate along the loops. The apparent propagation speeds roughly vary between 65 and 150 km s^{-1} which is close to the expected sound speed of the coronal loops. The measured propagation speeds seem to suggest that the transients are sound (or slow) wave disturbances.

I INTRODUCTION

In this paper we report on the study and detection of propagating disturbances (PDs) along coronal loops. Our study is focused on active region AR8218, using image sequences of JOP80 [1]. This JOP was performed on May 13, 1998 and produced simultaneously high cadence image sequences of the same active region in different wavelenghts. The analysis in this paper is based on observations of two instruments in two complementary wavelengths EIT (195 Å) [2] and TRACE (171 Å). The extremely high temporal cadence enabled us to study the dynamics on time-scales of the order of minutes.

Many observations of time variability and velocities of active region loops exist. Kjeldseth-Moe and Brekke [3] investigated loops at temperatures ranging from

---

¹⁾ Research Associate of the Fund for Scientific Research in Flanders (FWO-Vlaanderen).

CP537, *Waves in Dusty, Solar, and Space Plasmas,* edited by F. Verheest, et al.
© 2000 American Institute of Physics 1-56396-962-9/00/$17.00

10^4 K to 2.7×10^6 K using the Coronal Diagnostic Spectrometer (CDS). They concluded time-variability on time scales of the order of 10-20 min to be a characteristic property, particularly of loops emitting at temperatures less than 1.5×10^6 K. At higher temperatures ($\geq 1.9 \times 10^6$ K) the variability is generally much smaller. By investigating line shifts, velocities from 20 km s^{-1} to even 300 km s^{-1} have been found. Still no clear answer to the question of the driving force exists. Berghmans & Clette [4] first discussed a new class of weaker footpoint brightenings that produce wave-like disturbances propagating along quasi-open field lines, which are further analysed in this report. The phenomenon of propagating disturbances along loops is very much comparable to moving features in solar polar plumes. DeForest & Gurman [5] report on quasi-periodic compressive waves in solar polar plumes which have been considered by Ofman et al. [6] to be slow magnetoacoustic waves. They measure speeds ranging from 75 to 150 km s^{-1} and intensity variations of 10-20% of overall intensity. Also more recently De Moortel et al. [7] reported on similar propagating oscillations in large diffuse coronal loops, using TRACE data. Not only longitudinal oscillations, but also transverse oscillations of active region loops have been observed directly [8] [9].

Nevertheless, the physical mechanisms behind oscillatory phenomena in the solar corona, which cover a wide range of wavelengths and periods, are still virtually unidentified, mainly because of poor spatial resolution and insufficient time cadence. Aschwanden et al. [8] give a very broad overview of spatial oscillations of coronal loops and coronal loop dynamics in general.

II JOP80, INSTRUMENTS AND DATA TREATMENT

On May 13, 1998, the EIT instrument on board SOHO (SOlar and Heliospherical Observatory) has produced a unique image sequence in the context of the multi-instrument campaign SOHO JOP 80 [1]. JOP80 is dedicated to the high-time resolution imaging study of coronal and transition region dynamics (EIT shutterless-mode campaign). The aim was to focus on the study of bright structures (active region loops, bright points) with the highest possible time resolution and wide spatial coverage. First results of this JOP80 have been reported by Berghmans & Clette [4]. In JOP80, EIT is the lead instrument, followed by several space-born instruments (SXT, TRACE, MDI, CDS, SUMER), as well as two ground based observatories (in La Palma and Sac Peak). The combination of these instruments allows for good statistics of many local events.

EIT achieved an exceptional 15 s cadence in the Fe XII bandpass at 195 Å by leaving EIT's shutter open for 1 hour and operating the CCD in frame transfer mode. EIT collected during the 1 hour JOP 80 run in total 229 images of 128×96 pixels (332×249 arcsec). Flat-field and grid pattern corrections were applied to the images. In the EIT data cosmic rays were identified as exceptional deviations (at the 5 σ level) in the pixel's light curves that appear in one image only. Their values were restored by linear interpolation between the neighbouring images. (see [10],

Instrument	Bandpass	Ion	Level	Temp	Cadence	Resolution
EIT	195 Å	Fe XII	Corona	1.6 MK	15 sec	2.59 arcsec/pix
TRACE	171 Å	Fe IX/X	Trans. Region	1.3 MK	25 sec	0.5 arcsec/pix

TABLE 1. Some characteristics of the instruments used: EIT and TRACE. As a function of temperature, the EIT Fe XII (195 Å) bandpass is relatively sharply peaked with a peak formation temperature of about 1.6×10^6 K, similarly the TRACE Fe IX-X (171 Å) bandpass is peaked around 1.3×10^6 K.

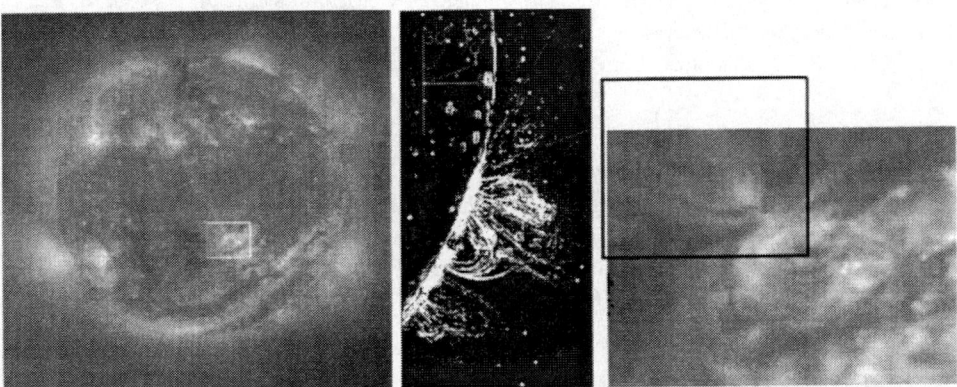

FIGURE 1. An overview of the EIT observations: *At the left,* a full disc image in the Fe XII (195 Å) bandpass, taken at 18h34m UT. The rectangle indicates the active region AR 8218. *In the middle,* a side view on AR 8218 when crossing the western limb. This image has been processed to enhance the edges of the loop structure. *At the right,* the field of view of AR8218, specified in the full disc at the left. Superimposed we have drawn the field of view of the TRACE image.

and [4] for all technical details). The TRACE images suffer more from radiation due to the radiation belt passages. To clean the TRACE images we ran twice the tracedespike IDL-routine (at the 6 σ level). The solar rotation during the 1 hr observation campaign (corresponding to about 3.7 EIT pixels) was compensated by simple correlation tracking.

III ACTIVE REGION 8218

The target of the campaign was a relatively small but highly dynamic active region (AR 8218). In Fig. 1(*left*), we show with an EIT synoptic image, the overall structure of this active region and zoom-in (Fig. 1, *right*) on the bundle of very long flux tubes that emanates from it to the NE. The width of the EIT field of view is about 2.3×10^8 m ($0.34 \times$ R$_\odot$). Taking profit of the solar rotation and assuming that the overall structure of this bundle remained constant during a week, we got a side view when the active region crossed the western limb (Fig. 1, *middle*). From

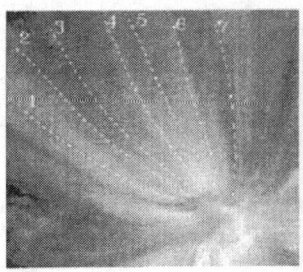  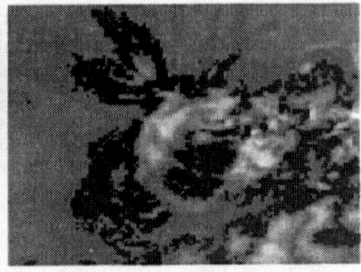

FIGURE 2. *At the left* TRACE image of the field of view as indicated on the right of Fig 1. In Table 2 we give the characteristics of the 7 loops tracked here. *In the middle*, the relative variance of the TRACE image is plotted and *at the right* the relative variance of the EIT image is plotted.

this we estimate that the flux tubes in the bundle make an angle α with the plane of the sky that ranges from roughly 30° up to 70°.

In a first inventory of the dynamics of this active region, a wide range in transients was discovered, ranging from a B3.5 flare producing a large plasma flow along pre-existing loops, to EUV versions of active region transient brightenings as previously observed by SXT on board YOHKOH [4].

IV   OBSERVATION AND INTERPRETATION

A   Tracking loops

Here we focus on an additional, new type of transients that was discovered propagating along the bundle of widely opening field lines in the NE of the active region. When investigating the region of the loops we have to take into account two facts. First we have to define the regions of solar variability. The variability we measure consists of solar variability, poisson noise (photon) and instrumental noise. Since the variance of the poisson noise equals the mean intensity, the variance of the total signal over mean should be constant when no solar variability was present in the data. We plotted the relative variance in Fig. 2 (*middle and right*). Clearly this relative variance is not constant. Moreover, our expectation of variability along the loops is confirmed.

In a second step we produce an image in which the loop structure becomes clearly visible. We make use of the variability in time along the loops to enhance the structure. We do this by calculating the standard deviation in each point (taken over time) divided by the median in that point. We identified oscillations in several loops during the 1998 May 13 time sequence, seven of which are selected visually for detailed analysis (Fig. 2, *left*). We tracked the loops on the TRACE image, and then converted the coordinates to EIT coordinates, to obtain the same loops in the EIT data. We number equidistant points along each loop from 0 (at the footpoint) to 9 (higher up in the loop) in the TRACE data. The path in the

EIT data is generally shorter, since the field of view is smaller at the North of the active region, where the loops are tracked. When assuming a semi-circular loop we visually estimate the length along an entire loop to be 8×10^8 m, and a path length (projection onto the f.o.v.) is estimated to be about 10^8 m.

In Fig. 3 we plot the relative standard deviation for the third loop selected, to get an idea of the variability along this loop. In this figure we also plotted the largest deviation from the median (to exclude cosmic rays), the second largest deviation (which is said to be representable for the variation in the loop). The variations are typically of the order of 2-10% w.r.t. the median. We find slightly more variability in the TRACE data than in the EIT data. For the 7 loops tracked this quantity is to be found in the first column in Table 2. In many cases, the relative variability is higher further up in the loops than at the footpoints. The fact that these curves only have small deviations from their linear fit indicates that the loops are well tracked.

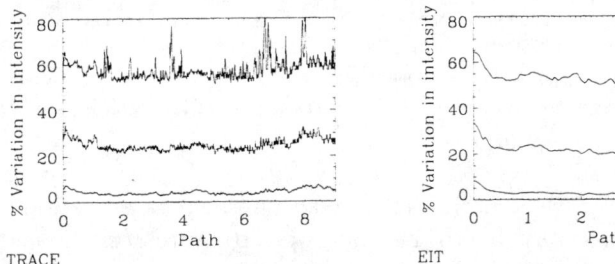

FIGURE 3. (*left:* TRACE, *right:* EIT) We plot from top to bottom, the largest and the third largest deviation from the median and the relative standard deviation of the third loop selected in Fig. 2 (*left*). These have to be interpreted as percentages of variability w.r.t. the median. The first two curves have been shifted over a percentage of 45 resp. 15. Notice that part of the path lies outside the EIT frame.

B   Variability along the loops

In Fig. 4 we show the temporal evolution (horizontal axis) along path 3 (vertical axis). A linear fit to each pixel's light curve was calculated. This linear fit is taken as the background EUV emission and in Fig. 4 we show the relative variations with respect to this background. In this diagram we see propagating disturbances as bright and dark diagonal ridges. The propagation speed along the loop of each disturbance can be determined by measuring the slope of the ridges. In this way we find an apparent propagation speed (because we only measure the component projected onto the plane of the sky). The observed propagation speeds measured in both data sets range from 65 to 150 km s^{-1}. However, the propagation speed could change by as much as 10 km s^{-1} from base to top without detection. When

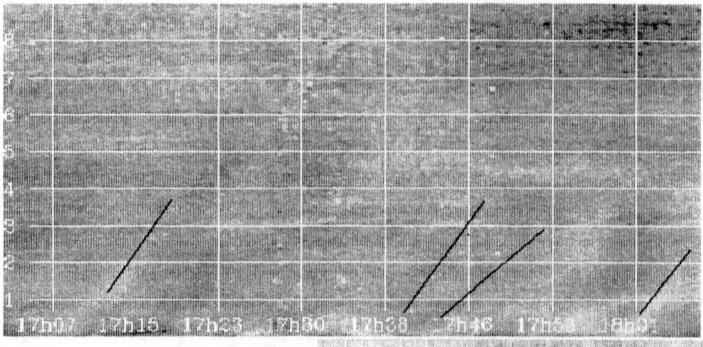
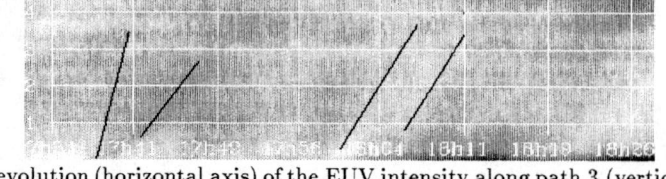

FIGURE 4. The temporal evolution (horizontal axis) of the EUV intensity along path 3 (vertical axis) in the TRACE data (*upper*) and in the EIT data (*lower*) tracked in Fig. 2. The inclined lines indicate the Propagating Disturbances. Notice that part of the path is lying outside the EIT frame.

tracking longer loops, we sometimes notice an apparent acceleration further up in the loops. This is probably due to the change of curvature of the loops. The speeds measured in the EIT data are higher than the speeds measured in the TRACE data. EIT measures at a higher temperature than TRACE. Hence the expected sound speed is higher for the EIT observations than for the TRACE observations (see Table 1). The sound speed C_S can be written as a function of temperature alone [12]. This allows us to derive a formal sound speed

$$
\begin{aligned}
C_S &= 152 \ \mathrm{T}^{1/2} \mathrm{m \ s^{-1}(T \ in \ °K)} \\
&= 173 \ \mathrm{km \ s^{-1}} \ \text{for TRACE data (171 Å)}, \quad (1) \\
&= 192 \ \mathrm{km \ s^{-1}} \ \text{for EIT data (195 Å)}. \quad (2)
\end{aligned}
$$

The apparent propagation speeds which we listed in table 2 as being V_p are of the order of this derived sound speed. This suggests that these PDs are sonic perturbations.

When comparing the velocity of the PD in the images, with the derived sound speed, we have to take into account the projection angle α between the loop direction and the plane of the sky. Assuming that the PDs indeed travel along the loop with the sound speed we derive the projection angle needed to bridge the gap between the apparent velocity and the sound speed to be:

$$
\alpha_{\exp} = \arccos\left(\frac{v_p}{C_S}\right). \quad (3)
$$

Path	σ_{rel} (in %)		V_p (in km s^{-1})		α_{exp}		Period (in min)
	EIT	TRACE	EIT	TRACE	EIT	TRACE	
P1	8.0	4.8	101	85	58 °	60 °	12, 38
P2	4.4	8.0	111	74	55 °	65 °	4-8
P3	8.1	8.7	94	86	61 °	60 °	4-8, 35
P4	2.9	4.03	60	53	70 °	72 °	10
P5	2.2	4.02	107	87	56 °	60 °	15
P6	4.2	7.1	97	83	60 °	61 °	17
P7	3.3	8.7	108	72	56 °	65 °	21
Average	4.7	6.5	97	77	60 °	64 °	
Cs			192	173			

TABLE 2. Characteristics of the propagating disturbances (PDs) along the paths outlined in Fig 2. σ_{rel} is the percentage of relative standard deviation with respect to the median as described above. V_p is the apparent speed (= speed projected onto the field of view), the. α_{exp} (see eq. 3) is the required correction angle which is needed to bring the apparent velocity v_p in agreement with the sound speed C_s which is 192 km s^{-1} for EIT and 173 km s^{-1} for TRACE. The period is the typical duration of a PD.

In the 4th column of Table 2 we give the angle α_{exp} that the loop is expected to make with the plane of the sky on the basis of this argument. These derived angles are in agreement with the angle derived from the side view of the active region (Fig. 1B). This means that the deviations of the measured propagation speed from the expected sound speed can be explained by the orientation of the magnetic bundle with respect to the plane of the sky. This further strengthens the sonic hypothesis.

V   RESULTS AND DISCUSSION

In Table 2 we give a short overview of the characteristics of the PDs along the paths outlined in Fig. 2 (*left*). We find outward propagating disturbances in virtually all loops within the opening bundle shown. Along some paths however we find more than one propagation speed, which we interpret as being due to the superposition of different loops. We draw the following conclusions:

1. Speeds roughly vary from 65 to 150 km s^{-1}, with an error of +/- 10 km s^{-1}.

2. We measure higher speeds in the EIT data than in the TRACE data. Since the sound speed is higher for features in the EIT dataset than for TRACE, this is in favor of the sound wave interpretation. We could only compare these speeds in the part of the xt-diagram which is common in time for both wavelengths. Since there is only an overlap of 35 min between the two data sets we need more data to confirm this.

3. The variations of the amplitude are of the order of 2-10 % in intensity or 1.4-3 % in density.

4. The absolute amplitude of a PD decays exponentially with a decay time of the order of 1.75 min. This means the PD vanishes before it reaches the other footpoint of the loop.

5. The expected projection angle lies in an acceptable range, such that the difference between the sound speed and apparent speeds can be explained by a projection angle.

6. We notice a good comparison between the angles measured in both wavelengths, which again points to the sound wave interpretation.

7. We find that the sound waves observed with EIT and TRACE along the same path have a common excitation source (see Fig. 4).

We aim to compare the observations to analytical models which model the propagation of sound waves in a coronal, isothermal loop structure, and study the feasibility of coronal diagnostics by such a method of comparison. [13].

The authors would like to thank Ronald Van der Linden *(Royal Observatory of Belgium)* for his continuous and patient support.

REFERENCES

1. Clette, F., *et al.*, *http://sohowww.~ nascom.nasa.gov/soc/JOPs/jop080.txt* (1998)

2. Robbrecht, E., Berghmans, D., Poedts, S., 'Slow magnetoacoustic waves in coronal loops?', in *Plasma Dynamics and Diagnostics in the Solar Transition Region and Corona*, edited by J. C. Vial & B. Kaldeich-Schürmann, Proceedings 8th SOHO Workshop, Paris, ESA SP-446, 1999, pp. 575-578.

3. Kjeldseth-Moe, O., Brekke, P., *Sol. Phys.* **182**, 73 (1998)

4. Berghmans, D., and Clette, F., *Sol. Phys.* **186**, 207B, (1999)

5. DeForest, C. E., and Gurman, J. B., *ApJ* **501**, L217 (1998)

6. Ofman, L., Nakariakov, V.M., DeForest, C. E., *ApJ* **514**, 441 (1999)

7. De Moortel, I., Ireland, J., Walsh, R.W., *Astron. Astrophys.* **355**, L23 (2000)

8. Aschwanden, M.J., Fletcher, L., Schrijver, C. J., Alexander, D., *ApJ* **520**, 880-894 (1999)

9. Ofman, L., Nakariakov, V.M., Deluca, E., Roberts, B., Davila, J.M. , *AAS* **194**, 7909 (1999)

10. Delaboudinière, J.-P. *et al.*, *Sol. Phys.* **162**, 291 (1995)

11. Berghmans, D., Clette, F. and Moses, D., *Astron. Astrophys.* **336**, 1039 (1998)

12. Priest, E., *Solar Magnetohydrodynamics*, D. Reidel Publ. Co., Dordrecht, Holland, (1984)

13. Nakariakov, V.M., Verwichte, E., Berghmans, D., Robbrecht, E., *Astron. Astrophys.*, **submitted**

Interaction of sound waves with inhomogeneous magnetized plasma in strongly nonlinear resonant slow wave layer

Michael S. Ruderman, Robert Erdélyi

Department of Applied Mathematics, University of Sheffield, Hicks Building, Western Bank, Sheffield S10 2TN, UK

Abstract. We consider slow resonant MHD waves in 1D planar equilibria with the unidirectional magnetic field. A nonlinear equation governing this waves in a slow resonant layer is derived. A periodic solution in the form of propagating wave with a permanent shape is found in the limiting case, where nonlinearity dominates dissipation. This solution is used to derive a connection formula that connects the values of the normal component of the velocity at two sides of the resonant layer. This connection formula is, in turn, used to study the interaction of an incoming sound wave with a slab containing an inhomogeneous magnetized plasma. The coefficient of the wave energy resonant absorption is calculated and compared with its counterpart obtained on the basis of linear theory.

INTRODUCTION

Resonant MHD waves have been intesively studied for a few decades in application to fusion plasmas ([1] and references therein), the magnetosphere ([2] and references therein), and the solar atmosphere (e.g., [3–7]). The remarkable property of resonant MHD waves is that their damping rate is independent of dissipative coefficients in weakly dissipative plasmas. This property makes resonant MHD waves very viable candidates for plasma heating.

There are two types of resonances: Alfvén and slow. In the case of Alfvén resonance the global plasma motion is in resonance with local Alfvén waves at a spatial position called the Alfvén resonant position. In the case of slow resonance it is in resonance with local slow waves at a spatial position called the slow resonant position. In this paper we concentrate on the slow resonance.

Until recently all analytical studies of resonant MHD waves were based on the linear description of plasma motion, and nonlinear effects were studied only numerically. Linear theory predicts that the amplitude of plasma motion in the vicinity of a resonant position can be very large even when it is small far away from the

CP537, *Waves in Dusty, Solar, and Space Plasmas*, edited by F. Verheest, et al.
© 2000 American Institute of Physics 1-56396-962-9/00/$17.00

resonant position. This observation implies a mixed description of resonant MHD waves, where the wave motion is described by the linear ideal MHD equations far away from the resonant position, and by the nonlinear dissipative MHD equations in a narrow resonant layer embracing the ideal resonant position.

Nonlinear effects are more pronounced in slow resonant layers than in Alfvén resonant layers. What is why the systematic analytical study of nonlinear effects in resonant layers has started from slow resonant layers. In [8] a nonlinear equation governing stationary motion in slow resonant layers in plasmas with isotropic viscosity and electrical resistivity has been derived. In [9] a similar governing equation has been obtained for plasmas with strongly anisotropic viscosity and heat conduction, as in the solar corona. In [10] the equation obtained in [8] has been generalized to include a stationary flow in the equilibrium state.

In this paper we derive a nonlinear equation governing non-stationary motion in slow resonant layers. Then we obtain the solution to this equation describing the stationary motion in a slow resonant layer in the case where nonlinearity dominates dissipation. This solution is used to derive a nonlinear connection formula that connects the normal component of the velocity to the left and the right of the resonant layer. This connection formula is then used to study the absorption of a sound wave, impinged on a slab containing inhomogeneous magnetized plasma, in a slow resonant layer.

DERIVATION OF GOVERNING EQUATION

We consider a plasma with isotropic viscosity and resistivity. Its motion is described by the standard system of MHD equations

$$\frac{\partial \rho}{\partial t} + \nabla \cdot (\rho \mathbf{V}) = 0, \quad \frac{\partial}{\partial t}\left(\frac{p}{\rho^\gamma}\right) + \mathbf{V} \cdot \nabla\left(\frac{p}{\rho^\gamma}\right) = 0, \tag{1}$$

$$\frac{\partial \mathbf{V}}{\partial t} + (\mathbf{V} \cdot \nabla)\mathbf{V} = -\frac{1}{\rho}\nabla p + \frac{1}{\mu\rho}(\nabla \times \mathbf{B}) \times \mathbf{B} + \nu\left(\nabla^2 \mathbf{V} + \tfrac{1}{3}\nabla\nabla \cdot \mathbf{V}\right), \tag{2}$$

$$\frac{\partial \mathbf{B}}{\partial t} = \nabla \times (\mathbf{V} \times \mathbf{B}) + \eta\nabla^2\mathbf{B}. \tag{3}$$

Here $\mathbf{V} = (V_x, V_y, V_z)$ is the velocity, $\mathbf{B} = (B_x, B_y, B_z)$ the magnetic induction, p the pressure, ρ the density, μ the magnetic permiability of vacuum, ν the kinematic coefficient of viscosity, η the coefficient of magnetic diffusion, and γ the ratio of specific heats. Note that, in spite of the presence of dissipation, we use the adiabatic equation (the second equation (1)). Numerical simulation [11] in the framework of linear MHD have shown that the account of dissipation in the second equation (1) practically does not change the behaviour of resonant MHD waves. We assume that this property remains valid in nonlinear theory.

We consider small-amplitude perturbations superimposed on a one-dimensional equilibrium state. In this equilibrium state all quantities depend on the x coordinate only in the Cartesian coordinates x, y, z. The equilibrium magnetic field is unidirectional and parallel to the yz-plane. There is an equilibrium flow parallel to the equilibrium magnetic field. In what follows the equilibrium quantities are labeled with the subscript '0'. The equilibrium pressure and magnetic induction satisfy the equation of the total pressure balance $p_0 + B_0^2/2\mu = $ const.

From linear theory we have the following physical picture of motion in a resonant layer. The global motion of the plasma is in exact resonance with slow waves at the ideal resonant position, and it is in quasi-resonance with slow waves in a narrow resonant layer. Since the resonant layer is thin, its inertia is very small. This implies that the perturbation of the total pressure only slightly varies across the resonant layer, and can be considered as an external driving force. If perturbations of all quantities are smaller than their background values, this picture remains valid in nonlinear regime. Hence, our aim is to derive a nonlinear equation governing slow waves in the resonant layer driven by the variation of the external pressure.

Let us introduce the dimensionless amplitude of perturbations far away from the resonant layer $\epsilon \ll 1$, and the total Reynolds number $R = (1/R_e + 1/R_m)^{-1}$. Here $R_e = V_h \ell/\nu$ and $R_m = V_h \ell/\eta$, V_h is the characteristic velocity that will be specified later, and ℓ is the characteristic scale of inhomogeneity. It is shown in [8] that the ratio of the largest nonlinear terms to the largest dissipative terms in the dissipative MHD equations is of the order of $\epsilon R^{2/3}$ in the resonant layer. This implies that nonlinearity is important in the resonant layer when $\epsilon R^{2/3} \gtrsim 1$. Since we intend to derive a governing equation that takes both nonlinearity and dissipation into account, we assume that $\epsilon R^{2/3} \sim 1$. However, this assumption is quite formal. In fact, the equation that we shall derive will be also valid in the two limiting cases, $\epsilon R^{2/3} \ll 1$ and $\epsilon R^{2/3} \gg 1$. In the first limiting case we can neglect the nonlinear term in this equation and arrive at linear description. In the second limiting case we can neglect the dissipative term and obtain the ideal nonlinear governing equation for the slow wave motion in the resonant layer.

In linear theory we can Fourier-analyze a perturbation and take the z-axis along the wavevector. Thus we obtain a two-dimensional problem. This approach is inapplicable in nonlinear analysis, and, in general, we have to deal with a three-dimensional problem. However, for the sake of simpicity, we restrict our analysis to a two-dimensional problem, and assume that perturbations of all quantities are independent of y.

The slow wave motion in the resonant layer is characterized by the property that the characteristic wavelength in the z-direction is much larger that the thickness of the resonant layer. This implies that the wave motion in the resonant layer can be represented by a superposition of infinite number of slow waves with the wavenumbers almost perpendicular to the resonant layer. These waves propagate with the phase velocity close to $c_T(x_c) + V_0(x_c)$ along the equilibrium magnetic field and, consequently, with the velocity $C = \mathcal{V} + V_0(x_c) \cos \alpha$ in the z-direction, where α is the angle between $\mathbf{B}_0$ and the z-direction, $x = x_c$ the ideal resonant position,

$V_0(x)$ the equilibrium flow velocity (recall that it is parallel to $\mathbf{B}_0$), $\mathcal{V} = c_T(x_c)\cos\alpha$, and c_T is the cusp speet given by

$$c_T^2 = \frac{c_S^2 v_A^2}{c_S^2 + v_A^2}, \quad c_S^2 = \frac{\gamma p_0}{\rho_0}, \quad v_A^2 = \frac{B_0^2}{\mu \rho_0}, \tag{4}$$

with c_S and v_A the sound and Alfvén speed. This observation inspires us to introduce the running variable $\theta = z - Ct$. Now we can take $V_{\mathrm{h}} = \mathcal{V}$.

The assumption $\epsilon R^{2/3} \sim 1$ implies $R \sim \epsilon^{3/2}$. Linear theory gives the estimate for the thickness of the resonant layer $R^{1/3}\ell \sim \epsilon^{1/2}\ell$. This estimate inspires us to introduce the stratching variable $\xi = \epsilon^{-1/2}(x - x_c)$. Once again using linear theory, we obtain the estimate that the ratio of the nonlinear terms to the linear terms in the MHD equations in the resonant layer is of the order of $\epsilon^{1/2}$. This estimate implies the following scenario of the evolution of an initial disturbance. Consider an initial disturbance either in the form of a puls with the characteristic duration T, or in the form of a periodic wave with the period T. Then the dominant motion of this disturbance will be the translation in the z-direction with the velocity $\mathcal{V}$, while its shape will change on the time-scale $\epsilon^{-1/2}T$. This observation inspires us to introduce the so-called "slow" time $\tau = \epsilon^{1/2}t$. Now we rewrite equations (1)–(3) in the new independent variables θ, ξ, and τ.

Since the characteristic scale of variation of equilibrium quantities, ℓ, is much larger than the thickness of the resonant layer, we use the approximation $f_0(x) = f_0(x_c) + \epsilon^{1/2}\xi f_0'(x_c)$, where f_0 represents any equilibrium quantity, and the prime indicates the derivative.

Linear theory predicts that the amplitudes of perturbations of p, ρ, $V_\parallel$, and $B_\parallel$ are of the order of $\epsilon^{1/2}$, while the amplitudes of perturbations of V_x, $V_\perp$, B_x, $B_\perp$, and the total pressure $P = p + B^2/2\mu$ are of the order of ϵ, where the subscripts $\parallel$ and $\perp$ indicate the components of a vector parallel and perpendicular to $\mathbf{B}_0$. These estimates inspire us to look for the solution to equations (1)–(3), rewritten in terms of θ, ξ, and τ, in the form of expansions with respect to $\epsilon^{1/2}$. We write these expansions as $f = f_0 + \epsilon^{1/2}f_1 + \epsilon f_2 + \ldots$ for p, ρ, $V_\parallel$, and $B_\parallel$, and as $g = g_0 + \epsilon g_1 + \epsilon^{3/2}g_2 + \ldots$ for V_x, $V_\perp$, B_x, $B_\perp$, and P.

Collecting terms of the lowest order with respect to $\epsilon^{1/2}$ in equations (1)–(3), we obtain the system of equation for the quantities of the first order approximation, which are the quantities with the subscript '1'. First of all, it follows from this system that P_1 is independent of ξ in the resonant layer, which is in complete agreement with the conclusion made previously on the basis of the qualitative analysis. Then, using this system, we express all quantities of the first order approximation in terms of u_1 and w_1, where $u_1 = V_{x1}$. In the case where perturbations of all quantities vanish as $|\theta| \to \infty$, $w_1 = V_{z1}$. When perturbations are periodic with respect to θ, w_1 is an oscillatory part of V_{z1}, determined as $w_1 = V_{z1} - \langle V_{z1}\rangle$, where $\langle\ldots\rangle$ denotes the mean value of a quantity over the period. And, eventually, we obtain the equation connecting w_1 and u_1

$$\frac{\partial u_1}{\partial \xi} + \frac{\mathcal{V}^2}{v_{Ac}^2} \frac{\partial w_1}{\partial \theta} = 0, \tag{5}$$

where the subscript 'c' indicates that a quantity is calculated at $x = x_c$.

In the second order approximation we obtain a linear inhomogeneous system of equations for the quantities with the subscript '2'. The right-hand sides of these equations are expressed in terms of w_1 and P_1. Their left-hand sides are obtained by the substitution of the quantites with the subscript '2' for the corresponding quantities with the subscript '1' in the equations of the first order approximation. Hence the homogeneous counterpart of the system of the second order approximation coincides with the system of the first order approximation. The system of the first order approximation has a non-trivial solution. This implies that the system of the second order approximation possesses a solution only if its right-hand side satisfies a compatibility condition. To obtain this condition we eliminate all quantities with the subscript '2' from the system of the second order approximation. As a result we arrive at

$$\frac{\partial w_1}{\partial \tau} - \frac{\Delta \xi}{2\mathcal{V}} \frac{\partial w_1}{\partial \theta} + \frac{\Lambda}{2\mathcal{V}} w_1 \frac{\partial w_1}{\partial \theta} - \lambda \frac{\partial^2 w_1}{\partial \xi^2} = \frac{-\mathcal{V}^2}{2\rho_{0c} v_{Ac}^2} \frac{\partial P_1}{\partial \theta}, \tag{6}$$

where

$$\Delta = \frac{d}{dx}[(C - V_0 \cos \alpha)^2 - c_T^2 \cos^2 \alpha]\Big|_{x=x_c}, \tag{7}$$

$$\Lambda = \mathcal{V}^5 \frac{(\gamma + 1)v_{Ac}^2 + 3c_{Sc}^2}{v_{Ac}^2 c_{Sc}^4}, \quad \lambda = \frac{\epsilon^{-3/2}}{2}\left(\nu + \frac{c_{Tc}^2 \eta}{v_{Ac}^2}\right).$$

Note that, in accordance with the assumption $\epsilon R^{2/3} \sim 1$, $\lambda \sim \ell \mathcal{V}$. Equation (6) is the governing equation for nonlinear slow waves in the resonant layer.

STATIONARY PERIODIC SOLUTION TO GOVERNING EQUATION AND CONNECTION FORMULAE

In this section we obtain a stationary ($\partial/\partial \tau = 0$) periodic ($w_1(\theta + L) = w_1(\theta)$) solution to equation (6). In general, such a solution can be found only numerically. To make analytical progress, we assume that nonlinearity strongly dominates dissipation in the resonant layer ($\epsilon R^{2/3} \gg 1$). This assumption enables us to neglect the dissipative term in equation (6), which is the last term on the left-hand side, in comparison with the nonlinear term, which is the third term on the left-hand side. Then integrating the resulting equation once with respect to θ, and taking into account that $\langle w_1 \rangle = \langle P_1 \rangle = 0$, we arrive at

$$\Lambda w_1^2 + 2\sigma w_1 - \Lambda \langle w_1^2 \rangle + 2\mathcal{V}^3 Q = 0, \tag{8}$$

283

where $\sigma = -\Delta\xi$ and $Q = P_1(\rho_{0c}v_{Ac}^2)^{-1}$. The solution to this equation is

$$w_1 = \Lambda^{-1}\left\{ -\sigma \pm \left[\sigma^2 + \Lambda^2\left\langle w_1^2\right\rangle - 2\Lambda\mathcal{V}^3 Q(\theta)\right]^{1/2} \right\}. \tag{9}$$

Since $w_1 \to 0$ as $|\sigma| \to \infty$, the sign in this equation coincides with the sign of the quantity σ. The condition $\langle w_1 \rangle = 0$ can be considered as the equation determining the unknown quantity $\langle w_1^2 \rangle$. When $Q(\theta)$ is fixed, it gives $\langle w_1^2 \rangle$ as a function of σ. It has been shown in [12] that for any fixed $Q(\theta)$ there is such a quantity $\sigma_0 > 0$ that $\langle w_1^2 \rangle$ is a sigle-valued even function of σ for $|\sigma| > \sigma_0$. This implies that w_1 is an odd function of σ for $|\sigma| > \sigma_0$. On the other hand, the equation $\langle w_1 \rangle = 0$, considered as an equation for $\langle w_1^2 \rangle$, has no solution for $|\sigma| < \sigma_0$. This implies that we have to use solutions with both signs to construct the solution to equation (8) in the strip $|\sigma| < \sigma_0$. This solution is discontinuous and contains a slow shock wave, which is not surprising at all because we have neglected the term with the highest derivative in equation (6). The intensity of this shock wave varies from zero at $\sigma = -\sigma_0$ to its maximum value at $\sigma = 0$, and back to zero at $\sigma = -\sigma_0$. It is shown in [12] that $\sigma^2 + \Lambda^2\langle w_1^2 \rangle = 2\Lambda\mathcal{V}^3 Q_M$, where Q_M is the maximum value of $Q(\theta)$ in the interval $[0, L]$ (recall that L is the period). Now it is straightforward to obtain the explicit solution in the strip $|\sigma| < \sigma_0$

$$w_1 = \Lambda^{-1}\begin{cases} -\sigma - \{2\Lambda\mathcal{V}^3[Q_M - Q(\theta)]\}^{1/2}, & -\sigma_0 < \sigma < \sigma_s(\theta), \\ -\sigma + \{2\Lambda\mathcal{V}^3[Q_M - Q(\theta)]\}^{1/2}, & \sigma_s(\theta) < \sigma < \sigma_0. \end{cases} \tag{10}$$

In accordance with the analysis in [12], $\sigma_s(\theta_M) = -\sigma_0$, $\sigma_s(\theta_M + L) = \sigma_0$, where θ_M is the point where $Q(\theta)$ takes its maximum value, $Q(\theta_M) = Q_M$. In addition, $\sigma_s(\theta)$ monotonically grows in the interval $[\theta_M, \theta_M + L]$, so it is more convenient to consider $\sigma_s(\theta)$ in this interval than in $[0, L]$. Since $\sigma_s(\theta)$ is periodic with the period L, it is discontinuous at $\theta = \theta_M + nL$, $n = 0, \pm 1, \ldots$. Introducing the inverse function $\theta_s(\sigma)$, $\sigma_s(\theta_s(\sigma)) = \sigma$, which is a monotonically growing function in the interval $[-\sigma_0, \sigma_0]$, and using equation (10), we write the condition $\langle w_1 \rangle = 0$ as

$$\sigma L = (2\Lambda\mathcal{V}^3)^{1/2}\left(\int_{\theta_M}^{\theta_s}\{Q_M - Q(\theta)\}^{1/2}\,d\theta - \int_{\theta_s}^{\theta_M+L}\{Q_M - Q(\theta)\}^{1/2}\,d\theta\right). \tag{11}$$

This equation determines $\theta_s(\sigma)$ in an implicit form. Or, vice versa, it gives the explicit expression for $\sigma_s(\theta)$, which is obtained simply by substituting σ_s for σ, and θ for θ_s. Hence, equations (10) and (11) give the explicit solution for w_1 in the strip $|\sigma| < \sigma_0$, written in terms of $Q(\theta)$.

Let us now obtain the connection formulae. We introduce the jump of a quantity $f(\sigma)$ across the resonant layer as $[f] = \lim_{\xi\to\infty}\{f(\xi) - f(-\xi)\}$. First of all, we note that, since P_1 is independent of ξ, $[P_1] = 0$. To calculated $[u_1]$, we just integrate equation (5) with respect to ξ from $-\infty$ to ∞. Since $w_1(\sigma)$ is odd for $|\sigma| > \sigma_0$, we have $\int_{-\infty}^{\infty} w_1\,d\sigma = \int_{-\sigma_0}^{\sigma_0} w_1\,d\sigma$. Since $w_1(\sigma)$ is a linear function of σ for $|\sigma| < \sigma_0$, the calculation of the latter integral is trivial. Using the approximation $u \equiv V_x \approx \epsilon u_1$, $P' \approx \epsilon P_1$, we eventually obtain

$$[u] = \frac{4\mathcal{V}^5}{\rho_{0c} v_{Ac}^4 |\Delta| L} \frac{\partial}{\partial \theta} \{P_M' - P'(\theta)\}^{1/2} \left(\int_{\theta_M}^{\theta} - \int_{\theta}^{\theta_M+L} \right) \{P_M' - P'(\vartheta)\}^{1/2} \, d\vartheta, \quad (12)$$

where $P_M' = P'(\theta_M)$ is the maximum value of the perturbation of the total pressure.

SOUND WAVE INTERACTION WITH INHOMOGENEOUS PLASMA

In this section we use the connection formulae, obtained in the previous section, to study the following problem. Consider an equilibrium state that consists of three regions. Region II is a slab, bounded by the planes $x = 0$ and $x = x_0$, that contains an inhomogeneous magnetized plasma. This region is sandwitched by regions I and III, which are semi-infinite in the x-direction. Region I contains a homogeneous magnetic-field-free plasma, and region III a homogeneous magnetized plasma. The equilibrium magnetic field is in the z-direction, and there is no equilibrium flow.

There is a sound wave incoming from $x \to -\infty$ in region I. This wave interacts with the inhomogeneous plasma in region II, and is partially reflected and patially absorbed in a slow resonant layer. The solution for u and P' in region I is

$$P' = \epsilon p_e \{\cos(k\Theta_+) + A(\Theta_-)\}, \quad u = \epsilon \frac{\chi p_e}{\rho_e \mathcal{V}} \{\cos(k\Theta_+) - A(\Theta_-)\}, \quad (13)$$

where p_e and ρ_e are the equilibrium pressure and density in region I, $\Theta_\pm = \theta \pm \chi x$, $(k\chi, 0, k)$ is the wavenumber, the first terms in curly brackets in both expressions describe the incoming wave, and the second terms the outgoing wave that exists due to partial reflection of the incoming wave from the inhomogeneous plasma. Our task is to determine the function $A(\theta)$ describing the outgoing wave. Note that this problem has been studied in the approximation of weak nonlinearity in the slow resonant layer in [13–15]. In this paper we use the approximation of strong nonlinearity in the slow resonant layer.

Here we only outline the procedure of solution of the problem, and present the main results. All details can be found in [12]. The analysis starts from deriving a governing equation for $A(\theta)$. The ideal slow resonant position in region II is determined by the condition $c_T^2(x_c) = c_{Se}^2(1 + \chi^2)$. We assume that $c_T(x)$ is a monotonically growing function, and $c_T^2(x_0) > c_{Se}^2(1 + \chi^2)$. Since $c_T(0) = 0$, this implies that there is exactly one slow resonant position in region II. Using the connection formula $[P'] = 0$, and the continuity of all quantities at $x = 0, x_0$, it is possible to obtain the solution to the linear ideal MHD equations in region III and in region II to the left and the right of the resonant layer in terms of $A(\theta)$. We use this solution to calculate $[u]$ in terms of $A(\theta)$. Then we compare the result with the expression given by equation (12). As a result we obtain a complicated nonlinear integral equation for $A(\theta)$. In spite of its complexity, this equation admits a very simple solution $A(\theta) = a\cos(k\theta + \varphi)$. This solution implies that the outgoing wave contains only the fundamental harmonic inspite that the motion in the slow resonant layer is strongly nonlinear. The amplitude a and the phase shift φ of the outgoing wave are expressed in term of equilibrium quantities, and the fundamental solution to the system of ordinary differential equations, describing the dependence of u and P' on x in region II outside of the resonant layer. These expressions are rather complicated and we do not write them down. They can be found in [12].

The most interesting quantity in the problem under consideration is the coefficient of the wave energy resonant absorption $K = (\Pi_{in} - \Pi_{out})/\Pi_{in}$. Here Π_{in} and Π_{out} are the energy fluxes in the incoming and outgoing wave respectively. It is straightforward to show that $K = 1 - a^2$. We have calculated this quantity numerically using the obtained expression for a, and compared the result with that given by linear theory. In what follows we use the notation K_{lin} and K_{non} for the coefficients of resonant absorption given by linear and nonlinear theory, respectively. When $kx_0 \ll 1$ (the long-wavelength approximation), $K_{non}/K_{lin} = 8/\pi^2 \approx 0.81$. In general, K_{non}/K_{lin} depends on the equilibrium quantities and kx_0. Our calculation has shown that, for $kx_0 \leq 10$ and a relatively large range of variation of the equilibrium quantities, $0.8 \leq K_{non}/K_{lin} \leq 1.2$. Hence, while linear theory fails to properly describe the wave motion in strongly nonlinear slow resonant layers, it gives fairly good approximation for the coefficient of resonant absorption.

REFERENCES

1. Vaclavik, J. and Appert, K., *Nuclear Fusion* **31**, 1945–1997 (1991).
2. Glassmeier, K.-H., Othmer, C., Cramm, R., Stellmacher, M. and Engelbretson, M., *Surveys in Geophysics* **20**, 61–109 (1999).
3. Ionson, J.A., *Astrophys. J.* **226**, 650–673 (1978).
4. Davila, J.M., *Astrophys. J.* **317**, 514–521 (1987).
5. Goossens, M., "MHD waves and wave heating in non-uniform plasmas" in *Advances in Solar System Magnetohydrodynamics*, edited by E.R. Priest and A.W. Hood, Cambridge University Press, 1991, pp. 137–172.
6. Hollweg, J.V., "Alfvén waves" in *Mechanisms of Chromospheric and Coronal Heating*, edited by P. Ulmschneider et al., Springer-Verlag, Berlin, 1991, pp. 423–434.
7. Poedts, S., "Waves in the transition region and corona – A theorist's view" in *Proceedings SP-448*, Florence, 1999, pp. 167–176.
8. Ruderman, M.S., Hollweg, J.V. and Goossens, M., *Phys. Plasmas* **4**, 75–90 (1997).
9. Ballai, I., Ruderman, M.S. and Erdélyi, R., *Phys. Plasmas* **5**, 252–260 (1998).
10. Ballai, I. and Erdélyi, R., *Solar Phys.* **180**, 65–79 (1998).
11. Poedts, S., Beliën, A.J.C. and Goedbloed, J.P., *Solar Phys.* **151**, 271–304 (1994).
12. Ruderman, M.S., *J. Plasma Phys.* **63**, 43–77 (2000).
13. Ruderman, M.S., Goossens, M. and Hollweg, J.V., *Phys. Plasmas* **4**, 91–100 (1997).
14. Ballai, I., Erdélyi, R. and Ruderman, M.S. *Phys. Plasmas* **5**, 2264–2273 (1998).
15. Erdélyi, R. and Ballai, I., *Solar Phys.* **186**, 67–97 (1999).

Nonlinear coupling of O- and X-mode radio emission and Alfven waves in the solar corona

O. Sirenko*, Yu. Voitenko†*, M. Goossens†, A. Yukhimuk*

*Main Astronomical Observatory, Holosiiv, Kyiv, 03680, Ukraine
†Centre for Plasma Astrophysics , K.U.Leuven, Celestijnenlaan 200B, B-3001 Heverlee, Belgium

Abstract. The nonlinear coupling of extraordinary and ordinary waves via kinetic Alfven waves(KAWs) is investigated on the basis of two fluid magnetohydrodynamics. The equation governing the time dependence of electric field of excited O-mode is found. We estimate the time of effective coupling between modes and corresponding interaction distance in solar corona. Our theoretical results show that the X- and O-mode couplings via Alfven waves can be efficient depolarization mechanism for the coronal radioemission.

I INTRODUCTION

The Sun's corona is a strong source of radio emission. The waves involved in the radio emission are the extraordinary (X-) and ordinary (O-) modes, the Z-mode, and the whistler mode. The waves of astrophysical interest are the X- and O-modes, because they can propagate from the place where they are generated in the solar atmosphere to infinity and hence can be observed on Earth. They are high frequency modes. In general they are elliptically polarized with opposite sense of the $\mathbf{E}$ vector rotation; the polarization of modes changes with the direction of the wavenumber $\mathbf{k}$ with respect to the external magnetic field $\mathbf{B_0}$. The polarization observations of these waves can be used to determine the strength and direction of the magnetic field in regions where no other information on the magnetic field is available. However, the interpretation of the polarization of solar radio emission is not always straightforward. One reason being that the apparent polarization of the radio emission can change as the emission propagates from its source to us - for example via mode coupling.

The theory of mode coupling in cold plasmas has been studied in detail([1] - [3]). The mode coupling occurs when the O- and X-modes cannot propagate separately, but are to exchange their energies. This effect may be caused by different reasons: local gradients in electron number density, magnetic field $\mathbf{B_0}$, or scattering on waves. The coupling between the two modes can be described with the parameter:

CP537, *Waves in Dusty, Solar, and Space Plasmas*, edited by F. Verheest, et al.
© 2000 American Institute of Physics 1-56396-962-9/00/$17.00

$Q = (\nabla k L)^{-1}$. ∇k is the difference in wavenumbers of the modes and L is the characteristic length over which the mode coupling is possible. If $Q \ll 1$ the modes do not overlap and hence are independent, but if $Q \gg 1$ the energy can be exchanged between them.

The coupling is detectable in the polarization, because in the case of weak coupling the O- and X-modes propagate like in an homogeneous medium. The polarization rotation direction should reverse whenever the longitudinal component of the magnetic field changes its direction. In the case of strong coupling, the strong energy exchange between the modes produces a frozen-in polarization, i.e. the polarization is constant; therefore the electromagnetic wave propagates as if the magnetic field was absent.

The theory of mode coupling has difficulties in explaining the observed polarization properties at meter wavelengths in quasi-transverse(QT) region. For example, the simple theory predicts that the emission at the fundamental frequency should be 100% polarized in the sense of the ordinary mode in solar radio burst of types II and III. But the observations show that the emission is only partially polarized, with a degree of polarization varying between 0 and 70%.

The type I storms, for which the sources are prevalent at wavelengths $\gtrsim 1m$, are highly circularly polarized (the degree of polarization is usually reduced for sources near solar limb). Theory predicts that the sense of polarization should reverse near meridian passage, but this reversal is not observed. So the observations show that mode coupling is strong in QT regions at meter wavelength, but, on the contrary, the theory suggests that it should be always weak.

Possible remedies for these discrepancies between theory and observation are given by Merlose [4], Bastian [5], Zheleznyakov [6]. However, so far not all discrepancies have been resolved.

In our work we study the mode coupling of the extraordinary and ordinary waves by kinetic Alfvén waves in a coronal plasma. Due to the kinetic properties these waves can more effectively interact with the plasma particles and other types of waves [7], [8]. We show that the presence of kinetic Alfvén waves in coronal loops leads to moderate mode coupling between ordinary and extraordinary waves. This process can be responsible for the depolarization of type II and III emission.

II   BASIC EQUATIONS

We consider a homogeneous plasma with a constant magnetic filed ($\mathbf{B_0} = B_0 \mathbf{e}_z$) and we use two-fluid magnetohydrodynamics:

$$\frac{\partial \mathbf{V}_\alpha}{\partial t} = \frac{1}{m_\alpha}(e_\alpha \mathbf{E} + \mathbf{F}_\alpha) + (\mathbf{V}_\alpha \times \omega_{B\alpha}) - \frac{T_\alpha}{m_\alpha n_\alpha}\nabla n_\alpha, \tag{1}$$

$$\frac{\partial n_\alpha}{\partial t} = -\nabla(n_\alpha \mathbf{V}_\alpha), \tag{2}$$

$$\nabla \times \mathbf{B} = \frac{4\pi}{c}\mathbf{j} + \frac{1}{c}\frac{\partial \mathbf{E}}{\partial t}, \nabla \times \mathbf{E} = -\frac{1}{c}\frac{\partial \mathbf{B}}{\partial t}, \nabla \cdot \mathbf{E} = 4\pi\rho, \tag{3}$$

where

$$\mathbf{j} = e(n_i \mathbf{V}_i - n_e \mathbf{V}_e), \quad \rho = e(n_i - n_e),$$

$$\mathbf{F}_\alpha = \frac{e_\alpha}{c}(\mathbf{V}_\alpha \times \mathbf{B}) - m_\alpha(\mathbf{V}_\alpha \nabla)\mathbf{V}_\alpha.$$

The index $\alpha = i, e$ corresponds to the ion and electron components of plasma respectively.

The electron density, electron speed, and the electric and magnetic fields are written as sums of background values and perturbations:

$$\begin{aligned}
n_e &= n_0 + n_{e2}, \\
\mathbf{V}_e &= \mathbf{V}_0 + \mathbf{V}_1 + \mathbf{V}_2, \\
\mathbf{E} &= \mathbf{E}_0 + \mathbf{E}_1 + \mathbf{E}_2, \\
\mathbf{B} &= B_0\mathbf{e}_z + \mathbf{B}_0 + \mathbf{B}_1 + \mathbf{B}_2,
\end{aligned} \tag{4}$$

where n_0 is the average value of plasma density, $B_0\mathbf{e}_z$ is the external magnetic field. The indexes 0, 1, and 2 refer to the ordinary, extraordinary and kinetic Alfvén waves respectively.

A   Extraordinary electromagnetic wave

We consider a region where the extraordinary wave propagates under an angle with respect to the ambient magnetic field. For frequencies $\omega_1 \gg |\omega_{Be}|$ the linear dispersion relation is:

$$\omega_1^2 = \omega_{pe}^2 + c^2 k_1^2 + \omega_{pe}^2 \frac{\omega_{Be}}{\sqrt{\omega_{pe}^2 + c^2 k_1^2}}\frac{k_{1z}}{k_1} \tag{5}$$

In this case wave polarization is elliptical. The electron velocity and magnetic field perturbations due to the extraordinary wave are determined by

$$\mathbf{V}_1 = \hat{A}_1 \mathbf{E}_1$$

and

$$\mathbf{B}_1 = \hat{b}_1 \mathbf{E}_1,$$

where matrices

$$\hat{A}_1 = -\frac{e}{m_e}\begin{pmatrix} \frac{i\omega_1}{\omega_1^2 - \omega_{Be}^2} & \frac{\omega_{Be}}{\omega_1^2 - \omega_{Be}^2} & 0 \\ \frac{-\omega_{Be}}{\omega_1^2 - \omega_{Be}^2} & \frac{i\omega_1}{\omega_1^2 - \omega_{Be}^2} & 0 \\ 0 & 0 & \frac{i}{\omega_1} \end{pmatrix}; \quad \hat{b}_1 = \begin{pmatrix} 0 & \frac{-ck_{1z}}{\omega_1} & 0 \\ \frac{ck_{1z}}{\omega_1} & 0 & \frac{-ck_{1x}}{\omega_1} \\ 0 & \frac{ck_{1x}}{\omega_1} & 0 \end{pmatrix}$$

The density perturbation produced by the extraordinary waves is negligible.

B   Ordinary electromagnetic wave

We consider the excitation of the ordinary wave propagating perpendicular to the ambient magnetic field. Since for this wave the perturbations of electric and magnetic field are parallel $E_0 \| B_0$, it affects the particle motions as if the magnetic field is absent. The linear dispersion is:

$$\omega_0^2 = k_0^2 c^2 + \omega_{pe}^2, \tag{6}$$

and the wave is linearly polarized. The non-zero components of electron velocity and magnetic field perturbations are

$$V_{0z} = \frac{e}{m_e} \frac{E_{0z}}{i\omega_0}, \qquad b_{0y} = -\frac{c}{\omega_0} k_{0x} E_{0z}.$$

The ordinary wave does not produce a density perturbation.

C   Kinetic Alfvén wave

In our model, the X- and O-modes couple because of the AWs in the corona. Low-frequency AWs can be excited at large scales at the base of corona by photospheric motions or higher up in the atmosphere by local reconnection events. During their propagation in the non-uniform corona, they develop short perpendicular length scales due to phase mixing. Since in the coronal plasma the effects of finite-ion-gyroradius prevail over the electron-inertia effects, we use the following linear dispersion equation for short-scale AWs:

$$\omega_2^2 = k_{2z}^2 V_A^2 \left(1 + k_{2x}^2 \rho_i^2 (1 + T_e/T_i)\right) \tag{7}$$

The polarization of these waves is almost linear. The electron velocity, density and magnetic field perturbations in the kinetic Alfvén wave are:

$$\mathbf{V}_2 = \hat{A}_2 \mathbf{E}_2,$$

$$\mathbf{B}_2 = \hat{b}_2 \mathbf{E}_2,$$

$$\frac{n_2}{n_0} = \frac{e}{m_e} \frac{i E_{02z}}{V_{Te}^2 k_{2z}},$$

where

$$\hat{A}_2 = \frac{e}{m_e} \begin{pmatrix} \frac{i\omega_2}{\omega_{Be}^2} & \frac{1}{\omega_{Be}} & -i\frac{k_{2x}}{k_{2z}}\frac{\omega_2}{\omega_{Be}^2} \\ -\frac{1}{\omega_{Be}} & 0 & \frac{k_{2x}}{k_{2z}}\frac{1}{\omega_{Be}} \\ -i\frac{k_{2x}}{k_{2z}}\frac{\omega_2}{\omega_{Be}^2} & -\frac{k_{2x}}{k_{2z}}\frac{1}{\omega_{Be}} & \frac{i\omega_2}{k_{2z}^2 V_{Te}^2} \end{pmatrix} ; \quad \hat{b}_2 = \begin{pmatrix} 0 & \frac{-ck_{2z}}{\omega_2} & 0 \\ \frac{ck_{2z}}{\omega_2} & 0 & \frac{-ck_{2x}}{\omega_2} \\ 0 & \frac{ck_{2x}}{\omega_2} & 0 \end{pmatrix}.$$

III EXCITATION OF THE O-MODE RADIATION BY THE FUSION OF X-MODE AND ALFVÉN WAVES

We expect that the ordinary waves are excited by the nonlinear interaction of finite-amplitude Alfvén and extraordinary waves. For the ordinary electromagnetic wave propagating along the X axis, the wave electric field is parallel to Z axis and obeys the equation

$$\frac{\partial^2 E_z(t,\mathbf{r})}{\partial t^2} - c^2 \frac{\partial^2}{\partial x^2} E_z(t,\mathbf{r}) + \omega_{pe}^2 E_z(t,\mathbf{r}) = -\frac{\omega_{pe}^2}{e} F_z(t,\mathbf{r}) + 4\pi e \frac{\partial (nV)_{NLz}}{\partial t}. \quad (8)$$

Here $F_z(t,\mathbf{r})$ and $e(nV)_{NLz}$ are the nonlinear force and current induced by the beatings of X-mode and AW.

In order to calculate the nonlinear right hand of equation (8) we decompose the electron velocity components, electron density perturbation, magnetic field perturbation related to KAW and extraordinary wave in a time-space Fourier integral

$$\left\{ \begin{array}{c} \mathbf{V}_{1,2}(t,\mathbf{r}) \\ n_2(t,\mathbf{r}) \\ \mathbf{B}_{1,2}(t,\mathbf{r}) \end{array} \right\} = \frac{1}{(2\pi)^{3/2}} \int_{-\infty}^{\infty} \left\{ \begin{array}{c} \mathbf{V}_{1,2}(\omega_{1,2},\mathbf{k}_{1,2}) \\ n_2(\omega_2,\mathbf{k}_2) \\ \mathbf{B}_{1,2}(\omega_{1,2},\mathbf{k}_{1,2}) \end{array} \right\} e^{i(\mathbf{k}_{1,2}\mathbf{r}-\omega_{1,2}t)} \, d\omega_{1,2} d\mathbf{r}_{1,2}. \quad (9)$$

We assume that the X-mode electric field is

$$\mathbf{E}_1(\mathbf{r},t) = E_{01} \exp\left[i(\mathbf{k}_1^w \vec{r} - \omega_1^w t)\right],$$

and AW electric field is

$$\mathbf{E}_2(\mathbf{r},t) = E_{02} \exp\left[i(\mathbf{k}_2^w \vec{r} - \omega_2^w t)\right],$$

where $E_{01} = E'_{01} \exp(i\delta_1)$ and $E_{02} = E'_{02} \exp(i\delta_2)$ are the complex amplitudes. Then their Fourier components are

$$\mathbf{V}_1(\omega_1,\mathbf{k}_1) = \hat{A}_1 \mathbf{E}_{01} \sqrt{2\pi} \delta(\mathbf{k}_1 - \mathbf{k}_1^w) \delta(\omega_1 - \omega_1^w)$$
$$\mathbf{V}_2(\omega_2,\mathbf{k}_2) = \hat{A}_2 \mathbf{E}_{02} \sqrt{2\pi} \delta(\mathbf{k}_2 - \mathbf{k}_2^w) \delta(\omega_2 - \omega_2^w) \quad (10)$$

$$\frac{n_2(\omega_2,\mathbf{k}_2)}{n_0} = \sqrt{2\pi} \frac{e}{m_e} \frac{iE_{02z}}{V_{Te}^2 k_{2z}} \delta(\mathbf{k}_2 - \mathbf{k}_2^w) \delta(\omega_2 - \omega_2^w) \quad (11)$$

$$\mathbf{B}_1(\omega_1,\mathbf{k}_1) = \hat{b}_1 \mathbf{E}_{01} \sqrt{2\pi} \delta(\mathbf{k}_1 - \mathbf{k}_1^w) \delta(\omega_1 - \omega_1^w)$$
$$\mathbf{B}_2(\omega_2,\mathbf{k}_2) = \hat{b}_2 \mathbf{E}_{02} \sqrt{2\pi} \delta(\mathbf{k}_2 - \mathbf{k}_2^w) \delta(\omega_2 - \omega_2^w) \quad (12)$$

Using the polarization of extraordinary and kinetic Alfvén waves and expression (10-12) we can rewrite equation (8) as:

$$\frac{\partial^2 E_z(t,\mathbf{r})}{\partial t^2} - c^2 \frac{\partial^2}{\partial x^2} E_z(t,\mathbf{r}) + \omega_{pe}^2 E_z(t,\mathbf{r}) = -\frac{\pi}{2} \frac{e}{m_e} \omega_{pe}^2 \times \tag{13}$$

$$\left[C_{11} e^{-i\left[\left(\omega_1^w + \omega_2^w \right) t - \left(\mathbf{k}_1^w + \mathbf{k}_2^w \right) \mathbf{r} + \delta \right]} + k.c \right],$$

where

$$\delta = \delta_1 + \delta_2,$$

$$C_{11} = i E_{01x} E_{02x} \times \left\{ \frac{k_{2z}^w}{\omega_2^w} \left(\frac{1}{\omega_1^w} + \frac{\omega_{Be}^2}{\omega_1^{w3}} \frac{\omega_{pe}^2}{\omega_1^{w2} - \omega_{pe}^2 - c^2 k_1^{w2}} \right) \left(1 + \frac{\mu_s}{1 + \mu_i} \right) - \tag{14}$$

$$\frac{\mu_s}{1 + \mu_i} \frac{\omega_1^w + \omega_2^w}{\omega_1^w} \frac{1}{V_{Te}^2 k_{2x}^w} \frac{c^2 k_{1x}^w k_{1z}^w}{\omega_1^{w2} - \omega_{pe}^2 - c^2 k_1^{w2}} \right\}.$$

We take the initial conditions for the excited wave as:

$$E_z(\mathbf{r}, 0) = 0; \quad E_{zt}(\mathbf{r}, 0) = 0. \tag{15}$$

We use the integral Fourier transformation with respect to x coordinate: we multiply the left and right hand sides of equation (13) with $\frac{1}{(2\pi)^{3/2}} e^{-i\mathbf{kr}}$ and integrate from $\int_{-\infty}^{\infty}$ on $d\mathbf{r}$. In this way we obtain the temporal equation for the spatial Fourier component of the electric field of ordinary wave :

$$\frac{\partial^2 E_z(t,\mathbf{k})}{\partial t^2} + E_z(t,\mathbf{k}) \left(\omega_{pe}^2 + c^2 k_x^2 \right) = A \sin \left[\left(\omega_1^w + \omega_2^w \right) t + \delta \right]$$

where

$$A = -\pi \frac{e}{m_e} \omega_{pe}^2 \delta(\mathbf{k} - (\mathbf{k}_1^w + \mathbf{k}_2^w)) C_{11}.$$

As a result, we obtain a simple equation for a harmonic oscillation with a periodic force. Its solution is:

$$E_z(t,\mathbf{k}) = -\frac{A}{\omega_{pe}^2 + c^2 k_x^2 - (\omega_1^w + \omega_2^w)^2} * \{ \sin \delta \cos \sqrt{\omega_{pe}^2 + c^2 k_x^2} t +$$

$$\frac{(\omega_1^w + \omega_2^w)}{\sqrt{\omega_{pe}^2 + c^2 k_x^2}} \cos \delta \sin \sqrt{\omega_{pe}^2 + c^2 k_x^2} t - \sin(\omega_1^w + \omega_2^w) t + \delta \} \tag{16}$$

Strong growth of the excited O-mode occurs when its frequency coincides with the sum of the frequencies of the KAW and X-mode $\sqrt{\omega_{pe}^2 + c^2 k_x^2} = \omega_1^w + \omega_2^w$:

$$E_z(t,\mathbf{k}) = A \left[t \cos(\sqrt{\omega_{pe}^2 + c^2 k_x^2} t + \delta) - \frac{\cos \delta}{\sqrt{\omega_{pe}^2 + c^2 k_x^2}} \sin \sqrt{\omega_{pe}^2 + c^2 k_x^2} t \right]$$

Using the inverse Fourier transform, we obtain the following solution for equation (13) for the resonant O-mode:

$$E_z(t, \mathbf{r}) = \frac{e}{m_e} \omega_{pe}^2 \frac{1}{8\sqrt{2\pi}} C_{11} * e^{i(\mathbf{k}_1^w + \mathbf{k}_2^w)\mathbf{r}} \{ t \cos(\sqrt{\omega_{pe}^2 + c^2(k_{1x}^{w2} + k_{2x}^{w2})} t + \delta) -$$

$$\frac{\cos\delta}{\sqrt{\omega_{pe}^2 + c^2(k_{1x}^{w2} + k_{2x}^{w2})}} \sin\sqrt{\omega_{pe}^2 + c^2(k_{1x}^{w2} + k_{2x}^{w2})} t \}, \tag{17}$$

The amplitude of the ordinary electromagnetic wave grows linearly in time.

In the non-resonant case $\sqrt{\omega_{pe}^2 + c^2 k_x^2} \neq \omega_1^w + \omega_2^w$ we simply have a beating of KAW and X-modes.

IV APPLICATION

In the previous section we obtained the expression for the dependence of the electric field of the excited ordinary electromagnetic wave from the electric field of the extraordinary and kinetic Alfvén waves. We focus on the resonant case, when the KAW and the extraordinary wave excite resonantly the ordinary wave.

Let us estimate the time of effective coupling of our waves in coronal plasma. Spectroscopic observations suggest the presence of Alfvén waves with magnetic field perturbations $B/B_0 \gtrsim 0.01$ in coronal loops [9]. Since these perturbations consist very probably of a spectrum of waves, we use $B/B_0 = 0.001$ for the effective field participating in the excitation of the particular O-mode. The corresponding electric field of the KAW is $E_x \approx 2.2$ V m^{-1}. The typical parameters of loops are: length $L = (2 - 5) \cdot 10^9$ cm, temperature $T \approx (2 - 4) \cdot 10^6$ K, magnetic field strength $B \approx 100 - 500$ G, number density $n \approx 0.5 \div 1 \cdot 10^{10}$ cm^{-3}. The characteristic scale of the inhomogeneity of the plasma across the loops is $L_\perp = (2 - 5) \cdot 10^7$ cm.

We consider the case when the X-mode with amplitude E_{1x} propagates across the magnetic loops. In the region where the matching condition $\omega_0 = \omega_1 + \omega_2$ is satisfied, the amplitude of the excited ordinary mode grows. From this resonant condition we estimate, that for coupling of the O- and X-modes at meter wavelengths, kinetic Alfven waves with $k_{2x} \approx 0.2 * 10^{-5}$ cm^{-1} are required in coronal loops.

The relative O-mode amplitude depends on time as

$$\frac{E_0}{E_{01x}} = \frac{1}{8\sqrt{2\pi}} \frac{e}{m_e} E_{02x} \frac{\omega_{pe}^2}{\omega_0^2} \frac{k_{2z}^w}{\omega_2^w \omega_1^w} \left[1 + \frac{t\mu_i}{1 + \mu_i} \frac{c^2}{V_{Te}^2} \frac{k_{1x}^w}{k_{2x}^w} \frac{\omega_0^w \omega_2^w}{\omega_1^{w2} - \omega_{pe}^2 - c^2 k_1^{w2}} \right]$$

$$\left[\tau \cos(\tau + \delta) - \frac{1}{2} [\sin(\tau - \delta) + \sin(\tau + \delta)] \right],$$

where the dimensionless time is $\tau = t/\omega_0$.

Investigation of this equation shows that the amplitude of the O-mode grows up to the amplitude of the X-mode after $t = 10^{-4}$ s. This corresponds to a characteristic distance for the effective interaction $L_{\text{int}} \approx 3 \cdot 10^6$ cm. Hence the effective interaction occur at a distance that is much shorter than the characteristic length scale of the nonuniformity across a loop, $L_{\text{int}} \ll L_\perp$.

It is interesting to note that here the efficiency of the excitation of the O-mode does not depend on the X-mode amplitude.

V   CONCLUSION

We considered the nonlinear excitation of O-mode radiation by the coupling of X-mode radiation and Alfvén waves. The equation governing the time dependence of the electric field of the excited O-mode is obtained on the basis of two-fluid magnetohydrodynamics. The solution (17) of this equation is valid up to the moment when the amplitude of the ordinary wave becomes equal to the amplitude of weakest pump wave; in our case this is the X-mode.

We applied our results to the coupling of X-mode and AW in coronal loops. Using values typical for coronal loops, we estimate the time of effective coupling and corresponding interaction distance.

As we considered the particular conversion of the X-mode into the O-mode ("moderate coupling"), our results can be used for the interpretation of depolarization of solar radio emission. As the X-mode converted into O-mode ("moderate coupling") this mechanism results in a depolarization of solar radio emission that is originally in the X-mode.

It is commonly accepted that the type II and III bursts are originally in O-mode. In this case depolarization happens via inverse coupling mechanism: O-mode + KAW$\Longrightarrow$X-mode. We suggest that this process is as efficient as the one that we considered here.

REFERENCES

1. Cohen, M.H., *ApJ*, **131**, 664-680 (1960).
2. Zheleznyakov, V. V., *Electromagnetic waves in space plasma*, Nayka, Moscow 1977(in Russian).
3. Bandiera, R., *Astron. Astrophys.* **112**, 52-60 (1982).
4. Melrose, D.B., *Solar Physics*, **119**, 143-156 (1989).
5. Bastian, T.S., *ApJ* **439**, 494-498 (1995).
6. Zheleznyakov, V. V., Kocharovsky, V. V., and Kocharovsky VI.V, *Astron.Astrophys*, **308**, 685-696 (1996).
7. Voitenko, Yu.M., *J. Plasma Physics*, **60**, 497-514 (1998).
8. Yukhimuk, A.K., Yukhimuk, V.A., Sirenko, O.K. and Voitenko, Yu.M., *J. Plasma Physics*, **62**, 53-64, (1999).
9. Hara, H. and Ichimoto, K., *ApJ*, **513**, 969-982 (1999).

Effects of physical parameters of solar structure on frequencies of helio-seismic modes

Petra Vanlommel* and Vladimir M. Čadež[†]

*Centre for Plasma Astrophysics, K.U.Leuven, Celestijnenlaan 200B, B-3001 Heverlee, Belgium
[†] Royal Observatory of Belgium and Belgian Institute for Space Aeronomy,
Ringlaan 3, B-1180 Brussels, Belgium

Abstract. We study the sensitivity of frequency eigen-spectra of solar global modes on physical parameters describing the applied model. Starting from a simple configuration that includes only two distinct regions, the solar interior and the solar atmosphere in the Cartesian geometry, we gradually add new features that improve the initial model and cause changes in eigen-frequencies. These frequency variations are computed numerically and plotted as functions of free physical parameters of a particular model which then gives us a better insight into how various elements of the model influence the eigen-spectra of solar $p-$, $f-$ and $g-$modes.

I INTRODUCTION

Many observational and theoretical studies of solar seismic global modes have been carried out since the sixties till the current SOHO space mission projects. These modes are detected by observing Doppler shifts of fluid velocities at the photospheric level and their modal frequencies are very accurately determined with relative errors as low as 10^{-5} [1,2]. For this reason, the seismic waves represent a solid tool for probing the solar interior: Its structural and dynamical properties. The high accuracy observations of eigen-frequencies, namely, requires a similar precision of knowledge of all other physical parameters that are relevant to a particular model chosen to describe the internal structure of the sun. Moreover, the presence of the structured solar atmosphere filled with magnetic fields may also noticeably contribute to values of global eigen-frequencies. It is therefore of basic importance to study and find out the sensitivity of frequency eigen-spectra on variations of each of the parameters, i.e. on the relative errors of their either known or estimated values, used in a chosen model.

In what follows, we consider an eigen-value problem for linear waves in composed models of solar interior topped by a structured atmosphere and analyze variations of computed eigen-frequencies with parameters such as the thickness of the chro-

CP537, *Waves in Dusty, Solar, and Space Plasmas*, edited by F. Verheest, et al.

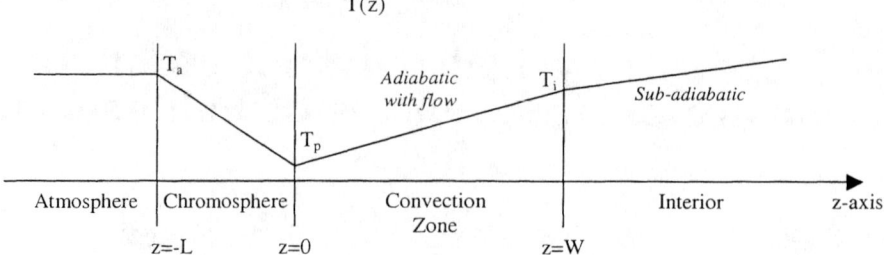

$$T(z)$$

	T_a		Adiabatic with flow	T_i	Sub-adiabatic
		T_p			
Atmosphere	Chromosphere		Convection Zone		Interior	z-axis
	z=-L	z=0		z=W	

FIGURE 1. The considered model of the unperturbed interior and atmosphere of the sun.

mosphere and the convection zone (parameters L and W respectively in Figure 1), temperature profiles in the atmospheric layers as well as in the solar interior below the convection layer (parameters τ and $\mathcal{A}$ respectively in Figure 1) and the speed profile of convective motions (parameters c and p in Eq (3)). Effects of magnetic field are left out from this paper. They are considered in detail in [3].

The modes we obtain are combinations of acoustic $p-$ and $f-$modes, together with gravity $g-$modes. The computed eigen-frequencies are generally in agreement with observations but their precise values strongly depend on values of parameters considered in the model. The strength of the dependence varies among the parameters and it turns out that the temperature gradient in the interior below the convection layer has the most pronounced influence on eigen-frequencies.

II   THE BASIC STATE AND EQUATIONS

Five-minute oscillations observed at the surface of the Sun are commonly described in terms of spherical harmonics $Y_l^n(\theta, \phi)$ of the co-latitude θ and the longitude ϕ, of degree l and order n. This description is most appropriate for global modes with low l when effects of spherical geometry are important.

In this paper we use the plane wave approximation, the modes are considered to be waves with sufficiently large values of l, which are not affected by the curved shape of the sun. Degree l is then related to the total horizontal wave number k on the surface of the sun by the expression $k = \sqrt{l(l+1)}/R_\odot$ where $R_\odot = 6.96\,10^2$ Mm is the solar radius. As there is no preferential horizontal direction, we treat the eigen-value problem in two dimensions, i.e. in the (x, z)-plane with the wavenumber k taken along the horizontal x-axis.

The initial unperturbed basic state is assumed in hydrostatic equilibrium implying the relation

$$\frac{d}{dz}\ln \rho_0 + \frac{d}{dz}\ln T_0 = \frac{g}{\mathcal{R}_* T_0} \tag{1}$$

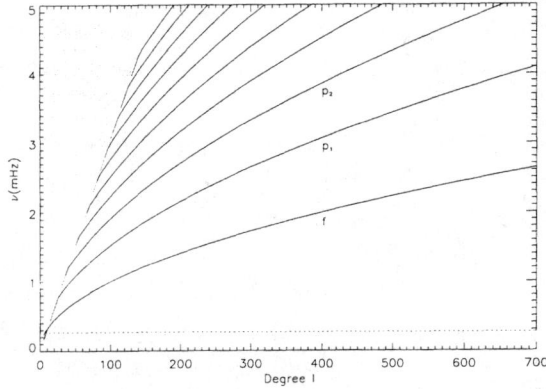

FIGURE 2. Dispersion curves showing the f-mode and acoustic p-modes in a stationary $(c = 0)$ reference model with $\tau = 400$; $L = 0$, $W = 0$; $\mathcal{A} = 1$.

where the vertical $z-$axis is oriented toward the interior of the sun, $\mathcal{R}_* = \mathcal{R}/\bar{\mu}$ is the individual gas constant, $\bar{\mu} = 1.3$ for the solar gas mixture, $\mathcal{R} = 8317$ J kg^{-1} is the universal gas constant. In the interior, we assume a temperature gradient which is not bigger than the adiabatic gradient $(dT/dz)_{ad}$:

$$\frac{dT}{dz} = \mathcal{A} \left(\frac{dT}{dz} \right)_{ad}.$$

The temperature profile $T_0(z)$ is schematically shown in Figure 1 where parameters τ and $\mathcal{A}$ are chosen according to the model.

The general model showing the structure of the atmosphere and the internal stratification of the sun is given in Figure 1. Such a basic state is disturbed by linear perturbations that are harmonic in time t and horizontal coordinate x while their $z-$dependence has to be calculated from the standard set of linearized equations of fluid dynamics. These equations can be reduced to a system of two coupled ordinary differential equations

$$D(z)\frac{d\xi_z}{dz} = C_1(z)\xi_z - C_2(z)P, \qquad D(z)\frac{dP}{dz} = C_3(z)\xi_z - C_1(z)P \qquad (2)$$

where ξ_z is the $z-$component of the Lagrangian displacement while P is the pressure perturbation.

The coefficients in Eq (2) are

$$D(z) = \rho_0(z)v_s^2(z)\Omega^2(z), \quad C_1(z) = \rho_0(z)g\Omega^2(z), \quad C_2(z) = \Omega^2(z) - \omega_s^2(z),$$

$$C_3(z) = \rho_0^2(z)\Omega^2(z)\left[v_s^2(z)\Omega^2(z) + gv_s^2(z)\frac{d}{dz}\log\rho_0(z) + g^2 \right].$$

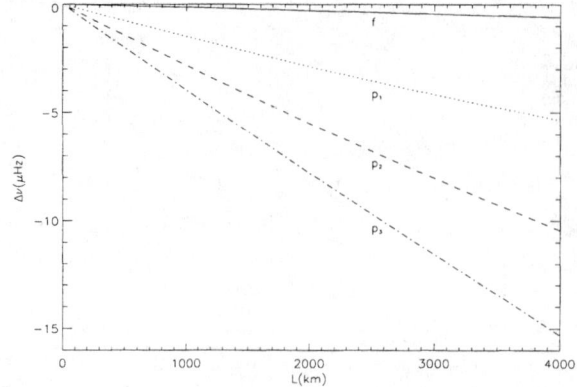

FIGURE 3. Shifts of eigen-frequencies $\Delta\nu = \nu(L) - \nu(0)$ of indicated modes at $l = 350$ in models that include the chromosphere of thickness $L \neq 0$. The remaining parameters are $\tau = 400$, $W = 0$ (no flow) and $\mathcal{A} = 1$.

The density distribution $\rho_0(z)$ is obtained analytically from Equation (1) with the profile $T_0(z)$ depending on the model according to Figure 1. The Doppler shifted wave frequency Ω and ω_s the sound frequency are defined as

$$\Omega \equiv \omega - kU_0(z), \quad \omega_s^2 \equiv v_s^2 k^2$$

respectively, where ω is the wave frequency and $U_0(z)$ is the profile of the horizontal flow in the convection zone.

We model the convective motion by

$$U_0(z) = cU_0 \left[\cos\frac{\pi z}{W} \right]^{2p+1}, \qquad \text{where:} \qquad p = 0, 1, 2, \ldots \tag{3}$$

The speed of the flow $U_0(z)$ thus changes its sign at $z = W/2$ and has maximal absolute values at $z = 0$ and $z = W$. Such a motion resembles a pattern of a convective cell of elongated extend in the horizontal direction. The integer parameter p in Equation (3) defines the concentration of the flow close to the boundaries of the cell while the parameter c is used for scaling the typical value for maximal flow speed of $U_0 = 2000$ km/s.

III MODELS AND RESULTS

Equations (1)-(3) are solved either analytically or numerically in each region of the model and the obtained solutions are matched through the boundary conditions. For details, see [4,5].

Requiring the solutions to be localized in the domain of finite $|z|$, we obtain the dispersion relation, i.e. the eigen-spectra $\omega = \omega(k)$ of global modes for a particular model.

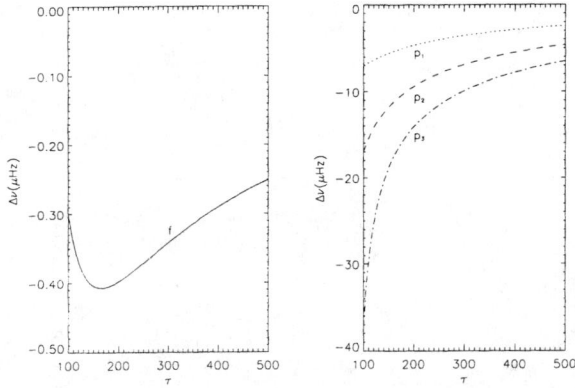

FIGURE 4. Frequency shifts in models with the chromosphere of thickness $L = 2000$ km for different values of the temperature ratio τ: $\Delta \nu = \nu(\tau) - \nu(0)$. The remaining parameters are $l = 350$, $\tau = 400$, $W = 0$ (no flow) and $\mathcal{A} = 1$.

A basic example of such a spectrum is given in Figure 2 showing frequencies $\nu(l) \equiv 2\pi/\omega$ for the f-mode and acoustic p-modes in a model with $\tau = 400$, $L = 0$, $W = 0$ and $\mathcal{A} = 1$. This is a simple case in which the sun is composed of two regions only: A stationary interior with a constant adiabatic temperature gradient and an isothermal atmosphere above the photospheric level $z = 0$ in which the temperature is τ times higher than at the photospheric level.

If a chromosphere, a layer of thickness L through which the temperature grows linearly, is added to the initial model, the eigen-frequencies change depending on L. As an example, Figure 3 shows such a dependence for waves with the degree $l = 350$. We see that the frequency of the f-mode changes by less than 1μHz at $L = 4000$ km while the p_n modes experience much larger frequency shifts depending on the order n. For example, the frequency of the p_2 mode which is about 3.5 mHz if $L = 0$, changes by 0.01 mHz or by some 0.3% of its value if $L = 4000$ km it taken.

Figure 4 shows the frequency shifts if the temperature ratio τ varies between 100 and 500 taking the chromosphere of the thickness $L = 2000$ km. The sensitivity of p mode frequencies to the variation of τ grows with the order of the mode and decreases with τ.

Models in which the temperature gradient in the interior of the sun takes different subadiabatic values yield frequency shifts as shown in Figure 5. The constant temperature gradients can range from zero ($\mathcal{A} = 0$) to the adiabatic gradient ($\mathcal{A} = 1$).

To model the convection zone, we add a new layer of the thickness $W = w\,R_\odot$ ($R_\odot$ is the solar radius) underneath the photosphere, with an adiabatic temperature gradient in it. The temperature gradient below the convection zone is assumed subadiabatic, with $\mathcal{A} = 0.8$. The frequency shifts due to variation of w are shown

299

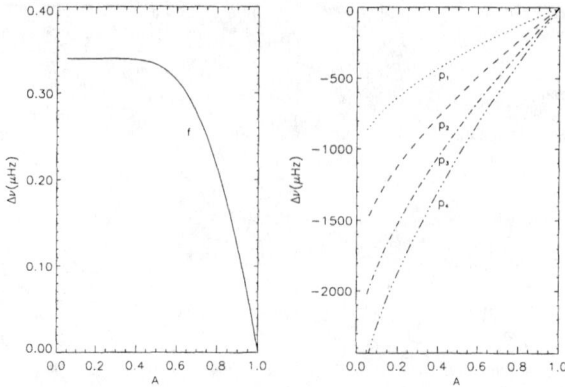

FIGURE 5. Frequency shifts caused by the variation of the temperature gradient in the interior of the sun: $\Delta\nu = \nu(\mathcal{A}) - \nu(1)$. The remaining parameters are $l = 350$, $L = 0$, $\tau = 400$ and $W = 0$ (no flow).

in Figure 6 The fluid is still assumed static i.e. in absence macroscopic motions. If now a horizontal flow given by Equation (3) is added to the convection layer, each of the three parameters c and p of the flow profile influences eigen-frequencies of the modes resulting in frequency shifts as shown in Figures 7-8.

We first consider flow profiles which differ in their intensity c only while the shape of the profile remains unchanged with $p = 0$. Figure 7 shows the related frequency shifts in comparison with the static case: $\Delta\nu = \Delta\nu(c) - \Delta\nu(0)$.

Finally, we fix the intensity of the flow to $c = 1$ and study consequences of varying the parameter p. The differences of the modal frequency for waves with $l = 350$ computed by taking $p = 0$ and an arbitrary integer value for p respectively, are plotted as continuous functions of p in Fig 8. We see that frequency shifts $\Delta\nu \equiv \nu(p) - \nu(0)$ tend to constant values at larger values of p. In this case, namely, the flow is already sufficiently localized to the region $z/W \approx 0$ and $z/W \approx 1$ and becomes much less sensitive to further increase of p.

IV   CONCLUSIONS

In this work we analyzed eigen-frequencies of solar global modes for various non-magnetic models of solar structure. We have shown that eigen-frequencies experience shifts if the characteristic parameters of the model are changing and that the relative shifts $\Delta\nu/\nu$ can easily be of the order of 10^{-2} for typical intervals of parameters applied in our calculations. These shifts are far above the observational accuracy that reaches 10^{-5}. So, a model that could explain eigen-frequencies with a high accuracy has to be based on values of physical parameters, such as the thickness of the convection layer, the chromosphere, temperature gradients, flow

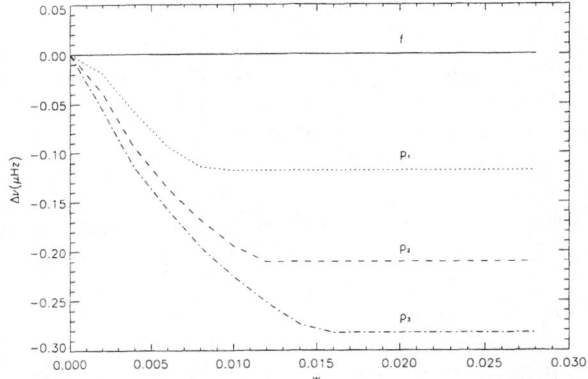

FIGURE 6. Frequency shifts caused by the variation of the dimensionless thickness $w \equiv W/R_\odot$ of a static convection zone: $\Delta\nu = \nu(w) - \nu(0)$. The remaining parameters are $l = 350$, $L = 0$, $\tau = 400$, $c = 0$ (no flow) and $\mathcal{A} = 0.8$.

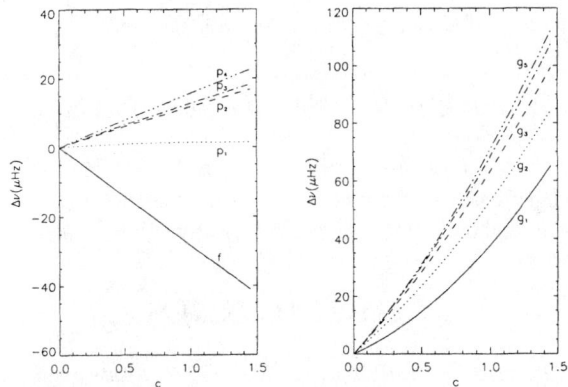

FIGURE 7. Frequency shifts $\Delta\nu = \nu(c) - \nu(0)$ arising from convection motions. The varying parameter of the flow is c while the other parameters of the model are $p = 0$, $\tau = 400$, $L = 2000$ km, $w = 0.01$ and $\mathcal{A} = 0.8$.

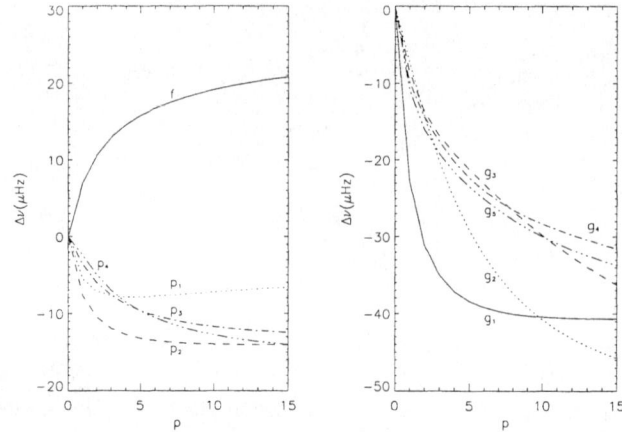

FIGURE 8. Frequency shifts $\Delta\nu = \nu(p) - \nu(0)$ arising from convection motions. The varying parameter of the flow is p while the other parameters of the model are $c = 1$, $\tau = 400$, $L = 0$, $w = 0.01$ and $\mathcal{A} = 0.8$.

profiles, etc, that should be determined with high precision. Currently, this cannot be achieved experimentally.

Our attention was to point out how each of the parameters entering our models influences frequencies of global modes in order to compose models that would explain characteristics of observed helio-seismic modes.

ACKNOWLEDGMENTS

V.M. Čadež acknowledges the support by the ESA PRODEX (Ulysses) and the Belgian Federal Services for Scientific, Technical and Cultural Affairs during his stay at the Belgian Institute for Space Aeronomy.

REFERENCES

1. Christensen-Dalsgaard, J., and Däppen, W, *Astron. Astrophys.*, **4**, 267 (1992).
2. Tomczyk, S., Schou, J., and Thompson, M.J., *Astrophys. J.*, **448** L57 (1995).
3. Vanlommel, P., and Goossens, M., *Solar Phys.*, **187**, 357 (1999).
4. Vanlommel, P., and Čadež, V. M., *Solar Phys.*, **182**, 263 (1998).
5. Vanlommel, P., and Čadež, V. M., *Solar Phys.*, submitted (2000).

Note on dispersive Alfven waves in the solar corona

Yu. Voitenko[*][†], M. Goossens[†], P. Shukla[¶] and L. Stenflo[‡]

[*] Main Astronomical Observatory, Holosiiv, Kyiv, 03680, Ukraine
[†] Centre for Plasma Astrophysics , K.U.Leuven, Celestijnenlaan 200B, B-3001 Heverlee, Belgium
[¶] Institut für Theoretische Physik-IV, Fakultät für Physik & Astronomie, Ruhr-Universität Bochum, D-44780 Bochum, Germany
[‡] Department of Plasma Physics, UmeàUniversity, S − 90187Umeà, Sweden

Abstract. The dispersive (kinetic) Alfvén waves in the ion-cyclotron frequency regime (ICKAWs), with frequencies close to the ion cyclotron frequency and with wavelengths close to the ion gyroradius, are studied in view of their ability to interact strongly with the plasma of solar corona. The combined action of finite ion cyclotron frequency and finite ion gyroradius effects has a profound influence on the Alfvén wave. The phase velocity of ICKAWs along the background magnetic field can be slower than the Alfvén velocity (due to finite ion cyclotron frequency effects), but also faster than the Alfvén velocity (due to finite ion gyroradius effects). The phase velocity depends on both the perpendicular and parallel wavenumbers, and spans a range of velocities from sub-Alfvénic to super-Alfvénic. Since these waves have a field-aligned component of electric field, they undergo Cherenkov resonant interaction and can be excited by the ion beam component set up in the low corona by the reconnection outflows from chromospheric reconnection events.

I INTRODUCTION

Kinetic and dispersive properties of short-scale (kinetic) Alfvén waves (KAWs) are of great interest [1]. Contrary to the ideal MHD Alfvén wave, the KAW can effectively interact with plasma via linear Cherenkov resonance, and with other waves via three-wave resonant interaction. As a result, KAWs can be excited and dissipated linearly and nonlinearly [2-5].

The commong approach so far was to study KAWs in the limit of weak parallel dispersion, $\omega/\Omega \ll 1$ (and often in the limit of weak perpendicular dispersion, $k_\perp \rho_i \ll 1$, also). However, KAWs with $k_\perp \rho_i \sim 1$ and $\omega/\Omega \lesssim 1$ may be excited in the corona by reconnection outflows [6]. In the case $\omega/\Omega \sim 1$ the wave dispersion can be different from that of low-frequency KAWs, and cyclotron damping may become important. The cyclotron absorption of parallel-propagating AWs excited by reconnection events in the solar corona has been proposed as a heating mechanism

CP537, *Waves in Dusty, Solar, and Space Plasmas*, edited by F. Verheest, et al.
© 2000 American Institute of Physics 1-56396-962-9/00/$17.00

[7].

In the present paper, kinetic theory is used to derive the linear equation that governs the excitation and damping of KAWs, taking into account the ion Larmor radius and finite ion cyclotron frequency effects. The linear response of the plasma is calculated for arbitrary values of the kinetic variable $\mu = k_\perp \rho_i$, and frequency ω/Ω. An important limiting case of KAWs in the ion-cyclotron frequency regime is considered and the results are applied to the problem of the excitation of KAWs by reconnection events in low corona.

II   DISPERSION RELATION

Let us consider a fully ionized uniform plasma in a background magnetic field B_0. We restrict our analysis to a plasma β in the range $m_e/m_i < \beta < 1$, typical for many laboratory and space plasmas (m_e/m_i is the electron/ion mass ratio, β is the plasma/magnetic pressure ratio). To study magnetically non-compressible ($B_z = 0$) AWs, we use two fluctuating electromagnetic wave potentials, scalar ϕ and (z-component) vector A_z [8]. These potentials induce in the j-th plasma component charge and current density perturbations q_j and j_j, obeying the quasi-neutrality equation

$$\sum_j q_j = \sum_j e_j n_j = 0;$$ (1)

and Ampere's law in the z direction:

$$\sum_j j_j = -\frac{c}{4\pi} \nabla^2 \mathbf{A_z}.$$ (2)

The difference between MHD and kinetic theories here is in the calculation of number density and current density perturbations. In order to take into account important kinetic properties of KAWs, we calculate n_j and j_j by use of the perturbed part f_j of the corresponding velocity distribution function ($F_j = F_j^0 + f_j$):

$$j = e \int d^3 \mathbf{V} V_z \mathbf{f};$$

$$q = e \int d^3 \mathbf{V} \mathbf{f}.$$ (3)

From here on we drop the subscript j for labelling the plasma specie unless it is really necessary to avoid misunderstanding. This has already been done in ([?]). The Fourier amplitude of the perturbed distribution function is [6]:

$$\int_0^{2\pi} d\vartheta f_k^L = 2\pi \frac{e}{m} \sum_n \frac{J_n^2(\mu\zeta)}{-\omega_k + k_z V_z + n\Omega} \times$$

$$\left[k_z(\phi_k - \frac{\omega_k}{k_z c} A_k) \frac{\partial}{\partial V_z} + n\Omega \left((\phi_k - \frac{V_z}{c} A_k) \frac{\partial}{\partial V_\perp \partial V_\perp} + \frac{1}{c} A_k \frac{\partial}{\partial V_z} \right) \right] F^0,$$ (4)

where $\zeta = V_\perp/V_T$, $\mu = k_\perp\rho = k_\perp V_T/\Omega$, that is $k_\perp V_\perp/\Omega = \mu\zeta$, and $J_n^2(\mu\zeta)$ is the Bessel function.

We use a factorized distribution function

$$F^0 = n_0 F_\parallel (V_z) F_\perp (V_\perp) \tag{5}$$

with a Maxwellian distribution in the perpendicular direction,

$$F_\perp (V_\perp) = \frac{1}{2\pi V_{T\perp}^2} \exp\left(-\frac{V_\perp^2}{2V_{T\perp}^2}\right). \tag{6}$$

The distribution function along the magnetic field $F_\parallel^0 (V_z)$ is a Maxwellian (with different, parallel temperature) for the main ion (proton) component ($V_0 = 0$), and a shifted Maxwellian distribution for the additional ion components:

$$F_\parallel^0 (V_z) = \frac{1}{\sqrt{2\pi} V_{T\parallel}} \exp\left(-\frac{(V_z - V_0)^2}{2V_{T\parallel}^2}\right). \tag{7}$$

Using this distribution function in the expression for the perturbed distribution function, we calculate current and charge perturbations. Then we solve the system of two equations: quasineutrality and Ampere's law. The mathematical analysis is given in another paper. The outcome of the analysis is the dispersion relation for low-frequency shear (magnetically incompressible) electromagnetic waves:

$$\left[1 + \sum_j \frac{n_j T_e}{n_0 T_j} \left(1 + \sum_n \Lambda_n (\mu_j^2) \sigma_{j0} Re Z_{jn}\right)(1 + i\delta_{ie})^{-1}\right] \frac{k^2 V_{Te}^2}{\omega_{pe}^2} +$$

$$-\sum_j \frac{n_j T_e}{n_0 T_j} \sum_n \Lambda_n (\mu_j^2) n\frac{\Omega_j}{\omega}\sigma_{j0} Z_{jn} \frac{V_k^2}{c^2} = 0. \tag{8}$$

Here

$$Z_n \equiv Z(\sigma_n) = \frac{1}{\sqrt{\pi}} \int_{-\infty}^{\infty} dy \left(\frac{1}{y - \sigma_n}\right) \exp\left(-y^2\right)$$

is the plasma dispersion function, and

$$\sigma_n = \frac{\omega_k - k_z V_0 - n\Omega}{\sqrt{2}k_z V_{T\parallel}}.$$

The expansions of Z_n are

$$Z(\sigma_n) = -2\sigma_n + \frac{4}{3}\sigma_n^3 - \ldots + i\sqrt{\pi}\exp\left(-\sigma_n^2\right)$$

for $\sigma_n^2 \ll 1$, and

$$Z(\sigma_n) = -\frac{1}{\sigma_n} - \frac{1}{2\sigma_n^3} - \frac{3}{4\sigma_n^5} - \ldots + i\sqrt{\pi}\exp\left(-\sigma_n^2\right).$$

305

for $\sigma_n^2 \gg 1$.

For weakly damped waves all $\sigma_{jn} > 1$ (except for $\sigma_{e0} < 1$ used before), and in a Maxwellian plasma with admixtures of minor species we have

$$\frac{V_k^2}{V_A^2} = \left[\left(1 - \Lambda_0 - \Lambda_1 \frac{2\omega_k^2}{\omega_k^2 - \Omega^2} \right)^{-1} + \frac{T_e}{T_p} \right] \mu^2 - i\frac{T_e}{T_j}\mu^2\delta_{ie} - i\frac{V_k^2}{V_A^2}\delta_1, \tag{9}$$

where the imaginary terms are:

$$\delta_{ie} = \sum_j \frac{n_j}{n_0}\frac{T_e}{T_j} \sum_n \Lambda_n\left(\mu_j^2\right) \sigma_{j0} Im Z_{jn};$$

$$\delta_1 = \sum_j \frac{n_j}{n_0}\frac{T_e}{T_j} \sum_n \Lambda_n\left(\mu_j^2\right) \frac{n\Omega_j}{\omega}\sigma_{j0} Im Z_{jn}.$$

III ANALYSIS

The real part of (9) gives the dispersion relation that includes both the finite ion Larmor radius corrections and the finite frequency corrections:

$$\frac{V_k^2}{V_A^2} = \left[\left(1 - \Lambda_0 - \Lambda_1 \frac{2\omega_k^2}{\omega_k^2 - \Omega^2} \right)^{-1} + \frac{T_e}{T_j} \right] \mu_j^2. \tag{10}$$

It is obvious from (10) that the effects of finite frequency (giving rise to parallel dispersion) become competitive with respect to the effects of finite ion gyroradius (giving rise to perpendicular dispersion) when the condition

$$\frac{1 - \Lambda_0}{2\Lambda_1} \lesssim \frac{\omega_k^2}{\omega_k^2 - \Omega^2} \tag{11}$$

is satisfied. This condition is not very restrictive. In a wide range of the kinetic parameter, $0 < \mu_j^2 \lesssim 1$, it is simply

$$\frac{\omega_k^2 - \Omega^2}{\omega_k^2} \lesssim 1. \tag{12}$$

As soon as the wave frequency approaches the ion cyclotron frequency, we have strong parallel dispersion:

$$\frac{V_k^2}{V_A^2} = \frac{\mu_j^2}{2\Lambda_1}\frac{\Omega^2 - \omega_k^2}{\omega_k^2} + \frac{T_e}{T_j}\mu_j^2. \tag{13}$$

In the opposite, low-frequency limit, $\omega/\Omega \ll 1$, the dispersion relation reduces to the usual KAW dispersion relation:

$$\frac{V_k^2}{V_A^2} = \left[\frac{1}{1 - \Lambda_0} + \frac{T_e}{T_j}\right] \mu^2. \tag{14}$$

In the limit of weak perpendicular dispersion, $\mu_j^2 < 1$, and for arbitrary frequency, we get

$$\frac{V_k^2}{V_A^2} = \frac{\Omega^2 - \omega_k^2}{\Omega^2} + \mu_j^2 \left[\frac{T_e}{T_j} + \frac{\Omega^2 - \omega_k^2}{\Omega^2}\left(\frac{3}{4}\frac{\Omega^2 - \omega_k^2}{\Omega^2} + \frac{\omega_k^2}{\Omega^2}\right)\right]. \tag{15}$$

In the low-frequency limit we obtain

$$\frac{V_k^2}{V_A^2} = 1 + \left(\frac{3}{4} + \frac{T_e}{T_j}\right)\mu_j^2. \tag{16}$$

In the ion-cyclotron frequency band we find

$$\frac{V_k^2}{V_A^2} = \frac{\Omega^2 - \omega_k^2}{\Omega^2} + \frac{T_e}{T_j}\mu_j^2. \tag{17}$$

From these expressions we see that the phase velocity of KAWs with $\mu_j^2 < 1$ can vary from super-Alfvénic at low frequencies to sub-Alfvénic in the vicinity of the ion-cyclotron frequency. The above consideration shows that the parallel dispersion is already as important as the perpendicular dispersion when the wave frequency is about half of the cyclotron frequency. The conditions for weak cyclotron damping, weak ion Landau damping, and strong ion cyclotron effects are compatible:

$$1 \ll \frac{\omega_k^2 - \Omega^2}{2k_z^2 V_T^2} = \frac{\omega_k^2 - \Omega^2}{\omega_k^2}\frac{\omega_k^2}{2k_z^2 V_T^2} \lesssim \frac{\omega_k^2}{2k_z^2 V_T^2}. \tag{18}$$

Allowing for an imaginary part of the wave frequency, $\omega \to \omega + i\gamma$, we find that the amplitude of KAW can change in time due to wave-particle interaction. The rate of this change is

$$\frac{\gamma}{\omega} = -i\frac{1}{2}\frac{T_e}{T_j}\frac{V_A^2}{V_k^2}\mu^2 \delta_{ie} - i\frac{1}{2}\delta_1. \tag{19}$$

The term δ_{ie} represents the ion and the electron Landau damping, as well as cyclotron damping, whereas δ_1 represents cyclotron damping only.

We can formally obtain the electrostatic limit by taking the limit $k_z^2\lambda_i^2 \to \infty$, then ICKAW should become a purely electrostatic acoustic IC wave. If we then drop in (10) the term originating from the electromagnetic part of the perturbation, $V_k^2/V_A^2 = 0$, we find that the other terms, originating from the electrostatic part of the perturbation, result in the well-known dispersion of the acoustic ion-cyclotron waves [9]:

$$\omega_k^2 = \Omega^2\left(1 + \frac{T_e}{T_j}\mu_j^2\right) = \Omega^2 + k_\perp^2 V_S^2. \tag{20}$$

However the phase velocity of the KAWs, V_k^2/V_A^2, cannot be arbitrary small. In fact, the electromagnetic term V_k^2/V_A^2 cannot be neglected for waves with a phase velocity lying in the range $V_{Ti}^2 \ll V_k^2 \ll V_{Te}^2$ and $\mu \neq 0$ if the Alfvén velocity is also in the same range.

The waves are essentially electromagnetic in this case and may be called ICK-AWs, ion-cyclotron kinetic Alfvén waves. Since the frequency of ICKAWs is close to the cyclotron frequency, $\omega_k \lesssim \Omega$, their phase velocity is determined by the parallel wavenumber,

$$\frac{V_k^2}{V_A^2} \lesssim \frac{1}{k_z^2 \delta_i^2},$$

and can be smaller than the Alfvén velocity when $k_z^2 \delta_i^2 \sim 1$, where δ_i is the ion inertial length. However, the requirement of weak cyclotron damping places a restriction on the parallel wavenumber, $k_z^2 \lambda_i^2 \ll V_A^2/V_{Ti}^2$, that is much more severe than $k_z^2 \lambda_D^2 < 1$. This condition is also in accord with the condition of weak ion Landau damping. As a result, as we already mentioned above, these waves are essentially electromagnetic.

IV APPLICATION: EXCITATION AND DAMPING OF ICKAWS IN CORONA

We apply our results to the model of coronal heating by Alfvén waves excited during short-scale reconnection events. There are strong observational indications that the bursty plasma heating events are closely related with magnetic reconnection events (nanoflares, microflares). High-frequency Alfvén waves, $\omega = 10 - 10^4$ s^{-1}, have been supposed to be generated by MHD perturbations which accompany this magnetic activity. These waves are able to heat the corona and to accelerate the solar wind [7, 10].

We consider here another, kinetic mechanisms for the wave excitation. Magnetic reconnection produces reconnection outflows with typical velocities equal to the Alfvén velocity based on the inflow conditions. Upward propagating outflows penetrate the coronal plasma and set up ion beams in it. We assume that the coronal plasma consists of background Maxwellian protons and electrons, and a V_z-shifted beam of Maxwellian protons. The excitation of KAWs by reconnection outflows has been investigated in [10]. This process requires super-Alfvén velocities of the beam and can be applied to the downward propagating beams. But upward propagating beams tend to become sub-Alfvénic rather than super-Alfvénic in the corona. Perpendicular dispersion varies the phase velocity of the Alfvén waves in the super-Alfvénic domain and is thus unable to built up Cherenkov resonance with sub-Alfvénic beams. However, as we showed in the present paper, the inclusion of the finite frequency effects in combination with finite gyroradius effects make KAW able to interact resonantly with the sub-Alfvénic beams as well.

For the model described above we have

$$\delta_{ie} = \sqrt{\frac{\pi}{2}} \frac{V_k}{V_{Te}} + \frac{T_e}{T_p} \sqrt{\frac{\pi}{2}} \frac{V_k}{V_{Tp}} \Lambda_0 \exp\left(-\sigma_0^2\right) \left[1 + \frac{\Lambda_1}{\Lambda_0} \exp\left(\sigma_0^2 - \sigma_1^2\right)\right] +$$
$$\frac{n_j}{n_0} \frac{T_e}{T_j} \sqrt{\frac{\pi}{2}} \frac{V_k - V_0}{V_{Tj}} \Lambda_{0j} \exp\left(-\sigma_{j0}^2\right) \left[1 + \frac{\Lambda_{1j}}{\Lambda_{0j}} \exp\left(\sigma_{j0}^2 - \sigma_{j1}^2\right)\right];$$

$$\delta_{i1} = i \left(\frac{V_k^2}{V_A^2} - \frac{T_e}{T_p} \mu^2\right) \frac{T_e}{T_p} \sqrt{\frac{\pi}{2}} \frac{V_k}{V_{Tp}} \Lambda_1 \frac{\Omega}{\omega} \exp\left(-\sigma_1^2\right). \tag{21}$$

We take a plasma β typical for the corona, $2V_{Ti}^2/V_A^2 = 0.04$, and we assume for simplicity $T_e = T_p = T_j$. We then find from (19) that a proton beam of relative number density $n/n_0 = 0.2$, and with a sub-Alfvénic velocity $V_0^2/V_A^2 = 1/2$ excites ICKAWs around the phase velocity $V_k^2 = 15V_{Ti}^2$ with maximum growth rate $\gamma/\omega = 0.05$.

The fate of these waves is obvious. Once they leave the unstable region where they are generated, they immediately meet the conditions of very strong damping. The damping rate, of about $\gamma/\omega = -0.2$, is mainly due to competing electron Landau and ion cyclotron damping. They dissipate quickly and cannot propagate far from their generation site. Therefore, if they are excited in the low corona, they contribute to the heating of the low corona and cannot directly contribute to the additional heating (presumebly due to cyclotron damping), found at higher heights. This issue concerns the waves excited by MHD disturbances as well: even a weak inhomogeneity across B_0 quickly develops high perpendicular wavenumbers, bringing about enhanced Landau damping, and the waves loose their energy before they reach the upper coronal levels where the signatures of cyclotron heating are observed.

A possible solution for this problem is that the waves undergo spectral transfer towards lower perpendicular wavenumbers, for example induced by nonlinear interaction [5]. There is also the possibility that the waves are excited already at higher coronal levels; the destabilising factor is uncertain in this case.

V CONCLUSIONS

We studied high-frequency kinetic Alfvén waves taking into account the combined action of the perpendicular (due to finite gyroradius) and the parallel (due to finite cyclotron frequency) effects. New analytical expressions reflecting important features of these waves are obtained.

We investigated the excitation of ICKAWs assuming that in the plasma of the low corona there is a beam-like population injected by reconnection events. The plasma outflowing from the reconnection site sets up neutralized proton beams in the surrounding plasma, providing free energy for the ICKAWs excitation. For the excitation of ICKAWs, the beam velocity should not be super-Alfvénic, as in the case of usual kinetic AWs. This fact makes the reconnection outflows evident drivers for unstable ICKAWs.

Depending on the wave frequency and parallel/perpendicular wavenumber ratio, the ICKAWs damp in the corona due to electron Landau and ion cyclotron damping. When the parallel wavenumber is increased, the phase velocity of ICKAWs shifts further into the region of ion thermal velocities, where ion Landau damping comes into play. Therefore, the ICKAWs that are excited in the low corona can heat the ion component of the coronal plasma in both the parallel and perpendicular direction, giving rise to an anisotropic heating of the ions, and the electrons.

This work was supported in part by INTAS grant 96-530. Yu.V. acknowledges the financial support by the KUL.

REFERENCES

1. Hollweg, J.V., *J. Geophys. Res.* **76**, 14811-14819 (1999).
2. Voitenko, Yu. M., Super-Alfvén Beams and Kinetic Alfvén Waves in Cosmic Plasma. *Preprint 89-9P of Institute for Theoretical Physics.* ITP, Kiev, Ukraine 1989 (in Russian).
3. Stenflo, L., & Shukla, P. K., *J. Geophys. Res.* A **100**, 17261-17263. (1995).
4. Yukhimuk, A. K., Voitenko, Yu. M., Fedun, V. N., Yukhimuk, V. A. *J. Plasma Phys.* **60**, 490-511 (1998). 5
5. Voitenko, Yu. M., Goossens, M., *Astron. Astrophys.* **357**, 1086-1092 (2000)
6. Voitenko, Yu. M., *Solar Phys.* **182**, 411-430 (1998).
7. Axford, W.I., McKenzie, J..F., *Cosmic Winds and the Heliosphere.* The University of Arizona Press, 31 (1997).
8. Hasegawa, A., Chen, L., *Phys. Fluids.* **30**, 1924-1934 (1976).
9. Brambilla, M., *Kinetic Theory of Plasma Waves,* Clarendon Press, Oxford, 1998.
10. Cranmer, S.R., Field, G.B., Kohl, J.L., *Astrophys. J.* **518**, 937-947 (1999).

Excitation of fast and slow magnetosonic waves by kinetic Alfven waves

A. Yukhimuk*, V. Fedun†, O. Sirenko*, Yu. Voitenko*‡

*Main Astronomical Observatory, Holosiiv, Kyiv-127, 03680, Ukraine
†Kyiv National University, Department of Astronomy and Space Plasma Physics, pr.Glushkova 6, Kyiv 252002, Ukraine
‡Centre for Plasma Astrophysics , K.U.Leuven, Celestijnenlaan 200B, B-3001 Heverlee, Belgium

Abstract. The nonlinear parametric interaction of Alfven waves with magnetosonic and ion-acoustic waves is considered on the basis of two-fluid magnetohydrodynamics. A nonlinear dispersion relation describing three-wave interaction, instability growth rate have been calculated and estimated. The analyses of theoretical results shows that kinetic effects in the Alfven waves (the finite ion Larmour radius) are essential for the parametric interactions of waves.

Nonlinear parametric processes studied in the paper could take place in the solar coronal loops, where plasma parameter is small. The products of the decay - magnetosonic and ion-acoustic waves, can effectively heat the coronal plasma in consequence of rapid dissipation.

I INTRODUCTION

Magnetohydrodynamic waves play an important role in the dynamics of Earth magnetosphere, solar wind and atmosphere of the Sun. The long-period geomagnetic pulsations observed in the Earth's magnetosphere have been considered as MHD Alfven waves ([1], [2]). The satellite observations show that the most part of turbulent pulsations are Alfven and magnetosonic waves in the solar wind [3]. They are thought to play an essential role in the solar wind heating. The MHD waves can provide a source of energy for corona heating. Alfven waves have enough energy for heating of coronal loops (see e.g. [4] and references therein). But they are slow-damped. This is the reson why Alfven wave can't heat coronal loops directly, and the problem arise: how the energy can be transferred from waves to plasma particles. The ion-acoustic and magnetosonic waves are rapid damped waves. Therefore, if the propagating Alfven wave decays into ion-acoustic and magnetosonic waves, they can transfer rapidly their energy to particles in consequence of dissipation. In the paper [5] was shown that processes of formation of acoustic waves from Alfven waves are effective in the region, where $V_A/V_s - 1/30 - 30$. The dissipation of wave energy arises on distances which are comparable with length of coronal

CP537, *Waves in Dusty, Solar, and Space Plasmas*, edited by F. Verheest, et al.
© 2000 American Institute of Physics 1-56396-962-9/00/$17.00

loops. Nonlinear interaction between Alfven and ion-acoustic waves was considered for the case of linear dispersion , i.e. without accounting of thermal effect (finite Larmor radius of proton). But in the magnetosheric and coronal plasma thermal effects play an important role. When the thermal effects are taken into account it leads to nonlinear dependence of frequency from wavenumber, which is essential for nonlinear wave interaction.

In the present work taking into account the thermal effect we consider nonlinear parametric interaction of Alfven wave with the ion-acoustic and magnetosonic waves.

II BASIC EQUATION

It is assumed that Alfven wave with finite amplitude propagates in homogeneous magnetized plasma ($\vec{B}_0 = B_0 \vec{e}_z$):

$$\vec{E}_0 = (E_{0x}\vec{e}_x + E_{0z}\vec{e}_z)\exp i(-\omega_0 t + \vec{k}_0 \vec{r}) + c.c., \tag{1}$$

where the frequency and wave vector are coupled by equality:

$$\omega_0^2 = k_{0z}^2 V_A^2(1 + t\mu_i), \tag{2}$$

Here $\mu_i = k_{0x}^2 \rho_i^2$, $\rho_i = V_{Ti}/\omega_{Bi}$ -proton Larmor radius, $\omega_{Bi} = eB/m_i c$ -ion cyclotron frequency, $t = T_e/T_i$. Our governing equations are a set of two-fluid MHD equation:

$$\frac{\partial \vec{v}_\alpha}{\partial t} = \frac{1}{m_\alpha}(e_\alpha \vec{E} + \vec{F}_\alpha) + (\vec{v}_\alpha \times \vec{\omega}_{B\alpha}) - \frac{T_\alpha}{m_\alpha n_\alpha}\nabla n_\alpha, \tag{3}$$

$$\frac{\partial n_\alpha}{\partial t} = -\nabla(n_\alpha \vec{v}_\alpha), \tag{4}$$

$$\nabla \times \vec{B} = \frac{4\pi}{c}\vec{j} + \frac{1}{c}\frac{\partial \vec{E}}{\partial t}, \tag{5}$$

$$\nabla \times \vec{E} = -\frac{1}{c}\frac{\partial \vec{B}}{\partial t}, \tag{6}$$

$$\nabla \cdot \vec{E} = 4\pi\rho, \tag{7}$$

where

$$\vec{j} = e(n_i \vec{v}_i - n_e \vec{v}_e),$$
$$\rho = e(n_i - n_e),$$

$$\vec{F}_\alpha = \frac{e_\alpha}{c}(\vec{v}_\alpha \times \vec{B}) - m_\alpha(\vec{v}_\alpha \nabla)\vec{v}_\alpha.$$

Index $\alpha = i, e$ corresponds to the ion and electron components of plasma respectively. As $F_{pi} = \frac{m_e}{m_i} F_{pe}$, influence of F_{pi} force is small, and can be neglected. In Ampere-Maxwell's equation we also neglect the displacement current.

We consider equations in Cartezian coordinates (x, y, z), supposing that all wave vectors are situated in OXZ plane. It is assumed that the wave synchronism conditions are satisfied:

$$\omega_0 = \omega + \omega_1, \quad \vec{k}_0 = \vec{k} + \vec{k}_1, \tag{8}$$

where $\omega_1, \vec{k}_1, \omega, \vec{k}$ -frequency and wave vector of magnetosonic and ion-acoustic wave respectively.

III   DISPERSION EQUATION FOR LOW-FREQUENCY PERTURBATIONS.

Since Alfven and ion-acoustic waves are ultralow-frequency waves, we can use the plasma approximation for obtaining of a dispersion equation:

$$\tilde{n}_e = \tilde{n}_i \tag{9}$$

where $\tilde{n}_e$ and $\tilde{n}_i$ are perturbations of ion and electron number densities. From the equation of motion and continuity equation, we can find expressions for $\tilde{n}_e$ and $\tilde{n}_i$:

$$\frac{\tilde{n}_e}{n_0} = \frac{e}{T_e}\left(\psi - \frac{k_x^2}{k_z^2}\frac{\omega^2}{\omega_{Be}^2}\phi - Q_{NL}\right), \tag{10}$$

$$\frac{\tilde{n}_i}{n_0} = \frac{e}{m_e}\left(\frac{k_z^2\psi}{\omega^2} - \frac{k_x^2\varphi}{\omega_{Bi}^2}\right) \tag{11}$$

Here

$$Q_{NL} = \frac{k_x\omega}{ek_z^2\omega_{Be}}\left(i\frac{\omega}{\omega_{Be}}F_x + F_y\right) + \frac{F_z}{iek_z}.$$

To find the relation between φ and ψ, we use z componet of the Ampere's law in two potential approximation and equation of charge conservation. Compare (10) to (11) and use relation between φ and ψ, we obtain dispertion equation for ultralow-frequancy perturbations:

$$\left[\frac{k_z^2 V_s^2}{\omega^2}\left(1 - \frac{\omega^2}{k_z^2 V_A^2}\right)\left(1 - \frac{\omega^2}{k_z^2 V_s^2}\right) - \mu_s\right]\varphi = -Q_{NL} \tag{12}$$

In the absence of pump wave ($Q_{NL} = 0$) we have:

$$\omega_A^2 = k_z^2 V_A^2 (1 + \mu_s),$$
$$\omega_s^2 = k_z^2 V_s^2 / (1 + \mu_s). \tag{13}$$

These expressions discribe dispersion laws for Alfven and ion-acoustic waves respectively. The special properties of these waves are their ability to propagate in xz plane, thus the wave energy can be transfered across the external magnetic field as well as along it. And since $\omega_s^2 \ll \omega_A^2$ and $\omega_A^2 \simeq k_z^2 V_A^2$, dispersion equation for ion-acoustic wave can be rewritten as:

$$\varepsilon \phi = \eta \psi_0 E_{1y}^* \tag{14}$$

where

$$\varepsilon = \omega^2 - k_z^2 V_s^2 / (1 + \mu_s),$$

$$\eta = -\frac{\omega^2}{\omega_{Be}} \frac{k_{0x}}{k_z} \frac{V_{Te}^2}{V_A} \frac{e}{T_e},$$

In the calculation of the coupling coefficient η, we have taken account of the pondermotive force created by interaction of the pump wave with the magnetosonic wave.

IV   DISPERSION EQUATION FOR MAGNETOSONIC WAVE

In order to obtain a dispersion equation for magnetosonic wave, which propagate across magnetic field, we also use plasma approximation. In this case ions and electrons velocity must coincide. From the electron equation of motion we find:

$$V_{ex} = \frac{e}{m_e \omega_{Be}} \left(E_{1y} + i \frac{\omega_1}{\omega_{Be}} E_{1x} \right) + \frac{1}{m_e \omega_{Be}} \left(i \frac{\omega_1}{\omega_{Be}} F_{1x} + F_{1y} \right), \tag{15}$$

where term proportional to (ω/ω_{Be}) describes the inertial drift, that must be taken into account for wave that propagates across magnetic field. The component of pondermotive force is determined by interaction of pump wave and ion-acoustic wave. From the ion equation of motion and Maxwells̀ equation we find expression for V_{ix}:

$$V_{ix} = \frac{k_1^2 V_A^2}{\omega_1^2} \frac{e E_{1y}}{m_e \omega_{Be}}. \tag{16}$$

Compare (15) to (16) we obtain dispersion equation for magnetosonic wave:

$$\varepsilon_1 E_{1y} = \eta_1 \varphi^* \phi_0, \tag{17}$$

where

$$\varepsilon_1 = \omega_1^2 V_A^2$$

and coupling coefficient is determined by

$$\eta_1 = -2V_A V_S \frac{m_e}{e} \left(\frac{V_{Te}}{V_A}\right)^2 \frac{k_{0x}}{\omega_{Be}} \omega_1^2 \omega_0 \mu_e \qquad (18)$$

$$\mu_e = k_x^2 \rho_e^2.$$

V NONLINEAR DISPERSION EQUATION AND DECAY GROWTH RATE

From ion-acoustic (14) and magnetosonic (17) dispersion equations we can find a nonlinear dispersion relation describing three-wave interaction:

$$\varepsilon \varepsilon_1^* = \eta \eta_1^* \mid \phi_0 \mid^2 . \qquad (19)$$

In the absence of the pump wave, $\phi_0 = 0$, two kinds of waves with frequencies $\omega = \omega_k$ and $\omega_1 = \omega_{1k}$ exist in the plasma. In the presence of the pump wave, $\phi_0 \neq 0$, energy transfer from the pump KAW to the other plasma modes occurs, and amplitudes increase with the growth rate. Putting $\omega = \omega_r + i\gamma$, $\omega_1 = \omega_{1r} + i\gamma$ ($\mid \gamma \mid \ll \omega_r, \omega_{1r}$) into (19) and expanding ε and ε_1 in Taylor series, we obtain an expression for the wave growth rate:

$$\gamma^2 = \left. \frac{\eta \eta_1^* |\phi_0|^2}{\frac{\partial \varepsilon}{\partial \omega} \frac{\partial \varepsilon_1}{\partial \omega_1}} \right|_{\omega=\omega_r, \omega_1=\omega_{1r}}, \qquad (20)$$

Where ω_r and ω_{1r} are defined from equations:

$$\varepsilon(\omega_r, \vec{k}) = 0, \qquad \varepsilon_1(\omega_1, \vec{k_1}) = 0.$$

substituting the expressions for the coupling coefficients η, η_1 and derivative $\frac{\partial \varepsilon}{\partial \omega}, \frac{\partial \varepsilon_1}{\partial \omega_1}$ we find:

$$\gamma^2 = \frac{W}{2} \frac{V_S}{V_A} \frac{\omega_0^2}{\omega_{Be} k_z V_A} \left(\frac{\omega}{\omega_{Be}}\right) \mu_e \omega_{pe}^2 \qquad (21)$$

where

$$W = \frac{|E_{0x}^2|}{4\pi n_0 T_e} \qquad (22)$$

VI CONCLUSIONS

Let us estimate the instability growth rate for the coronal plasma. Spectroscopic observations suggest the presence of Alfven waves with amplitudes $B/B_0 = 0.01$ in coronal loops [6]. To examine existence of Alfven waves, these authors used the method first proposed by [7]. Typical parameters for such loop are length $L = (2 - 5) \times 10^9$ cm, $n = (0.5 - 1) \times 10^{10}$ cm^{-3}, $B = 100 - 500G$, $T = 4 \times 10^6 K$, $\omega_{pe}/\omega_{Be} \approx 10$, $\mu_e \approx 1$, $\omega_1 \approx 10^{-1}\omega_{Bi}$. Substituting these values of the plasma parameters and a KAW intensity of $W \approx 10^{-5} - 10^{-6}$ into (21), we obtained $\gamma \approx (10^2 \div 10^3) c^{-1}$ and the time of instability development respectively $\tau = \gamma^{-1} \approx 0.01$ c. The nonlinear interaction of Alfven waves with magnetosonic and ion-acoustic waves presented in this paper could take place in the Earth's magnetosphere as well as in the solar magnetosphere, where a value of the plasma parameter β is low. In solar atmosphere the sources of Alfven waves can be 5-min. chromospheric oscillation, that swing the basis of magnetic loops and lead to propagation of alfven waves into corona. Alfven waves will transform into kinetic Alfven waves during the propagation along magnetic field. The kinetic Alfven wave can interact more effective with other kinds of waves because of the presence of a non-zero longitudinal component of the electric field E_z. When KAW reaches the region with low plasma parameter β it will decay into magnetosonic and ion-acoustic waves. Due to their rapid dissipation, these waves are more effectiv for heating of the coronal plasma. The exitation of fast magnetosonic waves by phase mixing of Alfven wave was considered in [8]. The relative efficacy exitation of slow and fast magnetosonic waves by Alfven waves need futher investigation.

REFERENCES

1. Nishida A., Mir, Moscov(in Russian) (1980).
2. Volokitin, A.S., and Dubinin, E.M., *Planet. Space Sci.* **37**,761-765 (1989).
3. Belcher, J.W., Devis, L., *J.Geophys.Res.*, **76**, 3539 (1971).
4. Goossens M. *Space Sci.Rev.* **68**, 51 (1994).
5. Uchida, Y., Kaburaki, O. *Solar Phys.* **35**, 451 (1974).
6. Hara, H. and Ichimoto, K., *ApJ*, **513**, 969-982 (1999).
7. McClements, et al. *Solar Phys.*, **131**, 41 (1991).
8. Nakarjakov, V.M., Roberts, B., Murawski, K., *Solar Phys.*, **175**, 93-105 (1997).

Alternating current generation in flux tubes by pressure fluctuations

Yuzef Zhugzhda*† and Marcel Goossens*

* Center for Plasma Astrophysics, K.U.Leuven, B–3001 Heverlee, Belgium
† IZMIRAN, Troitsk, Moscow region, 142092 Russia

Abstract. Alternating current in flux tubes is one of the possible contributors to the heating of the solar corona. The mechanism of the generation of alternating currents in flux tubes by turbulent pressure fluctuations in convective zone is proposed. The thin-flux-tube approximation shows, that the linear and weakly nonlinear torsional waves in twisted flux tube, are accompanied by tube cross section fluctuations. Consequently, external pressure fluctuations generate not only slow MHD waves but torsional waves as well. The torsional waves are nothing else than alternating current.

INTRODUCTION

Solar corona heating has been the focus of attention for decades. Heating by direct currents and magnetohydrodynamic waves are the two main competing ideas. Direct currents in the corona flow along coronal arches, which can be considered as flux tubes imbedded in a current-free magnetized plasma. Low frequency alternating currents in flux tubes are torsional Alfvén waves. The reflection of Alfvén waves on their way from the solar photosphere to the corona is overcome by the resonance filtering in the coronal loops and arcades [1]. Torsional Alfvén waves can be excited by random vortex motions in convection zone [2,3].

Torsional waves in current flux tubes deserve our attention, because they are accompanied by pressure fluctuations [4] and, consequently, can be excited by external pressure fluctuations in the convection zone. The treatment of torsional waves in current (twisted) flux tubes has become possible due to thin-flux-tube approximation for current (twisted) flux tubes [4]. The paper is organized as follows. The properties of linear and weakly nonlinear torsional waves are outlined. The possibility of the generation of alternating currents due to modulation of the direct current in flux tubes is discussed.

ALTERNATING CURRENTS IN FLUX TUBES

Alternating currents or, in other words, torsional waves in twisted or, in another way, current thin flux tubes are explored in the frame of the thin-flux-tube ap-

CP537, *Waves in Dusty, Solar, and Space Plasmas*, edited by F. Verheest, et al.
© 2000 American Institute of Physics 1-56396-962-9/00/$17.00

proximation. To obtain the thin-flux-tube approximation, all physical quantities are expanded in a Taylor series in the tube radius r with coefficients depending on z and t. When the expansions are sustituted into the set of MHD equations and terms of the same power in r are collected, an infinite set of equations for the coefficients of expansion is obtained. The reason of an infinite set is that the equations have to be satisfied for all values of the radial coordinate r. The thin-flux-tube approximation is the result of the truncation of the infinite set of equations. The dynamics of thin tube is governed by the following set of the approximate equations [4]

$$\rho\left(\frac{\partial u}{\partial t} + u\frac{\partial u}{\partial z}\right) + \frac{\partial p}{\partial z} = 0,$$

$$\frac{\partial}{\partial t}\left(\frac{\rho}{B}\right) + \frac{\partial}{\partial z}\left(u\frac{\rho}{B}\right) = 0,$$

$$\frac{\partial}{\partial t}\left(\frac{\Omega}{B}\right) + u\frac{\partial}{\partial z}\left(\frac{\Omega}{B}\right) = \frac{B}{4\pi\rho}\frac{\partial}{\partial z}\left(\frac{J}{B}\right),$$

$$\frac{\partial}{\partial t}\left(\frac{J}{B}\right) + \frac{\partial}{\partial z}\left(u\frac{J}{B}\right) = \frac{\partial\Omega}{\partial z}, \tag{1}$$

$$\frac{d}{dt}\left(\frac{p}{\rho^\gamma}\right) = 0,$$

$$\frac{\partial A}{\partial t} + u\frac{\partial A}{\partial z} - 2Av = 0,$$

$$\frac{\partial B}{\partial t} + u\frac{\partial B}{\partial z} + 2Bv = 0,$$

$$p + \frac{B^2}{8\pi} - \frac{A}{2\pi}\left[\rho\left(\frac{\partial v}{\partial t} + u\frac{\partial v}{\partial z} + v^2 - \Omega^2\right) + \frac{1}{4\pi}\left(J^2 + \frac{B}{2}\frac{\partial^2 B}{\partial z^2} - \frac{1}{4}\left(\frac{\partial B}{\partial z}\right)^2\right)\right] = p_{\text{ext}},$$

where ρ, A, Ω, u, are the density, cross section, angular velocity, longitudinal velocity and magnetic field in the tube, while ϕ-component of magnetic field is a linear function of radius $B_\phi = Jr$. Thus, the variable J is a measure of the current in the tube $(\nabla \times \mathbf{B})_z = 2J$. The gas pressure and longitudinal magnetic field are $p = p + p_2 r^2$, $B_z = B + B_{z2}r^2$. To simplify the problem the variables p_2 and B_{z2} are excluded from the set of equations (see details in [4]). But they have to be taken into account in the equilibrium model of the tube.

The linear and nonlinear waves are studied for the following equilibrium model of a twisted flux tube

$$B_{z2}^{(0)} = 0, \quad \Omega_0 = 0, \quad p_0 + \frac{B_0^2}{8\pi} - \frac{R_0^2}{8\pi}J_0^2 = p_{\text{ext}}, \quad p_2^{(0)} + \frac{1}{4\pi}J_0^2 = 0, \tag{2}$$

where R_0 is the tube radius.

Linear torsional waves

The dispersion equation for linear waves in a twisted tube has been obtained by (1) in [5] and reads, as follows,

$$(\omega^2 - C_T^2 k^2)(\omega^2 - C_A^2 k^2) - \frac{A_0}{4\pi} \frac{(\omega^2 - C_A^2 k^2)^2(\omega^2 - C_S^2 k^2)}{C_A^2 + C_S^2} -$$
$$- \frac{J_0^2 A_0 C_A^2}{2\pi B_0^2} \frac{(\omega^2 - C_A^2 k^2)(\omega^2 - C_S^2 k^2) + 2\omega^2 k^2 C_S^2}{(C_A^2 + C_S^2)} = 0, \tag{3}$$

where C_S, C_A, C_T are the sound, Alfvén and tube velocities, k is the longitudinal wavenumber. The first term in the dispersion equation corresponds to wave propagation in an infinitely thin untwisted tube ($J_0 = 0$, $A_0 = 0$). The second term reflects the influence of the finite cross-section. The third term is due to the effect of the twist of the manetic field. It is clear that dispersion appears only due to the second term. Introducing the fast and slow speeds, $C_\pm$, we can rewrite the dispersion equation (3) as

$$(C_A^2 + C_S^2 - K C_A^2)(\omega^2 - C_+^2 k^2)(\omega^2 - C_-^2 k^2) + \tfrac{A_0}{4\pi}(\omega^2 - C_A^2 k^2)^2(\omega^2 - C_S^2 k^2) = 0, \tag{4}$$

where
$$C_\pm^2 = C_A^2 \frac{C_A^2 + 2C_S^2 + K(C_S^2 - C_A^2) \pm \sqrt{S}}{2(C_A^2 + C_S^2) - 2C_A^2 K}, \tag{5}$$

$$S = C_A^4 + 2K(3C_A^2 C_S^2 + 4C_S^4 - C_A^4) + K^2(C_S^4 - 6C_S^2 C_A^2 + C_A^4). \tag{6}$$

$$K = \frac{B_\phi^{(0)}(R_0)}{2B_0} = \frac{J_0^2 A_0}{2\pi B_0^2} = \frac{A_0 \alpha^2}{8\pi} = \frac{\alpha^2 R_0^2}{8}, \tag{7}$$

where $\alpha = J_0/B_0$ and R_0 is the radius of the flux tube. If the parameter K tends to zero, the fast speed C_+ tends to the Alfvén speed C_A, and the slow speed C_- tends to the tube speed C_T. In the case of an untwisted tube, the dispersion relation (5) splits into the dispersion relations for the slow and torsional Alfvén waves

$$\omega^2 = C_T^2 k^2 + \frac{A_0}{4\pi(C_A^2 + C_S^2)}(\omega^2 - C_S^2 k^2)(\omega^2 - C_A^2 k^2), \qquad \omega^2 = C_A^2 k^2. \tag{8}$$

In the untwisted case, the torsional mode is not dispersive. The Alfvén torsional and slow magnetosonic sausage waves are modified by the twisting of the tube. The speeds C_+ and C_- can be considered as the *modified Alfvén speed* and *modified tube speed*, respectively.

In the case of weak dispersion (small $k^2 A_0$), when $\omega \approx C_\pm k$, the dispersion equation (5) can be approximated by

$$\omega^2 \approx C_\pm^2 k^2 \left(1 \pm \frac{A_0}{4\pi} \frac{(C_\pm^2 - C_A^2)^2(C_\pm^2 - C_S^2)}{C_A^2 \sqrt{S}} k^2\right). \tag{9}$$

For the slow body waves in an untwisted tube, (8) approximately gives

$$\omega^2 \approx C_T^2 k^2 \left(1 + \frac{A_0 C_T^4}{4\pi(C_A^2 + C_S^2)} k^2\right). \tag{10}$$

Except for the factor ~ 1 in front of A_0 this dispersion equation coincides with the dispersion equation for the slow body sausage waves obtained without using the thin-flux-tube approximation. As shown in [4], this is due to the truncation of the infinite set of equations, which replaces the magnetohydrodynamic equations when the dependent variables are expanded over the tube radius. The truncation of this set of equations is, at the same time, the truncation of the radial eigenfunction of the problem as well. The value of the first root of the truncated radial eigenfunction slightly differs from the root of the Bessel function, which is an exact eigenfunction of the problem. Zhugzhda [4] proposed to introduce an "effective" cross section and an "effective" radius of the tube

$$\mathcal{A} = 4A\xi_1^{-2}, \quad \mathcal{R} = 2R\xi_1^{-1}, \tag{11}$$

where $\xi_1 = 2.4$ is the first root of the zeroth order Bessel function. The "effective" cross section makes the approximation not only an exact one but is valid for *non-thin flux tubes* as well. We replace A by $\mathcal{A}$ in the following analysis. In the case of a weakly twisted flux tube, it is assumed that the same effective cross section can be used in (10).

The propagation of torsional waves in a twisted flux tube is accompanied by fluctuations of the tube cross section, which are proportional to current fluctuations

$$\frac{A}{A_0} = \frac{\mathcal{A}}{\mathcal{A}_0} = \frac{(C_+^2 - C_S^2)(C_+^2 - C_A^2)}{C_+^2 C_S^2 - (C_+^2 - C_S^2)(C_+^2 - C_A^2)} \frac{J}{J_0}. \tag{12}$$

In the limit of a current-free tube $J_0 \to 0$, the cross section fluctuations tend to zero, because $C_+^2 - C_A^2 \sim J_0^2$. The ratio of the relative fluctuations of cross section to the relative current fluctuations $F = (A/A_0)/(J/J_0)$ is plotted against the twist parameter K (7) on Fig.1. In the case of $\beta > 1.1$ the fluctuations of the tube cross section and current are in phase, that is to say, the tube cross section increases with current. In the case of $\beta < 1.1$, they are out of phase. The fluctuations of tube cross section due to torsional waves in a twisted tube are negligible for $\beta \sim 1.1$. Thus, external pressure fluctuations excite both slow and torsional waves. The wavelength of slow and torsional waves are different for a fixed frequency. Consequently, the kind of waves, that are excited, depends on the time and space scales of the external pressure fluctuations. To tackle the problem the external pressure has to be replaced by a source term. Then the treatment is similar to the problem of the generation of sound by turbulence [6]. The equations (36-39) from [5] can be rewritten as follows

$$(C_A^2 + C_S^2 - KC_A^2)\mathcal{D}_+\mathcal{D}_-B + \frac{\mathcal{A}_0}{4\pi}\mathcal{D}_S\mathcal{D}_A{}^2 B = \frac{B_0}{\rho_0}\mathcal{D}_S\mathcal{D}_A P_{ext}(t, z), \tag{13}$$

where the D'Alembert's operators are defined as

$$\mathcal{D}_A \stackrel{\text{def}}{=} \frac{\partial^2}{\partial t^2} - C_A^2\frac{\partial^2}{\partial z^2}, \quad \mathcal{D}_S \stackrel{\text{def}}{=} \frac{\partial^2}{\partial t^2} - C_S^2\frac{\partial^2}{\partial z^2}, \quad \mathcal{D}_\pm \stackrel{\text{def}}{=} \frac{\partial^2}{\partial t^2} - C_\pm^2\frac{\partial^2}{\partial z^2}, \tag{14}$$

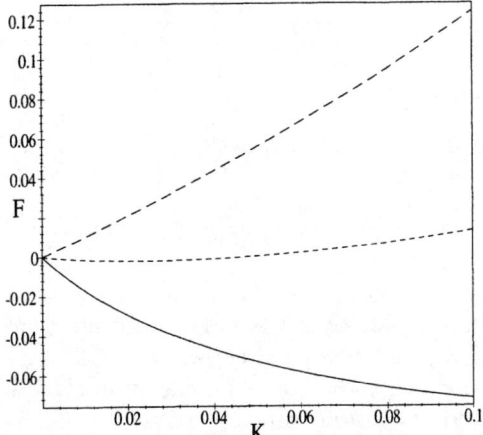

FIGURE 1. The dependence of $F = (A/A_0)/(J/J_0)$ as a function of twist parameter K (7) for linear torsional waves in the twisted flux tube with $\beta = 0.5$(dashed), 1.1((dotted), 2(solid).

and the external pressure is a given function. Equation (13) is akin of Lighthill equation [6] and the external pressure can be replaced by the so-called Lighthill stress tensor, T_{ij}, is given by

$$T_{ij} = \rho u_i u_j + (p' - C_{Se}^2 \rho')\delta_{ij} - \tau_{ij}, \tag{15}$$

where τ_{ij} is the viscous stress, and the prime denotes a fluctuating quantity. The excitation of both torsional and slow waves by turbulent pressure fluctuations is described by equation (13).

The generation of linear waves only is presumed in this case. The generation of nonlinear waves is discussed in the following section.

Weakly nonlinear torsional waves

Zhugzhda and Nakariakov [5] showed, that weakly nonlinear torsional waves are governed by the KdV equation

$$\frac{\partial B}{\partial \tau} - \delta \frac{\partial^3 B}{\partial \xi^3} - \varepsilon B \frac{\partial B}{\partial \xi} = 0, \tag{16}$$

where
$$\delta = \mp \frac{A_{\prime}(C_+^2 - C_S^2)(C_+^2 - C_A^2)^2}{8\pi\sqrt{S}C_A^2 C_+}, \tag{17}$$

$$\varepsilon = \mp \left(2\sqrt{S}C_A 2C_+\right)^{-1} \frac{C_S^2}{B_0} \left\{ \frac{C_+^2 - C_S^2}{C_S^2} \left[-2KC_A^2(C_+\tilde{N}_4 + \tilde{N}_1) - \right.\right.$$
$$\left.\left. - (C_+^2 - C_A^2)\left(C_A^2(1 + 2KC_A^2\tilde{N}_8) + \frac{KC_A^2}{C_+}(\tilde{N}_6 + \tilde{N}_7)\right)\right] + \right.$$

$$+ \left(C_+^2 - C_A^2 - 2KC_A^2\right)\left(C_+\tilde{N}_2 + \tilde{N}_1 + \tfrac{C_+}{C_S^2}\tilde{N}_5\right)\Big\}, \tag{18}$$

$$\tilde{N}_1 = \frac{C_+^2 C_S^2}{C_+^2 - C_S^2}, \quad \tilde{N}_2 = -\frac{2C_+^3 C_S^2}{\left(C_+^2 - C_S^2\right)^2}, \quad \tilde{N}_4 = -\frac{2C_+^5 C_S^2}{\left(C_+^2 - C_S^2\right)^2\left(C_+^2 - C_A^2\right)},$$

$$\tilde{N}_5 = -\frac{C_+^5 C_S^2(\gamma - 1)}{\left(C_+^2 - C_S^2\right)^2}, \quad \tilde{N}_6 = \frac{C_+\left(2C_S^2 - C_+^2\right)}{C_+^2 - C_S^2}, \quad \tilde{N}_7 = -\frac{C_+\left(C_+^2 + C_S^2\right)}{2\left(C_+^2 - C_S^2\right)},$$

$$\tilde{N}_8 = \frac{C_+^6 - C_+^4 C_A^2 - 2C_+^4 C_S^2 + 2C_+^2 C_A^2 C_S^2 - C_S^2 C_A^4}{C_A^2\left(C_+^2 - C_A^2\right)\left(C_+^2 - C_S^2\right)^2}. \tag{19}$$

which is valid for long wavelength, weakly nonlinear, slow body and torsional Alfvén waves in twisted flux tubes. Moreover, the KdV equation is valid for torsional waves of arbitrary wavelength in weakly twisted tubes. The KdV equation (16) is valid for a tube with free boundaries ($p_{\text{ext}} = const$ on the tube boundary). In what follow, our analysis is restricted to soliton solutions of KdV, which are a zero frequency limit of a more general cnoidal wave solution. In the case $\beta < 1$ the solution reads

$$B = B_a \cosh^{-2}\left(\tfrac{\xi + V\tau}{L}\right), \tag{20}$$

where
$$L \approx \sqrt{6}\mathcal{R}_0 K\beta(1 - \beta)\left(\tfrac{B_0}{B_a}\right)^{1/2}, \quad V \approx \frac{C_A B_a}{3(1 - \beta)B_0}, \tag{21}$$

are the wavelength and speed of the soliton in the reference frame, which is moving with velocity C_+. Thus, the Alfvén soliton is a propagating *widening* of the tube. The speed of the soliton *exceeds* the modified Alfvén velocity C_+. The Alfvén soliton could produce strong twist in a weakly twisted flux tube, when the full ϕ-component of the magnetic field on the surface of the flux tube is approximately

$$B_\phi \approx B_\phi^{(0)} - \frac{4BB_0}{(1 - \beta)B_\phi^{(0)}}, \tag{22}$$

where $B_\phi^{(0)}$ is the ϕ-component of the field on the surface of the undisturbed tube and B is defined by solution (24). The soliton produces the strongest twist of the tube, when the equilibrium twist of the tube is very small $B_\phi^{(0)} \ll B_0$. The effect increases when $\beta \to 1$. In the soliton, the twist is accompanied by a fast rotation of the plasma. The azimuthal component of the plasma velocity on the boundary of a weakly twisted tube is approximately

$$V_\phi \approx -\frac{4C_A}{1 - \beta}\frac{B}{B_\phi^{(0)}}. \tag{23}$$

The temperature, density and gas pressure decrease in the soliton.

When $\beta > 1$, both parameters δ and ε change sign and the solution of (16) has to be rewritten as

$$B = -B_a \cosh^{-2}\left(\tfrac{\xi - V\tau}{L}\right), \tag{24}$$

where
$$L \approx \sqrt{6}\mathcal{R}_0 K\beta(\beta - 1)\left(\tfrac{B_0}{B_a}\right)^{1/2}, \quad V \approx \frac{C_A B_a}{3(\beta - 1)B_0}. \tag{25}$$

In this case, the Alfvén soliton appears as a narrowing of the tube, where twist and rotation increase according to (22) and (23), while both the pressure and density decrease. However, the changes of sign of δ and ε do not occur exactly at $\beta = 1$ and, moreover, do not occur for the same values of β. This means, that there is a rather small range of parameters β and K, where the coefficients δ and ε have different signs. Obviously, in the case of nonlinear wave propagation along the tube with longitudinally varying parameters β and K, the crossing of the region of $\beta \approx 1$ has to be accompanied by some special effects. This is the region, where the twist by the soliton could increase strongly and the wave corresponding to a narrowing is transformed in a widening wave or vice versa. This effect is not described by the quadratic nonlinear theory and higher order nonlinear theory has to be applied.

The torsional solitons can be exited by external pressure fluctuations of finite amplitude. The expression (25) can be rewritten as

$$LR \approx \sqrt{6}K\beta(\beta - 1)R_0^2, \tag{26}$$

where R is the dicrease of the tube radius at the neck of tube narrowing due to the torsional soliton. It has been speculated,that, if external pressure fluctuations produce tube disturbance whose scales correspond to (26), the torsional soliton appears only. A similar condition for slow body soliton in twisted flux tube [5] reads

$$LR \approx \beta\sqrt{\frac{3(1 + 4K)}{1 - 3K}}R_0^2 \tag{27}$$

For $K \ll 1$ and $\beta > 1$ in the convection zone of the Sun, the scale products LR for torsional and slow solitons differ widely.

DISCUSSION

There are reasons to believe, that the photospheric flux tubes are legs of chromospheric and coronal loops. The direct current in the flux tubes appears due the effect of large scale shear motions on the legs of the loop in the solar photosphere. The time variations of the transverse motion produce transverse Alfvén waves in the loop only. Small scale vortex plasma movements can induce the torsional waves in tubes [2,3]. However, the point arises of whether there are small scale vortexes in photosphere, which rotate sufficiently fast. Another way to generate alternating current in the tube is to modulate the direct current by the tube diameter variations. But, it is evident, that not all tube cross section variations modulate alternating current. For example, a very slow change of the tube diameter affects current density but does not change the full current in the tube. Alternating currents in a flux tube are torsional Alfvén waves, which can be excited only by disturbances of definite time and (longitudinal) space scales. In the case of linear alternating current, the scales are defined by the wavelength and period of linear torsional waves. In the case of a weak impulse current, the scales have to correspond to torsional soliton parameters.

To make an estimate of the effect we consider typical flux tubes in the photosphere with field strength $1000 - 1700\,G$ and tube diameter $100\,km$. If $B_\phi^{(0)} = 0.1B_0$ is assumed, the ratio $K \sim 0.05$ and, consequently, $F \sim 0.05$. The relative external pressure fluctuations due to turbulence in the photosphere are about $\rho_0 v^2/p_0 = \gamma v^2/C_S^2 \sim 0.1$, where $v \sim 1 \div 2\,km\,sec^{-1}$ and $C_S = 6.7$ $km\,sec^{-1}$. If $\beta \lesssim 1$, the relative fluctuations of magnetic pressure $2B/B_0 \sim 0.025$, because they are about one-half of the relative fluctuations of the external pressure. The relative fluctuations of the cross section is also about 0.0125, because $AB = const$. Given these estimates for F and A/A_0, the relative fluctuations of the current $J/J_0 = A/FA_0 \sim 0.25$ and, the fluctuations of $B_\phi \sim 25 \div 40\,G$. We arrive at an estimate of the energy flux of the torsional waves in photospheric tubes due to the effect of the turbulent pressure fluctuations. The value of the energy flux is about $(1/4\pi)B_\phi C_a \sim 2.5 \div 6.4 \cdot 10^7 erg\,cm^{-2}\,sec^{-1}$. Due to spreading of the flux tubes in the corona the cross section of tube can increases $200 - 400$ times and the energy flux drops by factor 10^5, that is comparable with the flux needed to heat the corona.

The above estimate provides reason enough to believe, that the generation of alternating currents in twisted flux tube, by pressure fluctuations in the solar convective zone has to be taken into account in connection with the dynamics and heating of the chromosphere and corona. Among the possible effects the appearance of the impulse of the current due to the tube collapse merits attention. The present paper calls attention to the generation of alternating currents in flux tubes by turbulent pressure fluctuations. So far it was expected that the only effect of the pressure fluctuations is a slow mode generation (see, [7] and references there). A more detailed treatment of the problem has to use the equations (13) and (16) as its basis.

Acknowledgments

Y. Z. acknowledges the financial support by 'Onderzoeksfonds K.U.Leuven' (senior fellowship F/97/095).

REFERENCES

1. Zhugzhda, Y.D., Locans, V., *Solar. Phys. A* **76** 77–108 (1982).
2. Ruderman, M.S., Berghmans, D., Goosens, M., Poedts, S., *A&A* **320** 305–318 (1997).
3. Ruderman, Goosens, M., Ballester, J.I., Oliver R., *A&A* **328** 361–370 (1997).
4. Zhugzhda, Y. D., *Phys. Plasmas.* **3**, 10–21 (1996).
5. Zhugzhda, Y. D. and Nakariakov, V.M., *Phys. Lett. A* **252**, 222–232 (1999).
6. Lighthill, J., *Proc. R. Soc. Lond.* **211**, 564–587 (1952).
7. Ulmschneider, P., Musielak, Z.E. *A&A* **338**, 311–321 (1998).

PART 3: SPACE PLASMAS

Waves and nonlinear structures in bi-ion plasmas

Konrad Sauer, James F. McKenzie
and Eduard Dubinin

Max-Planck-Institut für Aeronomie, 37191 Katlenburg-Lindau

Abstract. The existence of a second ion population in addition to (mostly) protons and electrons is a common signature of laboratory and space plasmas, and even a small admixture may modify the plasma properties significantly. A bi-ion fluid approach is used to describe the collisionless coupling between the two ion populations. It is shown that the appearance of additional wave modes leads to new types of stationary nonlinear structures, such as bi-ion solitons and a magnetic pile-up boundary (or ion composition boundary). We highlight effects observed at comets and Mars.

INTRODUCTION

A strong motivation to conduct theoretical studies of bi-ion plasmas comes from ground-based and in-situ observations of pronounced density structures at comets, and Venus and Mars where a direct interaction of the solar wind with cometary/planetary ionospheres takes place. The paper laid out as follows. Firstly, bi-ion effects occurring in laboratory and space are briefly discussed. Then, the basic fluid model which takes into account the electromagnetic coupling between the two ion fluids is described. This bi-ion fluid model forms the basis for further developments namely dispersion relations, solitons and instabilities. Bi-ion structures which arise from 2D simulations of solar wind mass loading are presented. Finally, a new type of plasma boundary, which exists only in bi-ion plasmas, the so-called 'magnetic pile-up boundary' is discussed.

BI-ION EFFECTS IN LABORATORY AND SPACE

Experimental and theoretical studies of bi-ion effects in laboratory plasmas were already initiated at the beginning of the seventies. The propagation characteristics of ion acoustic waves in an Argon-Helium plasma, for example, were analyzed by Fried et al. [1] in 1971. Two waves modes were predicted. Experimental results [2] are shown in Fig.1a where the wave pattern and the dispersion relation are plotted for different abundance ratios from 5% to 50% Ar admixture in a He plasma. The interference pattern in the middle left panel is due to the superposition of the two wave modes. The right panel on the left shows the variation of the phase velocity of both

CP537, *Waves in Dusty, Solar, and Space Plasmas*, edited by F. Verheest, et al.
© 2000 American Institute of Physics 1-56396-962-9/00/$17.00

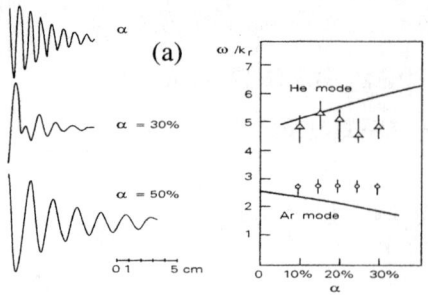
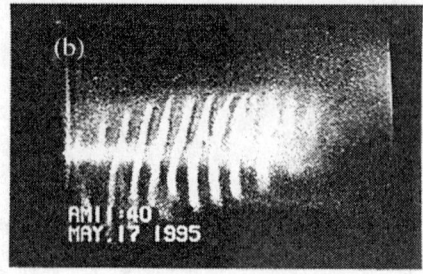

FIGURE 1 a) Wave pattern and dispersion relation of ion-acoustic waves in a He-Ar plasma [2]. b) Propagation of dust-acoustic waves in a Q machine [4].

waves versus the abundance ratio. It has been known for a long time that charged dust particles can act as a second ion population with the result that additional wave modes can arise in a dusty plasma. Dust acoustic waves were first discussed by Rao et al. (1990) [3]. Later, these waves were observed in a Q machine [4], using a video camera as shown in Fig.1b.

Another interesting feature of bi-ion plasmas is the way in which α-particles behave in the solar wind. From the Helios measurements [5] in the eighties it was found that the alphas move faster than the protons by about the local Alfven velocity. The difference velocity increases with decreasing distance to the Sun. Combining results of Helios and Ulysses [6], the measurements follow a $1/r$ dependence.

Pronounced plasma structuring at comets can be seen from ground. The most impressive structures are the cometary tail rays which may extend over several millions of kilometers. As an example, Fig. 2 (left) shows a picture of Hale-Bopp during its largest activity in 1997. Using the bi-ion fluid model, we were able for the first time to explain tail rays as multiple Mach cones [7] This is a subject of later discussion, see also the middle left plot in Fig. 8.

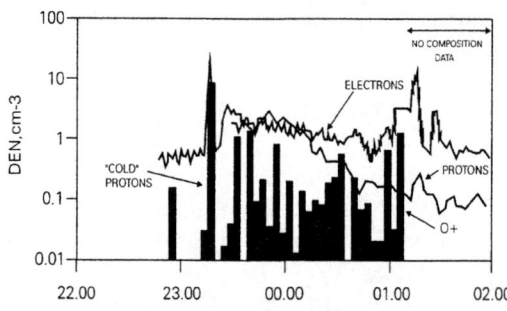

FIGURE 2. Plasma structuring at comet Hale-Bopp (left) and Mars (right). The plasma measurements in the magnetosheath of Mars were made by the Phobos spacecraft in 1989 , after [8] Dubinin et al. (1996).

Measurements by the Russian spacecraft Phobos-2 in 1989 revealed strong plasma structuring in the magnetosheath of Mars. Figure 2 (right) shows the density of protons and oxygen ions within the bow shock. A periodic increase and decrease of both densities is observed with the general trend that a transition from a proton dominated outer sheath to a inner region populated mainly by planetary ions takes place. We shall return to this point later.

THE BI-ION FLUID MODEL

A tractable plasma model for describing different bi-ion effects is the bi-ion fluid model developed e.g. in [9], Sauer et al. (1994). Its main characteristics are that protons (p) and heavies (h) are considered as separate fluids which are coupled by electromagnetic forces. Electrons are massless. Further, charge neutrality is assumed and no collisions are included. Thus, the basic system of equations consists of two continuity and momentum equations for protons and heavies (p↔h), Faraday's law and an electron energy equation (isothermal conditions are taken for the ions):

$$\frac{\partial}{\partial t} n_p + \nabla \cdot (n_p \mathbf{v}_p) = q_p \tag{1}$$

$$\frac{\partial}{\partial t}(n_p \mathbf{v}_p) + \nabla \cdot (n_p \mathbf{v}_p \mathbf{v}_p + P_p/m_p) =$$

$$\frac{1}{m_p} \frac{n_p}{n_e}\left[e n_h(\mathbf{v}_p - \mathbf{v}_h)\mathbf{x}\mathbf{B} - \nabla\left((P_e + \frac{B^2}{2\mu_0})\mathbf{I} - \frac{\mathbf{B}\mathbf{B}}{\mu_0}\right)\right] \tag{2}$$

$$\frac{\partial \mathbf{B}}{\partial t} + \nabla \mathbf{x}\left[\frac{1}{n_e}(n_p \mathbf{v}_p + n_h \mathbf{v}_h)\mathbf{x}\mathbf{B}\right] - \frac{\mathbf{B}\cdot\nabla\mathbf{B}}{e n_e \mu_0} = 0 \tag{3}$$

$$\frac{\partial}{\partial t} P_e + \nabla \cdot(\mathbf{v}_e P_e) + (\gamma - 1)P_e \nabla \cdot \mathbf{v}_e = 0 . \tag{4}$$

with $\quad n_e = n_p + n_h, \quad$ and $\quad \mathbf{v}_e = \frac{n_p \mathbf{v}_p + n_h \mathbf{v}_h}{n_e} - \frac{\nabla \mathbf{x}\mathbf{B}}{n_e \mu_0} . \tag{5}$

The momentum equations directly follow from the common bi-ion fluid equations by eliminating the electric field from the inertia-less electron equation of motion. The main signature is the appearance of a Lorentz-type force term, which becomes important if protons and heavies move with different velocities. Generally, in the continuity equation for the heavy ions a cometary-type source term was taken. Equations (1) – (5) form the governing system of equations for our theoretical studies in the subsequent sections. In the numerical simulations the equations were solved by flux-corrected codes in one and two dimensions.

BI-ION WAVES AND SOLITONS

Before considering stationary one-dimensional solutions of the full non-linear system, we first consider low-frequency electromagnetic waves in a bi-ion plasma. The first such studies were already done in the sixties [10, 11]. But new aspects arising from looking for soliton-type solutions require a more detailed analysis. The procedure is very common. The linearized equations (1) – (5) yield a dispersion relation $D(\omega,k) = 0$. The parameters are θ (the angle between $\mathbf{k}$ and $\mathbf{B}$), β_e (the electron plasma beta), the density ratio $\alpha = n_h/n_p$ and the mass ratio $\mu = m_h/m_p$. The solutions as (ω,k) plots are shown in Figure 3 for $\theta = 30^0$ (a) and for $\theta = 88^0$ (b). The upper left panel shows the dispersion of LF waves in a cold plasma of protons and electrons. There are two modes, the right-hand polarized fast (f) and left-hand-polarized Alfven mode (A). The inclusion of warm electrons leads to a modification of the Alfven mode (intermediate mode = i) and the appearance of an additional mode which is the slow mode (s) [12]. The admixture of heavy ions creates a new mode (marked by hi) and a splitting occurs. The arrow marks the so-called bi-ion cutoff frequency. For quasi-transverse propagation (Fig.3b) the slow mode disappears leaving behind only the magneto-acoustic mode (m-a). But, the signatures, caused by the second ion population, of the existence of a bi-ion cutoff frequency and the heavy-ion mode remain.

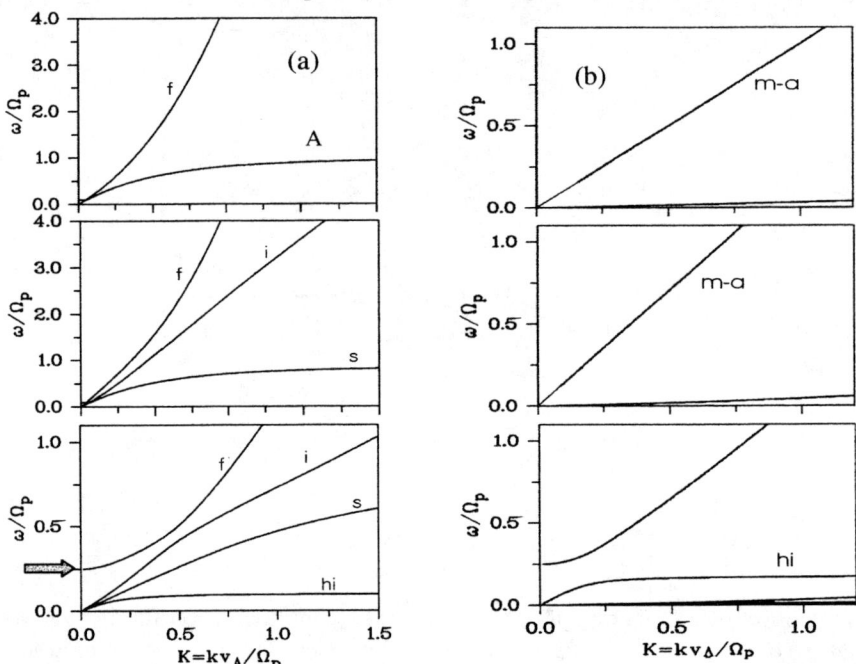

FIGURE 3. Dispersion relation of LF electromagnetic waves for two directions of propagation: a) $\theta = 30^0$, b) $\theta = 88^0$. Upper panel: plasma of cold protons and cold electrons; middle panel: plasma of cold protons and warm electrons ($\beta_e = 3.0$); lower panel: admixture of cold heavy ions. The arrow marks the bi-ion cutoff frequency.

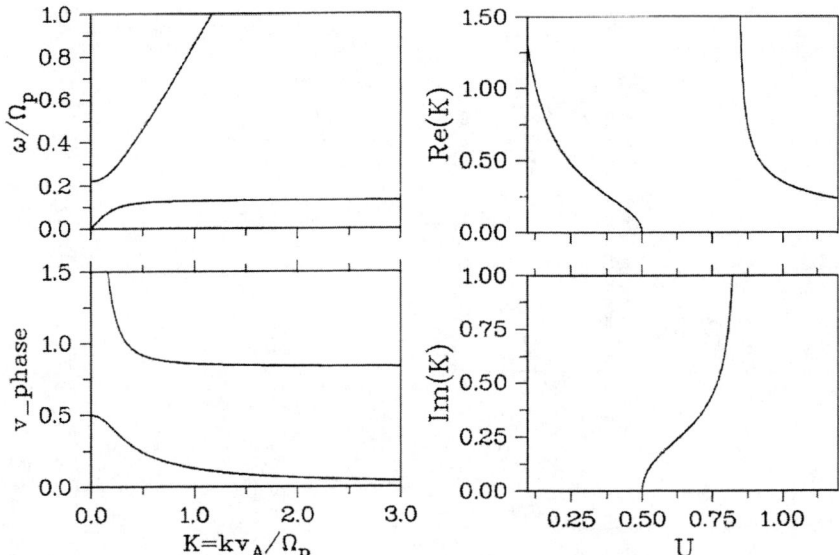

FIGURE 4. Dispersions relation of transverse LF bi-ion waves: Left) Frequency and phase velocity versus k; right) wave number k (real and imaginary part) versus speed of moving frame (stationary waves). The gray-dashed area marks the region of evanescent waves.

In Fig. 4a, both the frequency and the phase velocity are plotted versus k for waves transverse to the magnetic field. Note that the splitting of the heavy-ion mode by the m-a mode leads to a gap in the phase velocity – k diagram. The dispersion relation for structures which appear stationary a frame with velocity U is obtained by putting $\omega = kU$ in the dispersion relation to obtain $k = k(U)$. The solutions are shown in Fig. 4b. The gray-dashed area marks the region where the waves are evanescent. As we will see, this is the region in which solitary solutions can occur.

The structure equations of stationary solutions are obtained from the governing one-dimensional equations using the ansatz $x = x' - M_A t$, where M_A stands for the Mach number of the moving structure ($M_A = U/v_A$, where v_A is the Alfven velocity of protons). In the case of structures moving transverse to **B** (which is in the z direction), the equations can be reduced to a relatively simple system of ordinary differential equations (not given here) for the x-component of the proton and heavy ion velocity and for the y-component of the proton velocity. If these three quantities are known, all other variables can easily determined by algebraic relations.

Examples for transverse bi-ion solitons are given in Figure 5 for two abundance ratios: (a) $\alpha = 0.2$ and (b) $\alpha = 1.5$. The mass ratio is $\mu = 15$ for both cases. The plots show the proton and heavy ion density (middle two panels), the x velocities of both species (lower two panels), the y- component of the proton velocity and the magnetic field (upper two panels). The bi-ion soliton moves with a uniform speed as a stable configuration. This was checked by using the non-stationary equations. Especially, it was seen that the two solitons may collide without changing their identity after the collision. An interesting new type of soliton, which may be of relevance to the Martian magnetosheath near the 'ion composition boundary', arise if the heavy-ion density begins to dominate (Fig. 5b). The most pronounced effect is the depression of the proton density in the center of the soliton, in regions where the heavy ion density

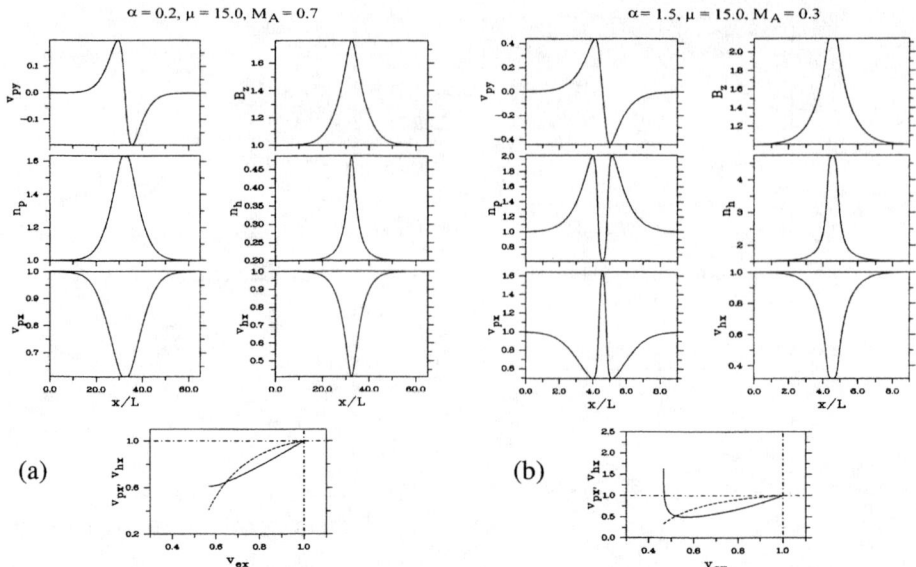

FIGURE 5. Spatial structure of transverse bi-ion solitons for two abundance ratios: (a) α=0.2 and (b) α=1.5. The mass ratio is μ=15 for both cases. Upper panels: y component of proton velocity and magnetic field B_z; middle panels: proton and heavy ion density; lower panels: x velocities of both ion species. The plot on bottom shows the 'momentum hodograph': v_{px} (solid line) and v_{hx} (dotted line) versus the electron velocity v_{ex}.

reaches its maximum. Such structures have not been described before, and we believe that they may explain some curious ion effects seen at Mars.

INSTABILITIES BY ION DRIFTS

Next we consider the excitation of LF waves by a relative drift parallel to the magnetic field. In most earlier papers a kinetic analysis was carried out using the Vlasov approach [13,14]. Recently, the dispersion relation of the multi-fluid theory was solved to determine the growth rate of unstable waves with respect to LF turbulence at comet Halley [15]. By using the bi-ion fluid approach, our analysis follows the same line. The motion of heavies with respect to the protons creates a beam-plasma configuration in which a resonant interaction of the beam with the right-hand polarized fast mode takes place. If we apply our fluid analysis to the conditions in [16], where a high-speed solar wind flow through a Barium cloud is considered, we obtain a complete agreement with the kinetic results. Thus, kinetic effects do not play an important role if the drift velocity is large enough to prevent wave-particle interaction.

New results were found for small-scale LF instabilities due to a relative drift *transverse* to the magnetic field [17-19]. We consider effects which arise on distances much smaller than the heavy-ion gyro radius. Thus, the heavies can be treated as non-magnetized and at rest. The model, therefore, considers moving protons, electrons and cold non-magnetized heavies at rest. The dispersion analysis shows that waves very close to the proton cyclotron frequency are excited. The mechanism is based on the resonant interaction between the beam and the fast mode.

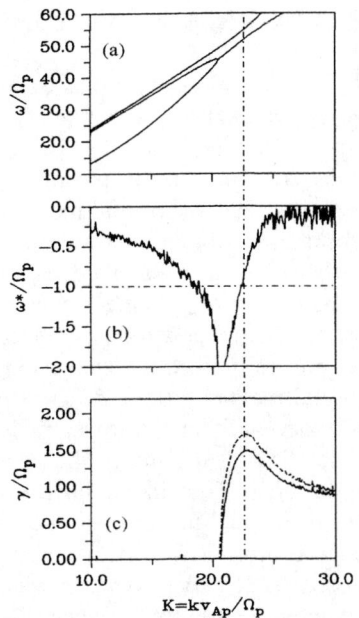

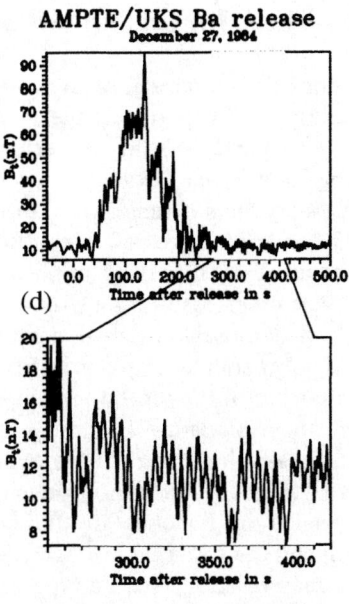

FIGURE 6. Dispersion relation for beam-excited whistlers ($M_A=3$, $\Theta=84^0$, $n_h/n_p=0.1$, $m_h/m_p=136$). The ω-k diagram arises from the intersection of the (heavy ion) beam mode with the fast mode of the background plasma; (a) in the proton reference frame, (b) in the beam frame. (c) shows the growth rate. Dashed and solid curves correspond to the cases with and without damping due to electron temperature effects ($\beta_e=0.1$). (d) Magnetic field during the barium release on December 27, 1984, in the solar wind [19]. The high-resolution plot (bottom) shows oscillations in two frequency bands, one is close to Ω_p.

The dispersion curves in the solar wind frame are shown in the left panels of Fig. 6. By resonant interaction a mode splitting arises (a), which is associated with an instability (c). Going back into the beam frame (b), that is the spacecraft frame, waves near the proton cyclotron frequency Ω_p are excited. This new type of wave excitation is supported by Phobos observations at Mars in 1989. A clear peak in the power spectrum of magnetic fluctuations at Ω_p was observed upstream of the bow shock [19,20]. Recent measurements with higher resolution aboard Mars Global Surveyor have confirmed that wave emission at Ω_p seems to be a general signature of the Martian magnetosphere indicating the presence of heavy ions of planetary origin. Evidence that this instability is really caused by the above mechanism and not by pick-up of protons also comes from the AMPTE barium release [19]. Fig. 6 (d) shows the magnetic field after Ba was released into the solar wind. First, there is a large compression up to a factor of about 10. In the later phase of the experiment, called the transition region, amplitude oscillations occur which result from the superposition of two waves. One emission is at the proton cyclotron frequency. This is clearly seen in the high-resolution plot inserted on the bottom of Fig. 6d. The lower frequency band is near $0.2\,\Omega_p$, which can be explained in a similar way as the Ω_p emission by resonant beam-plasma interaction in the range of the bi-ion cutoff frequency [17,19].

2D SIMULATIONS OF SOLAR WIND MASS LOADING

We now consider numerical results from our 2D bi-ion fluid simulations of solar wind mass loading [21]. This means there is a cometary-type source of heavy ions (located at $x = 0$, $y = 0$) and we study the effect of solar wind interaction. After filling the box with the solar wind at $t = 0$, the integration was continued up to the point where a quasi-stationary state is reached. In order to check that the bi-ion fluid model is able to describe the main features of mass loading, hybrid code simulations were also carried out. In contrast to the fluid approach, here the ions are described as individual particles. All other assumptions are the same, namely massless electrons and quasi-neutrality. A comparison between both approaches for solar wind interaction with a weak heavy-ion source is given in Fig. 7. From top to bottom the heavy ion density, the x-component of the proton velocity and the total magnetic field are shown as color plots. The main signature is the cycloidal trajectory seen in the upper panels along which the heavy ions begin to move due to the ExB force, as in the case of test particles. Even if the heavy ion density is small a clear fluid-like behavior arises, as the occurrence of bunches in the heavy ion flow due to resonant beam-plasma interaction and the formation of Mach cones in the proton flow. It is interesting to note that Pluto may resemble such a type of weak source [22] which would produce a very asymmetric plasma configuration around the planet. At first glance it may be surprising that there is a good agreement between the kinetic and bi-ion fluid simulations, but it supports the general philosophy that the crucial effects are caused by the electromagnetic coupling between both fluids, and not by single particle dynamics.

Fig. 8 shows plasma structures from bi-ion fluid simulations which are formed by solar wind interaction for three mass loading regimes, beginning with a weak source $(Q \sim 10^{25}\ s^{-1})$ on top to a moderate source $(Q \sim 10^{27}\ s^{-1})$ on bottom. A general feature of the interaction is plasma structuring and also tail rays are clearly visible. The structures are moving downstream and new ones arise due to the continuous production of new ions. We have to emphasize the interesting fact that tail rays resemble here multiple Mach cones which have nothing to do with magnetic field draping as suggested by Alfven [23] long ago. Another remarkable property which came out of the simulations is the 'bow shock splitting' [24]. Looking on the middle and upper two panels, one can see that the jump in the proton density at the bow shock does not coincide with the first sharp increase of the heavy ion density which occurs further inside of the proton shock. The filamentary structure within the bow shock becomes more pronounced with a further increase of the source strength. On approaching the center of the obstacle, the heavies begin to dominate over the protons. But, as shown in several papers [9, 25] the change in ion composition is, at least, not a gradual process. There is a 'critical point' where the flow velocity reaches the local phase velocity of a bi-ion wave mode. This marks a discontinuity (protonopause = ion composition boundary) at which the proton flow is stopped leaving behind a 'proton cavity' [9], seen in the lower right plot of Fig. 8 as a 'dark area'. How the discontinuity above is related to the magnetic pile-up boundary seen at comets and Mars will be discussed subsequently.

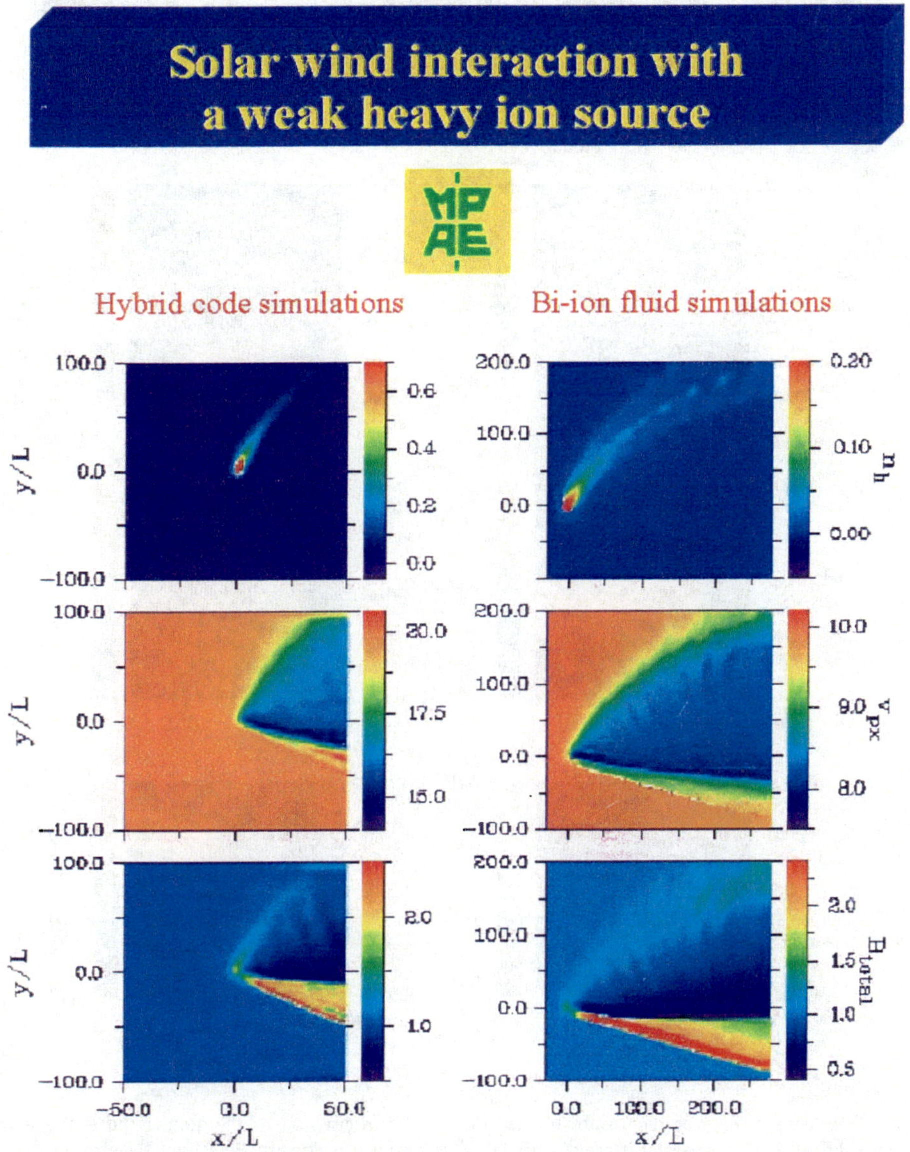

Solar wind interaction with a weak heavy ion source

Hybrid code simulations   Bi-ion fluid simulations

FIGURE 7. Solar wind interaction with a weak heavy ion source, comparison between 2D hybrid code and bi-ion fluid simulations [21]. The magnetic field is out of the simulation box. Although the flow parameters are not the same (left: $M_A=20$, right: $M_A=10$), the overall interaction pattern, showing the heavy-ion cycloid and multiple Mach cones, coincides well. Weak plasma structuring is already visible.

Plasma structures for three mass-loading regimes

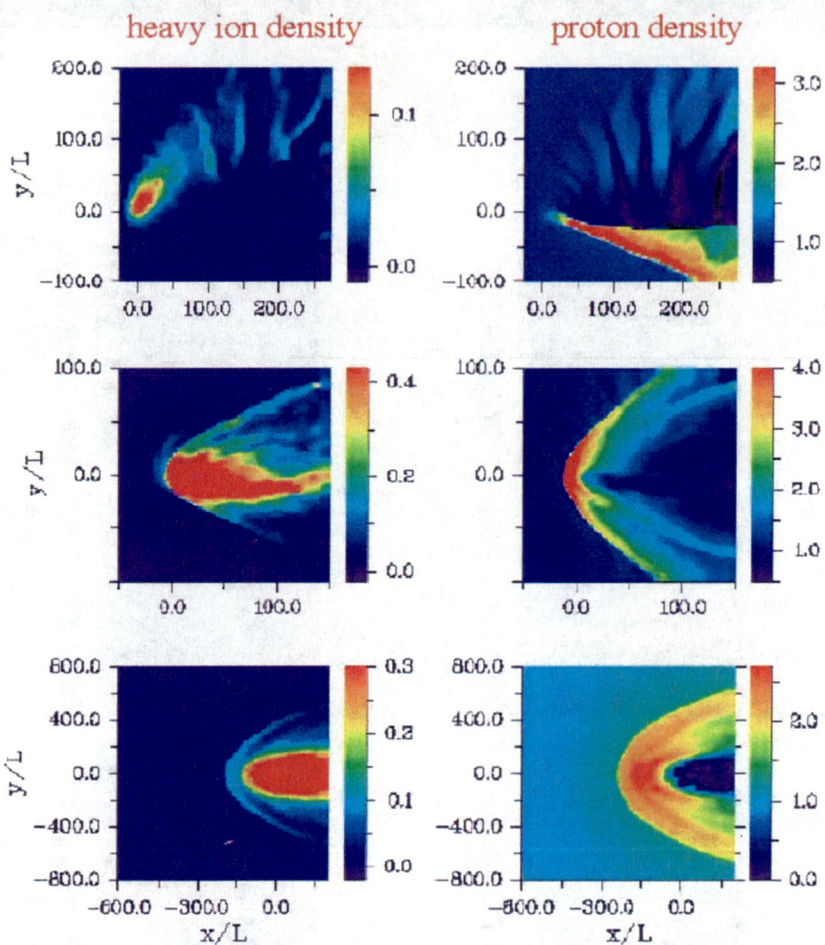

Figure 8. Plasma structures which are formed due to the interaction of the solar wind (SW) with a cometary heavy-ion source of different activity. The heavy ion density (left) and the proton density (right) are shown for three production rates: $Q \sim 10^{25}$ s^{-1} (on top), $Q \sim 10^{26}$ s^{-1} (in the middle) and $Q \sim 10^{27}$ s^{-1} (at bottom). In the first two examples (top and middle pannels) the SW magnetic field is oriented out of the simulation plane. In the bottom case, the SW magnetic field is in the y direction [21].

336

THE MAGNETIC PILE-UP BOUNDARY

Finally, we briefly discuss the structure of the Martian magnetosphere. From recent MGS measurements [26] it has been confirmed that Mars has only a very weak intrinsic magnetic field. Therefore, the exosphere/ionosphere is not shielded by a magnetic field with implication that the solar wind interaction has cometary-type character. A characteristic variation of the magnetic field along the MGS elliptical orbit (without crossing magnetic anomalies) is shown in the top panel of Fig. 9. Three plasma boundaries have been observed: bow shock (BS), magnetic pile-up boundary (MPB) and ionopause (IP). The nature of the MPB is still under ongoing debate, since one-fluid MHD models are not able to explain it. The same problem exists at comets where a similar boundary was detected during the flybys of comet Halley, Giacobini-Zinner and Grigg-Skjellerup. The bottom panel of Fig. 9 shows the observation at comet Halley [27].

To solve the problem about the physical nature of the MPB, the former ion measurements at Mars aboard the Phobos spacecraft in 1989 are of great importance. They showed the existence of a 'proton boundary' which is related to a relatively abrupt change in the ion composition [28, 29]: protons are replaced by planetary ions. Now there is a general agreement that the magnetic pile-up, the change in ion composition and the drop of the electron temperature are different signatures of one and the same plasma boundary, which obviously cannot be described by one-fluid MHD models. As discussed in the previous section, an additional plasma boundary (protonopause [9]) was obtained in our bi-ion fluid simulations which posses all the main signatures of the unusual discontinuity observed at comets and Mars [21, 25].

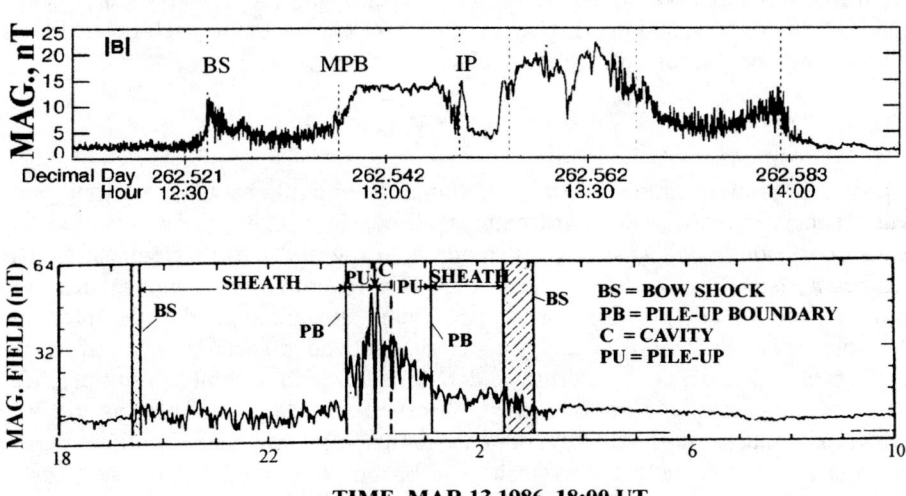

FIGURE 9. Top) Plasma boundaries observed at Mars by MGS [26]: BS = bow shock, MPB = magnetic pile-up boundary; IP = ionopause. Bottom) Similar boundaries were observed at comet Halley [27] on much larger scales.

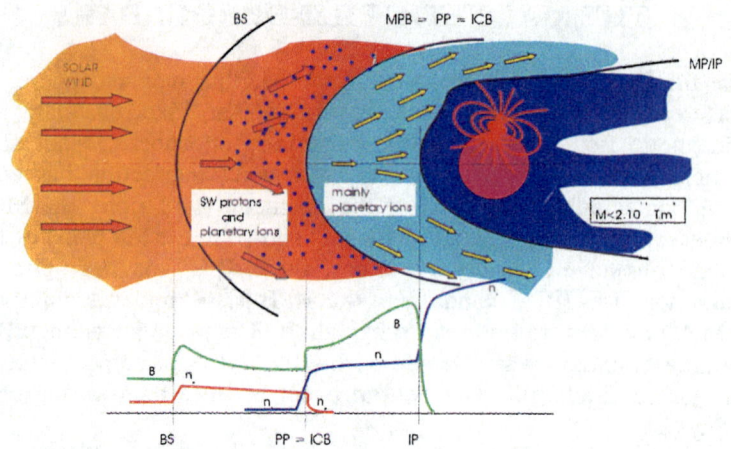

Figure 10. Our present view of the Martian magnetosphere, s.a. [25]. Its main signature is the presence of three plasma boundaries: bow shock (BS), magnetic pile-up boundary (MPB) = protonopause (PP) = ion composition boundary (ICB) and magneto/ionopause (MP/IP). The variation of the plasma parameters along the Sun-Mars line is schematically drawn at the bottom.

Fig. 10 shows our present view of the Martian magnetosphere. It was constructed by combining essential experimental results from both the Phobos and MGS mission with the outcome of our numerical simulations described before. The main characteristics is the presence of a third plasma boundary between bow shock and ionopause with following signatures: magnetic pile-up (MPB), stop of proton flow (PP), change in ion composition (ICB). This boundary arises due to the electromagnetic coupling between the both ion fluids and was studied in several papers. But, more analysis is necessary to find out how the up- and downstream parameters are related to each other, like the Rankine-Hugoniot relations for usual shocks.

SUMMARY

The bi-ion fluid description of the collisionless coupling between protons and a heavier species reveals interesting features absent in a one-fluid description. The appearance of additional wave modes, mode splitting and cut-off frequencies gives rise to beam instabilities and new types of nonlinear stationary structures, such as bi-ion solitons and ion composition boundaries. The results are applied to the interaction of the solar wind with comets and weakly magnetized planets such as Mars. The general feature is plasma structuring and tail rays which resemble multiple Mach cones. The magnetic pile-up boundary observed at Mars and comets finds an explanation. Finally, it should be pointed out that a lot of other situations exist in space where one can expect new results if bi-ion fluid models are used instead classical MHD approaches. The ongoing interpretation of the actual measurements near the moons of Jupiter, e.g. [30] (and hopefully later for the moons of Saturn) may profit from the experience gained from the studies of plasma-plasma interaction at Mars and comets.

ACKNOWLEDGMENTS

The work of Eduard Dubinin was partially supported by the guest program of the Deutsche Forschungsgemeinschaft.

REFERENCES

1. Fried, B. D., White, R. B., and Samec, T. M., *Phys. Fluids*, **14**, 2388 (1971).
2. Tran, M. Q., and Coquerand, S., *Phys. Rev. A*, **14**, 2301 (1976).
3. Rao, N. N., Shukla, P. K., and Yu, M. Y., *Planet. Space Sci*, **38**, 543-546 (1991).
4. Barkan, A., D'Angelo, N., and Merlino, R. L., *Advances in Dusty Plasmas*, World Scientific, Singapore, 1997, pp. 30-40.
5. Marsch, E., et al., *J. Geophys. Res.*, **87**, 35, 1982.
6. Neugebauer, M., Goldstein, B. E., Smith, E. J., and Feldman, W. C., *J. Geophys. Res.*, **101**, 17,047, 1996.
7. Sauer, K., and Dubinin, E., *Earth, Moon and Planets*, 77, 271-278, 1999.
8. Dubinin, E., Sauer, K., Lundin, R., Baumgärtel, K. and Bogdanov, A., *Geophys. Res. Lett.*, 23, 785 - 788, 1996.
9. Sauer, K. Bogdanov, A., and Baumgärtel, K., *Geophys. Res. Lett.* 21, 2255-2258, 1994.
10. Buchsbaum, S.J., *Phys. Fluids*, **3**, 418, 1960.
11. Smith, R. L., and Brice, N., *J. Geopyhs. Res.*, **69**, 5029, 1964.
12. Krauss-Varban, D., Omidi, N., and Quest, K. B., *J. Geophys. Res.*, **99**, 5987, 1994.
13. Gary, S. P., Smith, M. A., Lee, M. A., Goldstein, M. L., and Forslund, D. W., *Phys. Fluids*, **27**, 1852, 1984.
14. Winske, D., and Gary, S. P., *J. Geophys.Res.*, **91**, 6825, 1986.
15. Verheest, F., Lakhina, G. S., and Tsurutani, B. T., *J. Geopyhs. Res.*, **104**, 24,863, 1999.
16. Wang, J. J., Gary, S. P., and Liewer, P. C., *J. Geopyhs. Res.*, **104**, 24,807, 1999.
17. Sauer, K., Dubinin, E., Baumgärtel, K., and Tarasov, V., *Earth, Planets, Space,***50**, 269, 1998.
18. Baumgärtel, K., Sauer, K., Dubinin, E., Tarasov, V., and Dougherty, M., *Earth, Planets, Space,***50**, 453, 1999.
19. Sauer, K., Dubinin, E., Dunlop, M., Baumgärtel, K., and Tarasov, V., J. Geophys. Res., **104**, 6763, 1999.
20. Russel, C. T., et al., Geophys. Res. Lett., **17**, 97, 1990.
21. Sauer, K., and Dubinin, E., MPAe Report, MPAE-W-100-99-02, 1999; to be published in *Physics of Mass Loaded Plasmas,* Kluwer Academic Publishers, 2001.
22. Sauer, K., Lipatov, A., Baumgärtel, K., and Dubinin, E., *Adv. Space Res.*, **20**, 295, 1997.
23. Alfven, H., *Tellur,* **9**, 12, 1957.
24. Sauer, K., Dubinin, E., Baumgärtel, K., and Bogdanov, A., *Geophys. Res. Lett.,* **23**, 3643, 1996.
25. Sauer, K., and Dubinin, E., *Adv. Space Res.*, **26**, 1633, 2000.
26. Acuna, M., et al., *Science*, **279**, 1676, 1998.
27. Neubauer , F., *Astron. Astrophys.* **187**, 73, 1987.
28. Breus, T. K., et al., J. Geophys. Res., **96**, 1165, 1991.
29. Dubinin, E., Sauer, K., Lundin, R., Baumgärtel, K., and Bogdanov, A., *Geophys. Res. Lett.*, **23**, 785 1996.
30. Khuruna, K. K., Kivelson, M. G., and Russel, C. T., *Geophys. Res. Lett.*, **19**, 2391, 1997.

Magnetohydrodynamic wave mode conversion at the Earth's magnetopause

J. De Keyser

Belgian Institute for Space Aeronomy, Ringlaan 3, B-1180 Brussels, Belgium
(Johan.DeKeyser@bira-iasb.oma.be)

Abstract. Broadband ULF waves are routinely observed throughout the magnetosheath. These waves may be of solar wind origin or they might be generated closer to Earth, for instance, in the foreshock. Magnetohydrodynamic (MHD) wave mode conversion is a mechanism capable of delivering the energy flux carried by these ULF MHD waves to the plasma at the magnetopause. We present numerical simulations for typical subsolar magnetopause configurations. Resonant coupling to Alfvén and slow-mode waves can occur. Energy is fed to the plasma in the resonant sheets, which can explain the enhanced ULF fluctuation level observed at the magnetopause.

INTRODUCTION

The magnetopause separates the magnetosheath from the magnetosphere. Although the magnetopause can be described globally by a tangential or rotational discontinuity equilibrium, it is strongly affected by transient phenomena. These are related to mass and energy transport mechanisms across the magnetopause, such as flux transfer events, impulsive plasma penetration, percolation and diffusion.

This paper focuses on the effect of the ever-present ultralow frequency (ULF) waves in the magnetosheath [1] on the magnetopause. These ULF waves exist in the broadband solar wind spectrum, but they can also be generated near the Earth, for instance as turbulence in the foreshock [2]. These waves are convected across the bow shock and through the magnetosheath [3–5] and can interact with the magnetopause, and in certain circumstances also with inhomogeneities deep within the magnetosphere [6,7]. The ULF fluctuation level peaks at the magnetopause and is suppressed in the magnetosphere [2,3,8–11]. The peak at the magnetopause could be due to wave mode conversion of the incident waves and the corresponding resonant amplification.

MHD wave mode conversion is the process by which MHD waves are transformed from one mode into another one. This occurs when fast-mode waves that propagate across magnetic field lines reach a point where their wave vector becomes parallel to the field lines. This happens in resonant sheets parallel to the field lines, where

CP537, *Waves in Dusty, Solar, and Space Plasmas*, edited by F. Verheest, et al.
© 2000 American Institute of Physics 1-56396-962-9/00/$17.00

the tangential wave vector of the incident wave matches the local Alfvén or cusp frequency. This produces Alfvén or slow-mode surface waves, which are damped due to dissipation. As energy flux across the field lines is diverted into the direction parallel to the field lines and dissipated, a jump in the normal energy flux is observed at the resonant sheets: the energy is resonantly absorbed. The wave amplitude at the sites of resonance becomes very large, hence the term resonant amplification.

SPECTRAL DESCRIPTION OF LINEAR MHD WAVES

We linearize the ideal MHD equations around a one-dimensional equilibrium configuration by writing every quantity (depending on position r and time t) as a sum of its equilibrium value and small-amplitude higher order contributions: $q(r, t) = q^{(0)}(x) + q^{(1)}(r, t) + \ldots$ The response of the system to a superposition of several small-amplitude waves is the superposition of the responses to each individual wave. We consider perturbations $q^{(1)}$ with frequencies $\omega_{\pm n} = \pm n \omega_{\text{base}} = \pm 2\pi n f_{\text{base}}$, where $n = 1, 2, \ldots, n_{\max}$, and with wave vectors $k = k_x(x)\mathbf{1}_x + k_t$, where the constant tangential wave vector $k_t = [0, k_y, k_z]$ is given:

$$q^{(1)}(r, t) = \sum_{n=1}^{n_{\max}} \hat{q}_n(x) e^{i(k_t r - \omega_n t)} + \hat{q}_{-n}(x) e^{i(k_t r + \omega_n t)}.$$

For each ω_n the perturbation $\hat{q}_n$ is computed from the linearized MHD equations in terms of the normal displacement $\hat{\xi}_x$ and the total pressure perturbation $\hat{\tau}$:

$$\frac{d}{dx}\hat{\tau} = C_1(x, k_t, |\omega_n|)\hat{\xi}_x, \qquad \frac{d}{dx}\hat{\xi}_x = C_2(x, k_t, |\omega_n|)\hat{\tau}.$$

Expressions for C_1 and C_2 are given in [12]. In uniform regions the equations become

$$\left(\frac{d^2}{dx^2} + K_x^2\right)\hat{\xi}_x = 0,$$

with $K_x^2 = -C_1 C_2$. If $K_x^2 > 0$ the solution is a superposition of a left- and a right-going wave, with $k_x = \pm K_x$ (propagating waves). If $K_x^2 < 0$ the solution consists of exponentially growing and decaying waves with $k_x = \pm i|K_x|$ (non-propagating waves). The resonance conditions reflect the coupling between the incident fast-mode and the Alfvén and slow-mode waves at the singular points:

$$|k_{\parallel}| = k_A = \frac{|\omega_n|}{v_A}, \qquad |k_{\parallel}| = k_S = \frac{|\omega_n|}{v_A}\sqrt{v_A^2 + c_s^2};$$

k_A and k_S are the local Alfvén and slow-mode wave numbers [12]. Each frequency component is characterized by its proper resonance conditions. Following *Belmont et al.* [13] we consider surface wave vectors $k_t + \epsilon_t i$ with $\epsilon_t \ll k_t$, corresponding to wave fronts that are slightly modulated in the tangential direction, rather than being exactly planar. This avoids the singularities at the points of resonance. It can be shown that the solution is independent of the precise value of ϵ_t if it is small enough. We solve the ordinary differential equations for $\hat{\xi}_x$ and $\hat{\tau}$ using an adaptive second-order implicit integrator. We have chosen $\epsilon_t/k_t \sim 10^{-6}$. Thanks to the adaptivity of the integrator we are able to resolve the resonant sheets, whose thickness scales with ϵ_t.

It is difficult to specify boundary conditions on the magnetosheath side, as both incident and reflected waves are present there. We therefore require (for each frequency component) that, at the magnetospheric side, only a left-going wave is present; then $\hat{\tau} = \pm \hat{\xi}_x C_1/C_2^{1/2}$ there, where the sign is chosen so as to select the left-going wave [12]. For each frequency the solution is computed with $\hat{\xi}_x(x_{\mathrm{msph}}) = 1$.

Total pressure variations in the magnetosheath may be due to dynamic, thermal, or magnetic pressure fluctuations. While we impose density perturbations in the magnetosphere, the model self-consistently computes the incident and reflected magnetosheath waves, which involve variations of all plasma and field parameters. The density perturbation at $x = x_{\mathrm{msph}}$, $y = z = 0$, has a Fourier expansion

$$s_{\mathrm{msph}}(t) = \sum_{n=1}^{n_{\max}} s_n e^{-i\omega_n t} + s_{-n} e^{+i\omega_n t}.$$

As s_{msph} is real, $s_n \equiv s_{-n}^*$ (* denotes the complex conjugate). Imposing $s_{\mathrm{msph}}(t) = \rho^{(1)}(x_{\mathrm{msph}}, t)$, or $s_n = \hat{\rho}_n(x_{\mathrm{msph}})$, produces the desired linear combination of the spectral components that matches the driver. We then have $\hat{q}_n \equiv \hat{q}_{-n}^*$ for any perturbed quantity, corresponding to real perturbations $q^{(1)}(x, t)$. We choose s_{msph} to have an arbitrarily small unit amplitude, as the solution is fixed only up to a scaling factor.

We consider frequencies f in the range $5 - 500$ mHz, below the ion gyrofrequency. The magnetopause layer was chosen to be several ion gyroradii wide, so that the MHD approximation does apply. The one-dimensional approximation is valid only when the wavelengths are smaller than the curvature radius of the magnetopause. The magnetosheath flow also cannot be ignored far from the subsolar point.

MONOCHROMATIC WAVES

We consider the subsolar magnetopause configuration used by *De Keyser et al.* [12], which is of the tangential discontinuity type without plasma flow, with a smooth density and temperature variation across the magnetopause and a unidirectional magnetic field. The magnetopause has a half-thickness $D = 300$ km [14] and is centered at $x = 0$ (x is the coordinate along the Earth–Sun line). This equilibrium is characterized by Alfvén speeds $v_{A,\mathrm{msh}} = 225$ km s^{-1} $< v_{A,\mathrm{msph}} = 1420$ km s^{-1} and sound speeds $c_{s,\mathrm{msh}} = 235$ km s^{-1} $< c_{s,\mathrm{msph}} = 525$ km s^{-1}.

Figure 1 (taken from [12]) shows some examples of monochromatic waves ($f = 0.5$ Hz) with various angles of incidence (different k_t). The left plots show the minimum and maximum values (over time) of the magnetic perturbation magnitude

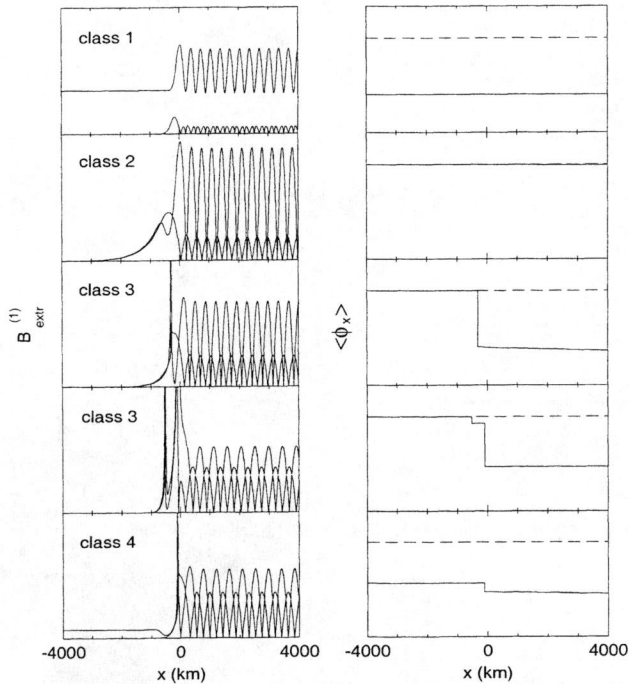

FIGURE 1. Classes of monochromatic wave solutions. The left plots give the magnetic field perturbation amplitude (in arbitrary units), the right ones give the normal time-averaged energy flux (in arbitrary units). The magnetosphere is to the left, the magnetosheath to the right in each plot. Class 1: Transmission and reflection. Class 2: Full reflection. Class 3: Absorption and reflection. Class 4: Transmission, absorption, and reflection.

$B^{(1)}$. The right plots give the time-averaged normal energy flux profiles; $\langle \phi_x \rangle < 0$ corresponds to earthward flux. The magnetosphere is to the left, the magnetosheath to the right in each plot. These examples show wave transmission (nonzero wave amplitude and nonzero flux in the magnetosphere, class 1 and 4), reflection (all classes), and/or absorption (narrow peaks in $B^{(1)}$, jumps in the flux, class 3 and 4). The resonances in the fourth example correspond to Alfvén and slow-mode waves.

INTERMITTENT WAVES

We now focus on the response of the magnetopause to density pulses, as an example of intermittent waves. We use a periodic sequence ($f_{\text{base}} = 5$ mHz, $n_{\text{max}} = 100$) of pulses of unit magnitude with alternating sign, with $\boldsymbol{k}_t = [0.5, 0]k_{A,\text{msph}}$. Each pulse lasts 10 s. The base frequency wave and its harmonics are all propagating in both the magnetosheath and the magnetosphere and no mode conversion occurs.

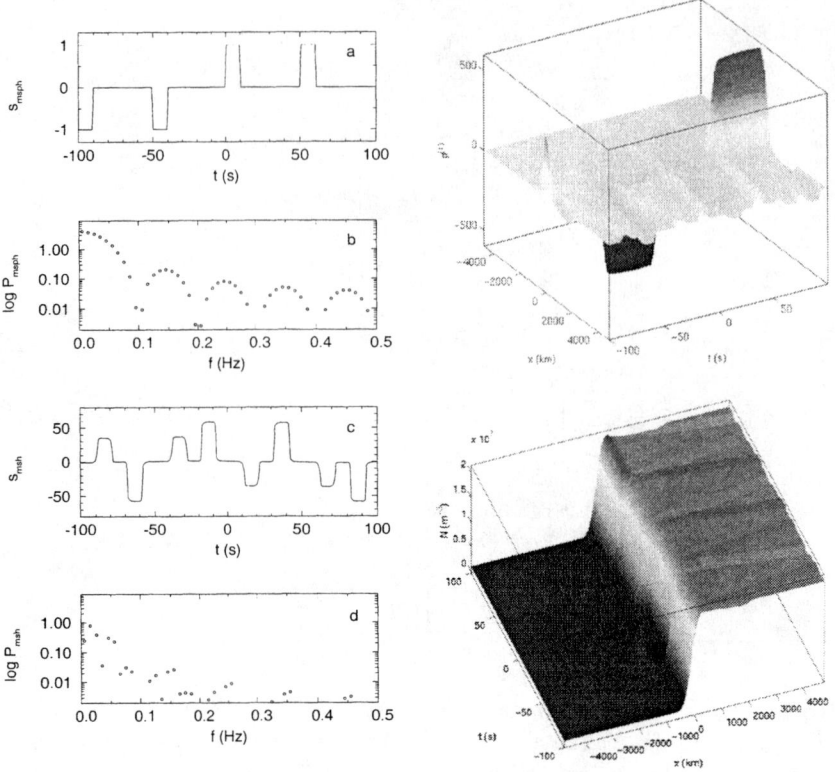

FIGURE 2. Density pulses incident on the magnetopause. (a) Density perturbation imposed as magnetospheric boundary condition. (b) Power spectral density. (c) Density perturbation computed at the magnetosheath edge. (d) Power spectral density. (e) Density perturbation as function of time. (f) Total density.

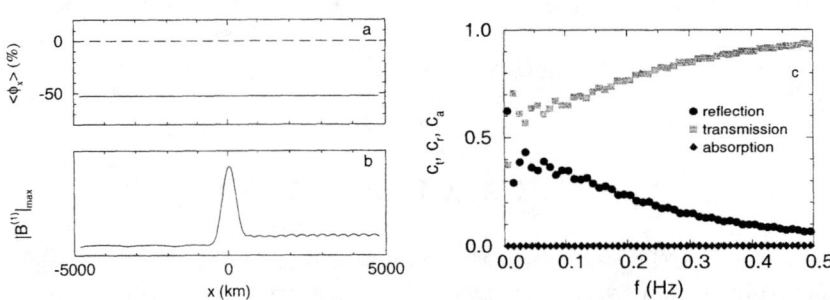

FIGURE 3. Density pulses incident on the magnetopause. (a) Energy flux. (b) Magnetic perturbation amplitude. (c) Spectral reflection, transmission, and absorption coefficients.

Figure 2a–2d (taken from [15]) show the density perturbation imposed on the magnetospheric side and its power spectrum, and the corresponding density perturbation on the magnetosheath side (showing the incident and reflected pulses) and its power spectrum. Figure 2e plots the density perturbations $\rho^{(1)}(x, y = 0, z = 0, t)$, revealing how the incident magnetosheath pulses produce reflections. The density perturbation is maximal in the magnetopause, where it has about 10 times the magnetosheath perturbation amplitude. The transmitted pulse amplitude is only ~1/50th of the incident one. Figure 2f illustrates the effect of the perturbation on the total density. We have superposed the perturbation on the equilibrium density, thereby choosing its amplitude small enough so as to guarantee positivity of density and pressure everywhere. The effect of the pulse train is an oscillating motion of the magnetopause over a distance of the order of the magnetopause thickness (larger displacements can only be produced in the nonlinear context).

Although there is no resonant absorption (see Figure 3a) the wave amplitude is enhanced at the magnetopause (Figure 3b). We define the transmission, reflection, and absorption coefficients c_t, c_r, and c_a as the transmitted, reflected, and absorbed time averaged flux relative to the incident time averaged energy flux ($c_t + c_r + c_a = 1$). The spectral transmission, reflection, and absorption coefficients are given in Figure 3c; the overall values (integrated over all frequencies) are $c_t = 53$ %, $c_r = 47$ %, and $c_a = 0$ %.

BROADBAND CONTINUUM WAVES

Observations often show sustained broadband ULF waves in the magnetosheath [4]. In order to simulate this, we choose $f_{\text{base}} = 5$ mHz, $n_{\text{max}} = 100$ and $k_t = [40, 40]k_{A,\text{msph}}$), corresponding to a tangential Alfvén wavelength of 5000 km (example taken from [15]). As the lowest frequency modes are non-propagating in the magnetosheath and the magnetosphere we remove them from the driver. There is a broad band of modes for which Alfvén and slow-mode resonances occur inside the magnetopause.

Figure 4a and 4b show the imposed magnetospheric perturbation (solid line) obtained by removing the lowest frequencies from the square pulse train (dashed line), and its power spectrum. The magnetosheath perturbation (Figure 4c and d) is irregular; the frequency components that do not propagate in the magnetosphere dominate the broadband ULF magnetosheath spectrum. The resonant absorption is evident in the energy flux profile (Figure 4e): the discrete frequency spectrum used here is so close to a continuous one that the profile is a smooth curve rather than a staircase line, at least in the earthward part of the magnetopause layer. The magnetic perturbation features closely spaced, narrow, high peaks (width and height depend on the dissipation level). Figure 4f shows the spectral transmission, absorption and reflection coefficients. Frequencies below 50 mHz are absent (non-propagating in magnetosphere and magnetosheath). Below 300 mHz, $c_t = 0$ as such waves do not propagate in the magnetosphere. Above 300 mHz the magnetosphere

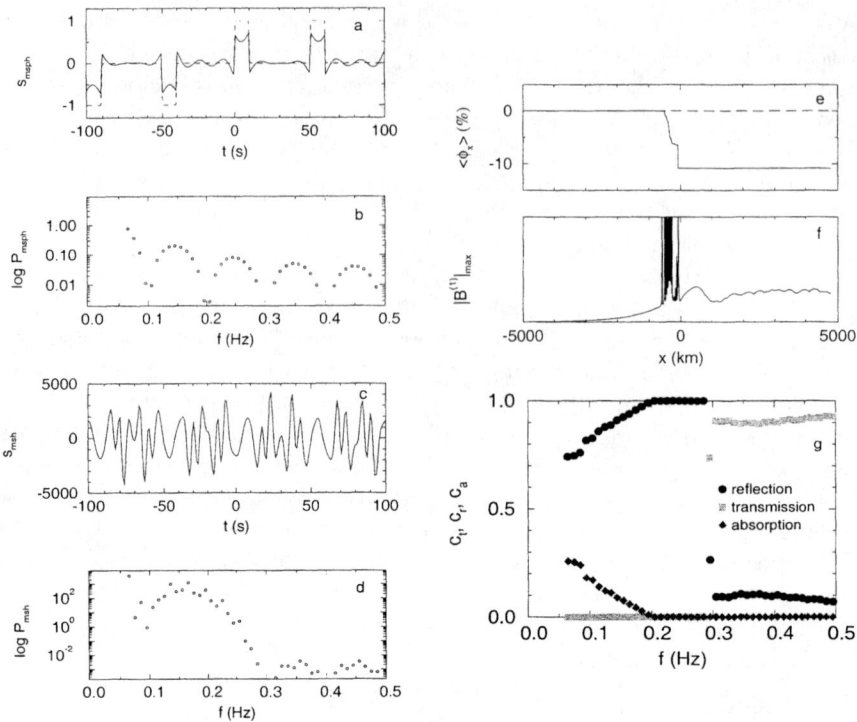

FIGURE 4. Broadband continuum waves incident on the magnetopause. (a) Density perturbation imposed as magnetospheric boundary condition. (b) Power spectral density. (c) Density perturbation computed at the magnetosheath edge. (d) Power spectral density. (e) Normal time-averaged energy flux. (f) Magnetic perturbation amplitude. (g) Spectral transmission, reflection and absorption coefficients.

becomes transparant; about 90 % of the incident flux is transmitted. The integrated transmission is about 0 %, the absorption amounts to 11 %, while 89 % is reflected. The small integrated transmission is due to the negligible contribution of the highest frequency components in the incident perturbation.

DISCUSSION

This paper evaluates a scenario for the transport of electromagnetic energy present in magnetosheath ULF fluctuations to the magnetopause. The calculations have been carried out in the framework of linear MHD. We have analyzed the behavior of monochromatic, intermittent, and broadband ULF waves at the magnetopause. Both intermittent, "spiky" perturbations, as well as broadband ULF continua have been identified in the magnetosheath [3–5,8,9].

These simulations can explain several observed phenomena at the magnetopause:

small-scale motion of the magnetopause, small fluctuations of the normal magnetic field at the magnetopause, the enhanced ULF fluctuation level throughout the magnetopause (not in just one or a few isolated layers), and the suppression of magnetic field fluctuations in the magnetosphere. In spite of the obvious limitations of linear MHD, the simulations corroborate the hypothesis that mode conversion does take place at the magnetopause. *Treumann et al.* [16] suggest that the ULF waves can provide sufficient diffusion to contribute significantly to the formation of the low latitude boundary layer.

ACKNOWLEDGMENTS

This work was supported by PRODEX (Cluster II) and by the Belgian Federal Office for Scientific, Technical and Cultural Affairs.

REFERENCES

1. LaBelle, J., and Treumann, R. A., *Space Sci. Rev.*, **47**, 175–202 (1988).
2. Engebretson, M. J., Lin, N., Baumjohann, W., Lühr, H., Anderson, B. J., Zanetti, L. J., Potemra, T. A., McPherron, R. L., and Kivelson, M. G., *J. Geophys. Res.*, **96**, 3441–3454 (1991).
3. Anderson, R. R., Harvey, C. C., Hoppe, M. M., Tsurutani, B. T., Eastman, T. E., and Etcheto, J., *J. Geophys. Res.*, **87**, 2087–2107 (1982).
4. Song, P., Russell, C. T., and Gary, S. P., *J. Geophys. Res.*, **99**, 6011–6025 (1994).
5. Lacombe, C., Belmont, G., Hubert, D., Harvey, C. C., Mangeney, A., Russell, C. T., Gosling, J. T., and Fuselier, S. A., *Ann. Geophysicae*, **13**, 343–357 (1995).
6. Southwood, D. J., *Planet. Space Sci.*, **22**, 483–491 (1974).
7. De Keyser, J., *J. Geophys. Res.*, (2000). In press.
8. Perraut, S., Gendrin, R., Robert, P., and Roux, A., "Magnetic pulsations ovserved on board GEOS-2 in the ULF range during multiple magnetopause crossings," in *Proceedings of the Magnetospheric Boundary Layers Conference*, ESA Science Publication 148, Noordwijk, 1979, pp. 113–122.
9. Rezeau, L., Morane, A., Perraut, S., Roux, A., and Schmidt, R., *J. Geophys. Res.*, **94**, 101–110 (1989).
10. Rezeau, L., Roux, A., and Russell, C. T., *J. Geophys. Res.*, **98**, 179–186 (1993).
11. Song, P., Russell, C. T., and Huang, C. Y., *J. Geophys. Res.*, **98**, 5907–5923 (1993).
12. De Keyser, J., Roth, M., Reberac, F., Rezeau, L., and Belmont, G., *J. Geophys. Res.*, **104**, 2399–2409 (1999).
13. Belmont, G., Reberac, F., and Rezeau, L., *Geophys. Res. Lett.*, **22**, 295–298 (1995).
14. Berchem, J., and Russell, C. T., *J. Geophys. Res.*, **87**, 2108–2114 (1982).
15. De Keyser, J., *J. Geophys. Res.*, (2000). Submitted.
16. Treumann, R. A., LaBelle, J., and Bauer, T. M., "Diffusion processes: An observational perspective," in *Physics of the Magnetopause*, edited by P. Song et al., AGU Geophysical Monograph Series 90, Washington, D.C., 1995, pp. 331–341.

Waves in non-Maxwellian plasmas with excess superthermal particles

M.A. Hellberg[1], R.L. Mace[2], F. Verheest[3]

[1] *School of Pure & Applied Physics, University of Natal, Durban, South Africa*
[2] *Physics Department, University of Durban-Westville, South Africa*
[3] *Sterrenkundig Observatorium, Universiteit Gent, Ghent, Belgium*

Abstract. Many space plasmas have an excess of superthermal particles and can be modelled by a κ-distribution. While $\kappa \to \infty$ yields the Maxwellian, a low value of κ describes a power-law distribution with a strongly accelerated, hard spectrum. Wave studies require the generalized plasma dispersion function, which may be expressed in terms of a hypergeometric function, making it easily usable. We review studies of waves in κ-distribution plasmas, focusing *inter alia* on applications to space plasmas. Wave behaviour for low-κ plasmas is significantly different from that found for Maxwellian plasmas.

INTRODUCTION

Although most studies of waves in plasmas are based on the assumption of a Maxwellian distribution function, non-Maxwellian distributions are common, both in natural and laboratory plasmas. In particular, in both space and stellar regions, power-law distributions are often observed, e.g. $4\pi v^2 f(v)dv \propto v^{-\alpha}dv$ for $|v| > v_{th}$. Examples are found in galactic cosmic ray distributions, solar flares, the solar wind plasma, the plasma sheet, the magnetotail, near plasma shock waves, etc. These distributions may have a significant high-energy tail arising from some external acceleration mechanism, i.e. a hard spectrum.

Vasilyunias [1] found that particle distributions measured in satellite-based experiments could be modelled by a generalized Lorentzian or "kappa" distribution, a family of modified power-law distributions, the real-valued parameter κ allowing one to fit to the actual distribution. This enables one to cover a wide range of distributions from the Lorentzian to the Maxwellian. Introducing a modified thermal speed, $\theta = \left(\frac{2\kappa-3}{\kappa}\right)^{\frac{1}{2}} v_{th}$, the normalized κ-distribution may be written as

$$f(v) = \left(\pi\kappa\theta^2\right)^{-3/2} \frac{\Gamma(\kappa + 1)}{\Gamma(\kappa - \frac{1}{2})} \left(1 + \frac{v^2}{\kappa\theta^2}\right)^{-(\kappa+1)} . \tag{1}$$

CP537, *Waves in Dusty, Solar, and Space Plasmas*, edited by F. Verheest, et al.
© 2000 American Institute of Physics 1-56396-962-9/00/$17.00

Here $v_{th} = (T/m)^{1/2}$ is the usual thermal speed. The definition of θ imposes a physical lower limit on κ, as it is real only for $\kappa > \frac{3}{2}$. For superthermal particles, $|v| \gg \theta$, the κ-distribution approaches a power-law, with $\alpha \simeq 2\kappa$. For $\kappa \to \infty$ the distribution becomes the Maxwellian, while low κ represents a hard spectrum with excess superthermal particles, arising from strong acceleration. There are numerous examples of satellite observations that have been modelled using κ-distributions, e.g. in the plasma sheet Christon *et al.* [2] found ions with $\kappa_i = 4.7$, while the electron distribution fitted $\kappa_e = 5.5$. In the distant magnetotail ion particle data matched $\kappa_i = 5.5$. Lemaire's group [3,4] have devised a Lorentzian ion exosphere model and a solar wind model with κ-distribution corona, using values in the range 2-6. A value of $\kappa_e = 4$ yielded agreement with electron observations in the solar wind. Fermi acceleration, arising from a collisionless shock wave, may produce power-law distributions with a spectral index, $\alpha = \frac{3r}{r-1}$, where $r = u_1/u_2$ is the shock compression ratio. In general, α is not an integer, and for strong shocks $\alpha \leq 4$, i.e. Fermi acceleration should yield $\kappa \leq 2$ near a shock [5].

For wave studies one needs to find a dielectric function. Based on the κ-distribution, one may define a generalized plasma dispersion function Z_κ for arbitrary real κ, which is found to be proportional to the Gauss hypergeometric function [5]. The latter is analytically well known, and may be calculated easily.

The function Z_κ may be used for a variety of wave studies, with applications in space and astrophysics. In this paper we shall discuss electron plasma waves, ion-acoustic and electron-acoustic waves, electrostatic fluctuations, electromagnetic wave propagation along the magnetic field, and a general dielectric tensor for a magnetoplasma. In all cases, the behaviour of κ-distributions with low values of κ differs importantly from what may be calculated for Maxwellian distributions.

PLASMA DISPERSION FUNCTION Z_κ

By analogy with the usual Fried & Conté [6] plasma dispersion function (Z-function), Summers and Thorne [7] introduced a modified plasma dispersion function, valid only for integer κ, with the normalized κ-distribution replacing the Maxwellian in the integrand. Their formulation may be expressed as a finite series of $\kappa + 1$ terms, which means that calculation is fast.

Following them, Mace & Hellberg [5] defined a generalized plasma dispersion function $Z_\kappa(\xi)$ for arbitrary real κ as

$$Z_\kappa(\xi) = \frac{1}{\pi^{1/2}\kappa^{3/2}} \frac{\Gamma(\kappa + 1)}{\Gamma(\kappa - \frac{1}{2})} \int_{-\infty}^{\infty} \frac{ds}{(s - \xi)(1 + s^2/\kappa)^{\kappa+1}} \quad \text{Im}(\xi) > 0. \quad (2)$$

We note that in general the integral has branch points at $z = \pm i\sqrt{\kappa}$. Evaluating the contour integral, Mace & Hellberg [5] reduced it via Pochhammer's integral to an expression proportional to the Gauss hypergeometric function. Hence

$$Z_\kappa(\xi) = \frac{i(\kappa + \frac{1}{2})(\kappa - \frac{1}{2})}{\kappa^{3/2}(\kappa + 1)} {}_2F_1[1, \, 2\kappa + 2; \, \kappa + 2; \, \frac{1}{2}(1 - \xi/i\sqrt{\kappa})]. \quad (3)$$

Clearly, this formulation is very useful as it allows for non-integer κ, and the well-known hypergeometric function can be easily manipulated and calculated, e.g. as an internal function `Hypergeometric2F1` in MATHEMATICA, or following an algorithm such as that of Press *et al.* [8]. For integer κ, the function Z_κ reduces to the expression of Summers & Thorne [7], and for $\kappa \to \infty$ it goes over into the Z-function. In practice, numerical values are very similar to those for the Maxwellian case for, say, $\kappa \geq 10$. Using the hypergeometric function one may deduce numerous relationships, including, for instance, the derivative [5], and hence that

$$Z'_\kappa(\xi) = -2\frac{(\kappa + \frac{1}{2})(\kappa - \frac{1}{2})}{\kappa^2}$$
$$\times \left\{ 1 + \frac{\kappa + 1}{\kappa + \frac{1}{2}} \left(\frac{\kappa + 1}{\kappa}\right)^{1/2} \xi\, Z_{\kappa+1}\left[\left(\frac{\kappa + 1}{\kappa}\right)^{1/2}\xi\right]\right\}$$

which reduces to the well-known expression $Z'(\xi) = -2\{1 + \xi\, Z(\xi)\}$ for $\kappa \to \infty$. Asymptotic expansions have also been generalized to real κ [9].

ELECTROSTATIC WAVE STUDIES

Using the generalized Z_κ, one may study waves in plasmas with κ-distributions, and hence interpret satellite-based experimental observations in space. One may easily show that the general dispersion relation for electrostatic waves in an unmagnetized plasma is (e.g. [7,10,11])

$$\epsilon(\mathbf{k}, \omega) = 1 + 2\sum_j \frac{\omega_{pj}^2}{k^2\theta_j^2}\left\{\frac{2\kappa_j - 1}{2\kappa_j} + \frac{\omega}{k\theta_j}Z_{\kappa_j}\left(\frac{\omega}{k\theta_j}\right)\right\} = 0. \tag{4}$$

Using the expression for $Z'_\kappa(\xi)$, this may also be written as

$$1 - \sum_j \frac{(\kappa_j - 1)^2}{\kappa_j(\kappa_j - \frac{3}{2})}\frac{\omega_{pj}^2}{k^2\theta_j^2}Z'_{\kappa_j - 1}\left[\left(\frac{\kappa_j - 1}{\kappa_j}\right)^{1/2}\frac{\omega}{k\theta_j}\right] = 0. \tag{5}$$

In general each plasma species j may possess its own value of κ, κ_j.

Electron plasma waves

Thorne & Summers [10] comprehensively investigated Langmuir waves in κ-distribution plasmas in space, for $\kappa = 2$ and 3. Many of their results, obtained for integer κ, have been generalized to arbitrary κ [5,12]. Making the usual high phase velocity assumption, and hence using asymptotic expansions of Z_κ, one finds, for small $|\omega_i/\omega_r|$ and small k, that the real part of the dispersion relation becomes

$$\omega_r^2 = \omega_{pe}^2 + 3\frac{\kappa_e}{2\kappa_e - 3}k^2\theta_e^2 = \omega_{pe}^2 + 3k^2v_e^2. \tag{6}$$

Thus, in this approximation, the dispersion relation is that for a Maxwellian plasma. The damping, on the other hand, shows significant dependence on κ [10,5], viz.

$$\gamma =\simeq -\pi^{1/2}\frac{\Gamma(\kappa_e + 1)}{\Gamma(\kappa_e - \frac{1}{2})}\omega_{pe}(2\kappa_e - 3)^{\kappa_e-1/2}(k^2\lambda_{De}^2)^{\kappa_e-1/2}. \tag{7}$$

Here $\lambda_{De} \equiv (T_e/4\pi n_e e^2)^{1/2}$ is the electron Debye length. Thus at long wavelengths, $|\gamma|$ decreases as a power-law, rather than exponentially. Hence damping increases with decreasing κ, as there are more resonant high-energy particles (see also [5]). Actually, the above approximate solutions also tacitly assume $2\kappa_e+2 \gg 5$. Numerical solution of the exact electrostatic dispersion relation shows that the Maxwellian-like dispersion obtained analytically is correct only for $k\lambda_{De} \lesssim 0.1$, especially for small κ_e. In fact, there is surprising dependence of ω_r on κ_e even for values of $k\lambda_{De} \sim 0.25$. One finds a monotonically decreasing phase speed with decreasing κ_e. This was found for integer κ [7] and subsequently confirmed more generally [5]. For very small wavenumber, damping of Langmuir waves increases for decreasing κ_e, but for $0.1 < k\lambda_{De} < 0.5$ there is significant (non-monotonic) variation with κ_e.

Ion-acoustic waves

The Summers-Thorne group [7,11,13] studied the ion-acoustic wave for integer κ, and their results have been generalized to arbitrary κ [12,14]. An important element in the physics of the ion-acoustic wave is the shielding by the electrons, which is governed by their mobility, and hence the Debye length. The shielding length for a κ-plasma [12,14] with its larger superthermal component, becomes

$$\lambda_{\kappa\alpha} \equiv \left[\left(\frac{\kappa_\alpha - \frac{3}{2}}{\kappa_\alpha - \frac{1}{2}}\right)\frac{\epsilon_0 T_\alpha}{n_\alpha q_\alpha^2}\right]^{\frac{1}{2}}. \tag{8}$$

This expression was also found independently by Bryant [15], and earlier from the wave dispersion relation [16]. The shielding distance is thus reduced from the usual Maxwellian Debye length by a factor $[(\kappa_\alpha - \frac{3}{2})/(\kappa_\alpha - \frac{1}{2})]^{\frac{1}{2}}$, which for, say, $\kappa = 1.8$, 2, 4, 10, means reduction by factors of 0.48, 0.58, 0.85 and 0.95, respectively. The dispersion relation is best expressed in terms of $\lambda_{\kappa e}$, rather than using λ_D [12],

$$\omega_r^2 = \frac{\omega_{pi}^2}{1 + 1/k^2\lambda_{\kappa e}^2} = \frac{k^2\lambda_{\kappa e}^2\omega_{pi}^2}{1 + k^2\lambda_{\kappa e}^2}, \tag{9}$$

which describes the usual physics, with $\lambda_{\kappa e}$ replacing λ_{De}.

Numerical solution of the full dispersion relation reveals a monotonic dependence on κ, with the Maxwellian yielding the largest phase speed, twice that for $\kappa = 1.666$. The expression for the damping is even more complicated than for the Langmuir wave [12]. Numerically, it yields three ranges of $k\lambda_{De}$: for $k\lambda_{De} < 1.5$ there is a monotonic decrease in damping with increasing κ, while for $k\lambda_{De} > 4$ the trend is reversed, the intermediate $k\lambda_{De}$ range providing the transition.

As distributions with small κ have excess superthermal particles, they will, compared to a Maxwellian, exhibit a reduction in particle number elsewhere in velocity space, thereby affecting slopes, and hence resonances. Hence the ion-acoustic instability driven by an electron drift may be affected variously as κ is lowered, depending on the value of the drift velocity relative to the details of the distribution. Meng *et al.* [11], assuming $\kappa_e = \kappa_i$, have shown that one cannot make a generalized statement as to whether a low κ distribution increases or decreases the instability. Critical factors include the electron-ion temperature ratio and the normalized drift speed ($V_{de}/v_{th,e}$). When the ions are very cold, the low κ distribution is destabilized by a smaller drift, as it also is for relatively hot ions ($T_i \geq \frac{1}{4}T_e$), while in the intermediate range the Maxwellian is more unstable.

Electron-acoustic waves

The electron-acoustic wave (EAW) [18], with a frequency between ω_{pe} and ω_{pi}, may be weakly damped in a two-electron-temperature plasma if the temperature ratio is large enough. For small k dispersion is acoustic-like, with a phase speed of the order of the hot electron thermal speed, v_h. The wave is typically strongly damped for both very small (hot electron Landau damping) and large wavenumbers (cool electron Landau damping), while being observable for intermediate k.

This mode, or the associated EA instability, has been used to explain such phenomena as broadband electrostatic noise (BEN) in the magnetotail [19] , electrostatic noise in the polar cusp region of the magnetosphere [20], and BEN and hiss in the cusp/cleft region [21]. Criteria for weak damping of EAW were studied, assuming cool Maxwellian electrons, hot κ-distributed electrons and cold ions [22],

$$1 - \frac{\omega_{pc}^2}{2k^2v_c^2}Z'\left(\frac{\omega}{\sqrt{2}\,kv_c}\right) - \frac{(\kappa-1)^2}{\kappa(\kappa-\frac{3}{2})}\frac{\omega_{ph}^2}{k^2\theta_h^2}Z'_{\kappa-1}\left[\left(\frac{\kappa-1}{\kappa}\right)^{1/2}\frac{\omega}{k\theta_h}\right] = 0.$$

The weakly-damped existence domains in the space of k and n_h/n_e, as a function of temperature ratio, T_h/T_c, were compared with the earlier bi-Maxwellian study [18]. The superthermal particles in the hot electron distribution affect both the dispersion and the damping rates. In particular, they tend to increase the hot ion Landau damping, and qualitatively a decrease in κ may be equated to an increase in temperature. As would be expected, these differences are most marked for low values of $\kappa(\simeq 1.6 - 5)$. A surprising feature is that for, say, a temperature ratio as low as 20, there is a large difference between the existence regimes found for $\kappa = 1.6$ and 2.0 [22].

Although the EAW plays an important role in space physics, it has not been studied much in the laboratory. However, we have recently revisited an earlier inconclusive high-frequency wave experiment, analyzing it by assuming κ-distributions for the two electron components [23,24]. As a result we have been able to confirm what was probably the first laboratory observation of the EAW, and to establish that the cool component was indeed Maxwellian, while the hot component had $\kappa \simeq 3.8$. Earlier attempts to understand the experimental dispersion and damping characteristics, by modelling with (a) a bi-Maxwellian, and (b) a cool Maxwellian with a hot waterbag distribution, had both been unsuccessful.

Electrostatic fluctuations

The discrete particles of a plasma are individually coupled to the electromagnetic field, with random motions inducing microscopic fluctuations in the average or macroscopic fields. The plasma parameter $g = 1/n\lambda_\kappa^3$ is a measure of discrete particle effects. As we have seen, for lower κ values, the Debye length is reduced, and thus g is increased considerably. As a result the level of electrostatic fluctuations is increased in a κ-distribution for low κ. The spectral density of fluctuations may be shown from test particle theory to depend importantly on the dielectric function $\epsilon(\mathbf{k}, \omega)$ and on the distribution $F_\alpha^{(0)}(\omega/k)$, both of which will be different for a κ-distribution plasma. Numerically [12,14] one finds that fluctuation levels are increased when there is an abundance of superthermal particles (small κ), and that fluctuations are more widely spread in (ω, k) space away from the normal modes than is the case for a Maxwellian distribution, where the fluctuation map is almost δ-function like in the vicinity of the plasma wave and the ion-acoustic wave. As may be expected, for $\kappa_e = \kappa_i \simeq 10$, the behaviour is essentially Maxwellian-like.

ELECTROMAGNETIC WAVES IN A MAGNETOPLASMA

The effects of κ-distributions on electromagnetic waves in a magnetoplasma were first studied by Leubner [25] (later corrected [26]), who used a bi-Lorentzian,

$$f(v_\parallel, v_\perp) = \frac{1}{\pi^{3/2}\theta_\parallel \theta_\perp^2} \frac{\Gamma(\kappa+1)}{\kappa^{3/2}\Gamma(\kappa - \frac{1}{2})} \left(1 + \frac{v_\parallel^2}{\kappa\theta_\parallel^2} + \frac{v_\perp^2}{\kappa\theta_\perp^2}\right)^{-(\kappa+1)}, \tag{10}$$

to study parallel propagation, in particular the R-mode, showing that in a plasma with low κ, growth was much stronger than for the Maxwellian case. Unlike the Maxwellian, the bi-Lorentzian is not separable. The Summers-Thorne group expanded this integer κ work and applied it to a number of space scenarios, considering waves such as the electron cyclotron mode, the R-mode, whistlers, etc. (see for example [13,27].

Mace [28] has explored the effect of a uniform magnetic field on wave propagation and obtained an integral expression for the full dielectric tensor in a uniform magnetoplasma, whose velocity distributions are isotropic kappa distributions. Although the result appears formally similar to Trubnikov's [29] dielectric tensor for an isotropic relativistic plasma, this connection is purely formal – Mace's calculation is non-relativistic. Having carried out the velocity integrals first, one is left with a complicated time-like integral. However, one must recognize that the result is a general formalism [28], and he has shown that from this formal expression one may recover the electrostatic dispersion relation, and that of the electromagnetic L and R modes for propagation along the magnetic field [30].

Subsequently, Mace [31] developed the formalism for the study of the parallel propagation of the R-mode in a plasma with each species having a bi-Lorentzian loss-cone distribution, expressing the dielectric function in terms of the Z_κ function for each species. In applying this to whistler waves, he neglected the loss-cone index, and hence explained the nearly field-aligned "1 Hz" whistlers (the terminology refers to the frequency observed at the satellite), seen in the Earth's electron foreshock.

CONCLUSION

The κ-distribution and the closely related bi-Lorentzian have been successfully used to model distributions observed in space and in astrophysics. Both the generalized plasma dispersion function Z_κ of Mace & Hellberg [5] and the modified plasma dispersion function of Summers & Thorne [7], applicable to integer κ, have been used to good effect in researching waves and instabilities in space. Electrostatic waves have been investigated, as has parallel electromagnetic propagation.

As the mathematical apparatus for the study of κ-distributions is now available, it seems that the time has come to move away from the ubiquitous Maxwellian and the Z-function of Fried & Conté [6] as the basis for exploring microinstabilities and high frequency waves in space plasmas. Experimentalists would be well-advised to use these new tools in interpreting their data when they find power-law distributions.

Acknowledgments

This work was supported by the Flemish Government (Department of Science and Technology) and the South African National Research Foundation in terms of the Flemish-South Africa Bilateral Scientific and Technological Cooperation on the Physics of Waves in Dusty, Solar and Space Plasmas. We thank our collaborators, especially Rudolf Treumann and Gareth Amery. MAH is grateful to Frank Verheest and the Universiteit Gent for their hospitality during a visit, during which this paper was written.

REFERENCES

1. Vasilyunias, V.M., *J. Geophys. Res.* **73**, 2839-2884 (1968).
2. Christon, S. P., Mitchell, D. G., Williams, D. J., Frank, L. A., Huang, C. Y., and Eastman, T. E., *J. Geophys. Res.* **93**, 2562-2572 (1988).
3. Pierrard, V., and Lemaire, J.F., *J. Geophys. Res.* **101**, 7923-7934 (1996).

4. Maksimovic, V., Pierrard, V., and Lemaire, J.F., *Astron. & Astrophy.* **324**, 725-734 (1997).

5. Mace, R.L., and Hellberg, M.A., *Phys. Plasmas* **2**, 2098-2109 (1995).

6. Fried, B.D., and Conté, S.D., *The Plasma Dispersion Function* Academic, New York, 1961.

7. Summers, D., and Thorne, R.M., *Phys. Fluids B* **3**, 1835-1841 (1991).

8. Press, W.H., Flannery, B.P., Teukolsky, S.A., and Vetterling, W.T., *Numerical Recipes: The Art of Scientific Computing*, 2nd ed. Cambridge University Press, Cambridge, 1992, pp. 208–211, pp. 271–273.

9. Grabbe, C., and Venturino, E., Phys. Plasmas **3**, 35-41 (1996).

10. Thorne, R.M., and Summers, D., *J. Geophys. Res.* **96**, 217-223 (1991).

11. Meng, Z., Thorne, R.M., and Summers, D., *J. Plasma Phys.* **47**, 445-464 (1992).

12. Mace, R.L., Hellberg, M.A., and Treumann, R.A., *J. Plasma Phys* **59**, 393-416 (1998).

13. Summers, D and Thorne, R.M., *J.Geophys. Res.* **97**, 16827-16832 (1992).

14. Mace, R.L., Hellberg, M.A. and Treumann, R.A., "Enhanced electrostatic fluctuations in plasmas", *International Conference on Plasma Physics*, edited by H. Sugai and T. Hayashi, Japan Society of Plasma Science and Nuclear Research, Tokyo, 1996, Volume 1, pp. 6-9.

15. Bryant, D.A., *J. Plasma Phys.* **56**, 87-93 (1996).

16. Chateau, Y.F., and Meyer-Vernet, N., *J. Geophys. Res.* **96**, 5825-5836 (1991).

17. Gary, S.P., and Tokar, R.L., *Phys. Fluids* **28**, 2439-2441 (1985).

18. Mace, R.L., and Hellberg, M.A., *J. Plasma Phys,.* **43**, 239-255 (1990).

19. Tokar, R.L., and Gary, S.P., *Geophys. Res. Lett.* **11**, 1180-1187 (1984).

20. Schriver, D., and Ashour-Abdalla, M, *J. Geophys. Res.* **92**, 5807-5819 (1987).

21. Mace, R.L., and Hellberg, M.A., *J. Geophys. Res.* **98**, 5881-5891 (1993).

22. Mace, R.L., Amery, G., and Hellberg, M.A., *Phys. Plasmas* **6**, 44-49 (1999).

23. Hellberg, M.A., Mace, R.L., Armstrong, R.J., and Karlstad, G., "Modelling of experiments on electron-acoustic waves", *International Conference on Plasma Physics & EPS Conference on Fusion and Plasma Physics*, edited by J. Stöckel et al., European Physical Society, Praha, 1998, Volume 1, pp 2065-2068.

24. Hellberg, M.A., Mace, R.L., Armstrong, R.J., and Karlstad, G., "Electron-acoustic waves in the laboratory: an experiment revisited", *J. Plasma Phys.: to appear* (2000)

25. Leubner, M.P., *J. Geophys. Res.* **88**, 469-473 (1983).

26. Summers, D., and Thorne, R.M., *J. Geophys. Res.* **95**, 1133-1135 (1990).

27. Xue, S., Thorne, R.M., and Summers, D., *J. Geophys. Res.* **98**, 17475-17484 (1996).

28. Mace, R.L., *J. Plasma Phys.* **55**, 415-429, (1996)

29. Trubnikov, B.A., *Plasma Physics and the Problem of Thermonuclear Reactions*, edited by M.A. Leontovich, Pergamon, Press, New York, 1959, Volume 3, p. 122.

30. Mace, R.L., *Phys. Scripta* **T63**, 207-210 (1996)

31. Mace, R.L., *J. Geophys Res* **103**, 14 643-14654 (1998)

Inertial Alfvén waves in the ionosphere: theoretical considerations and experimental constraints

N. Ivchenko*, G. Marklund* and Y. Khotyaintsev[†]

*Alfvén Laboratory, Royal Institute of Technology, 10044 Stockholm, Sweden.
[†]Swedish Institute of Space Physics, 75591 Uppsala, Sweden.

Abstract. Perturbations of electric and magnetic fields with periods of the order of 1 second are commonly observed in the auroral region by satellites and sounding rockets. The events are often accompanied by magnetic field aligned electron precipitation. The observations have been interpreted as inertial Alfvén mode waves. A variety of theories, some of which are surveyed here, have been suggested to describe such events. Recent observations of Alfvénic activity by sounding rockets and satellites are presented and their implications for the theoretical models are discussed.

INTRODUCTION

Since the beginning of sounding rocket and satellite measurements in the auroral ionosphere a wide variety of phenomena at different scales have been observed. Fluctuations in both electric and magnetic fields with frequencies in the range of 0.1-10 Hz are frequently observed in the auroral region. Frequencies below the local ion gyrofrequency and correlation of the transverse components of δE and δB suggest that the fluctuations are due to the presence of magnetohydrodynamic waves at the probe location. Another possible interpretation is in terms of stationary field aligned current structures closed by transverse currents in the ionosphere, which also produce correlated signatures in δE and δB. To distinguish between the two interpretations, the $\delta E/\delta B$ ratio is often used [1,2]. For spatial structures the ratio is $1/\mu_0\Sigma_P$, where Σ_P is altitude-integrated Pedersen conductivity, while for Alfvén waves the ratio is roughly the Alfvén speed.

A number of experiments suggest that fluctuations occuring on scales of 10-100 km (corresponding to frequencies below about 0.5 Hz in the satellite frame) are often current structures, and smaller scale (higher frequencies in the satellite frame) disturbances are likely to be of Alfvénic nature. Density depletions and accelerated electrons are found to be associated with the small scale waves, and several theories have been developed, addressing various aspects of the phenomena. For an extensive review on the subject see [3].

CP537, *Waves in Dusty, Solar, and Space Plasmas*, edited by F. Verheest, et al.
© 2000 American Institute of Physics 1-56396-962-9/00/$17.00

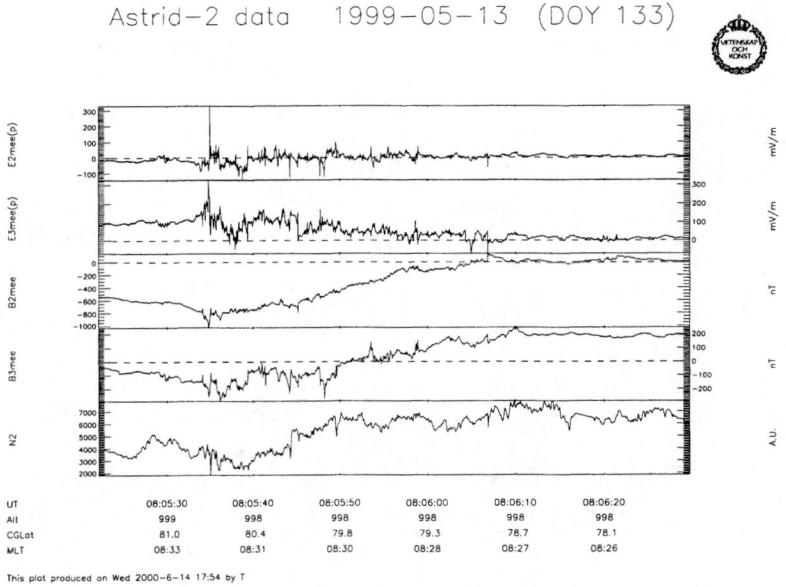

Astrid−2 data   1999−05−13   (DOY 133)

FIGURE 1. An example of Alfvénic activity in the cleft observed by Astrid-2 microsatellite. The panels present from top to bottom the eastward and southward components of δE and δB fields respectively and the Langmuir probe current being proportional to the plasma density.

Figures 1 and 2 present two typical examples of Alfvénic phenomena observed by the Astrid-2 Swedish microsatellite [4] at 1000 km altitude. The first example is a crossing of a large scale region of broadband electromagnetic turbulence in the pre-noon highlatitude ionosphere at a typical location of the dayside cleft. Correlated disturbances in δE and δB are observed with amplitudes of up to 200 mV/m and 100 nT respectively, together with strong density fluctuations. In some cases deep density depletions are collocated with the strongest spikes in the electric field. The second example presents two rather localized perturbations seen in both δE and δB. The first perturbation reaches a maximum electric field of over 100 mV/m and shows irregular variations in both the electric and the magnetic field. A 10 % density depletion is associated with the perturbation. The second perturbation is of smaller amplitude, has a quasiperiodic nature and lacks any density variation.

Satellite measurements have an intrinsic time-space ambiguity, as it is impossible to distinguish between temporal variations ($\partial/\partial t$) and spatial gradients crossed by the spacecraft ($V_{sat}\nabla$). This makes interpretations and comparisons with models ambiguous. Careful consideration of the combined sets of observations together with some theoretical insight can though yield some information to solve the ambiguity. We present here some theoretical considerations and recent observations of the Alfvénic phenomena. Implications of the observations are discussed.

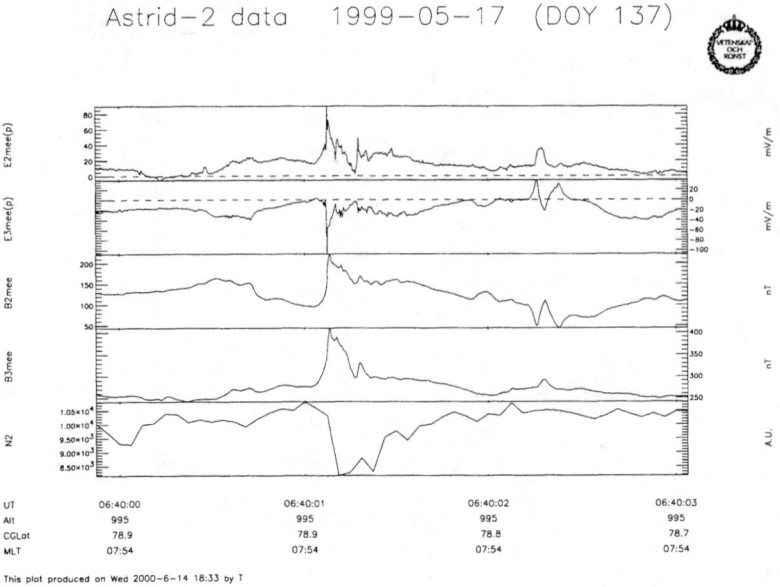

FIGURE 2. An example of "solitary" Alfvénic structures. The format is the same as in Fig. 1.

THEORETICAL CONSIDERATIONS

Alfvén waves

Magnetohydrodynamic waves below the ion gyrofrequency can exist in a plasma [5]. The shear mode, were both δE and δB are perpendicular to the ambient magnetic field, has a frequency of $\omega = k_\parallel V_A$, where $V_A = B_0/\sqrt{\mu_0 \rho}$ is the Alfvén speed, $k_\parallel$ is the wave vector component along the magnetic field. In the auroral ionosphere typical values of V_A are 10^3-10^5 km/s depending on the altitude. Electric and magnetic field disturbances are perpendicular to each other and to the ambient magnetic field, and $\delta E_x/\delta B_y = V_A$.

Effect of finite electron inertia

In the ideal MHD approximation the wave frequency does not depend on the perpendicular components of the wavevector. In the two-fluid approximation, though, the dispersion relation changes due to the contribution of the electron inertia and pressure effects [6]. For a cold plasma ($\beta \ll m_e/m_i$) (the regime encountered below ca. 10^4 km altitude) the dominant effect is due to the inertia of the electron. The equation of motion for electrons along the magnetic field is:

$$m_e \frac{\partial V_{e\parallel}}{\partial t} = -eE_\parallel \tag{1}$$

Thus, the parallel electric field $E_\parallel$ is no longer zero and is coupled to the variation of the parallel current of the wave. The dispersion relation for the waves becomes:

$$\omega = \frac{k_\parallel V_A}{\sqrt{1 + k_\perp^2 \lambda_e^2}} \tag{2}$$

here $\lambda_e = c/\omega_{pe}$. The wave has a perpendicular dispersion and a parallel electric field $E_\parallel = \frac{k_\perp k_\parallel \lambda_e^2}{1 + k_\perp^2 \lambda_e^2} E_\perp$. This field is significant for $k_\perp \lambda_e \geq 1$.

Models of the Alfvén waves in the ionosphere

The electric field often shows elliptical polarization in regions of Alfvénic activity. An Alfvén vortex description [7] was developed to account for these observation. Plasma depletions observed together with localized waves have been suggested in models of Alfvénic shocks [8] and resonance cones [9] to play an essential role in modifying the propagation characteristics of the waves, which in their turn contribute to a further deepening of the plasma cavity. A way to modulate the electron energies by stationary Alfvén structures in a flowing plasma has been described [10].

Freja observations of short-scale (of order of 0.1 s) disturbances in δE and δB stimulated the development of the solitary kinetic Alfvén wave model [11,12]. The terms SKAW is also used to refer to events like those in Fig. 2. The SKAW model invokes the time evolution of obliquely propagating nonlinear waves in one dimension. In the long time limit the shape of the nonlinearly steepened waves was similar to waveforms observed on Freja.

Ionospheric Alfvén resonator

Some of the theories referred to above were developed for a homogenious infinite (along the magnetic field) plasma. For the ionosphere this approximation is not strictly correct. Firstly, the flux tube is bounded from below by the conductive ionospheric layer, which provides for reflection of the Alfvén waves. Secondly, the density of plasma in the ionosphere decreases rapidly with altitude, from $10^4 - 10^6$ cm^{-3} in the F layer maximum to below 100 cm^{-3} above 1-2 R_E. The value of the Alfvén velocity increases with altitude up to about 1-2 R_E, above which it decreases due to the decrease of the Earth's magnetic field. The gradient in the Alfvén speed acts so as to partially reflect upward propagating waves.

The boundary conditions taken together lead to the formation of the ionospheric Alfvén resonator [13]. The resonator can sustain a set of standing waves with discrete frequencies. The basic frequency is about 0.5-1 Hz for typical conditions. The resonator can become unstable to the feedback instability [14], arising due to the modulation of the ionospheric conductivity by the field-aligned current. In the presence of a large-scale convection electric field driving a Pedersen current in the ionosphere a localized fluctuation of the conductivity polarizes and launches an upward Alfvén pulse. The pulse reflects due to the Alfvén speed gradient and returns to the ionosphere, where if can further increase the initial fluctuation.

RECENT OBSERVATIONS

$\delta E/\delta B$ ratio

It has been observed that the ratio $\delta E/\delta B$ is often frequency dependent and is generally increasing with frequency. The part of the spectrum where $\delta E/\delta B$ roughly equals the local Alfvén velocity was usually interpreted as due to Alfvén waves and the higher frequency part as due to some electrostatic mode.

On the other hand it was shown on the basis of Freja observations that observed $\delta E/\delta B$ ratio may correspond to the theoretical dependence for inertial Alfvén waves

$$\left|\frac{\delta E_\perp}{\delta B_\perp}\right| = v_A\sqrt{(1 + k_\perp^2 \lambda_e^2)}, \qquad (3)$$

in the whole frequency range from 1 to 500 Hz, representing low frequency ($\approx$1 Hz) turbulence of inertial Alfvén waves with a wide range of transverse scales being Doppler-shifted by the motion of the spacecraft (7 km/sec) [2]. Correlation of simultaneous measurements of density variations by two separated Langmuir probes confirms this conclusion. Time shifts between the similar features observed by the probes were found to be consistent with the satellite crossing a density structure, which does not change in shape or move during the crossing.

Quasiperiodic oscillations at the edge of an auroral arc

Alfvénic activity is often observed at the edges of large-scale auroral structures, where it is associated with field-aligned accelerated electrons. Quasiperiodic narrow-band oscillations of E and B were observed by sounding rockets on several occasions [15,16]. Recent measurements by the Auroral Turbulence II sounding rocket reveal the temporal nature of the oscillations.

The Auroral Turbulence II rocket was launched at 8:36:41 UT on Feb 11, 1997 from Poker Flat, Alaska, into a premidnight aurora. Two subpayloads had separated early in the flight, providing three-point measurements with separations of several kilometers. The payloads reached an apogee of about 500 km, crossing several auroral arcs during the flight. At the poleward edge of the arc close to the apogee electric and magnetic field oscillations with a period of about 2 s were observed by all three payloads. The oscillations were accompanied by field aligned energy-dispersed electron bursts, occurring in phase with the waves. The source tracing of the electrons gave an altitude of 10^4 km [17].

At the crossing of the arc the velocity of the payloads was about 1 km/s, and the arc itself moved in the same direction with a velocity of 500 m/s. The slow motion was fortunate to reveal the extent of the oscillation region. Since for some time all the payloads observed the oscillations in phase, the extent must have been larger than the payload separation (around 3 km at the time of the crossing). By considering the times when the payloads entered and exited the oscillation region, it was possible to estimate its scale length across the arc being about 7 km, and thus much smaller than the arc thickness. The value of the frequency and the almost harmonic appearance of the oscillations suggest the waves were captured in the ionospheric Alfvén resonator.

A satellite would have crossed the region in 1 s, which is less than the oscillation period. Thus the whole region would have been observed as a "solitary" disturbance, and the temporal structure of the waves would have been unresolved.

Occurrence of Alfvénic turbulence

As can be seen from figures 1 and 2 the Alfvénic phenomena may have a number of various appearances. Whereas a prolonged period (corresponding to several hundred km) of broadband turbulence is observed in the cleft region in the first example, the two perturbations in the second case are rather localized, one of which has a much narrower spectrum. Observations reported in the previous subsections suggest in one case a broad wavevector spectrum of relatively narrow band low frequency fluctuations, and in the other case coherent quasiperiodic oscillations observed in a region several kilometer wide.

This suggests that several mechanisms may be operating, probably leading to different classes of events. A statistical study of the occurrence of various types of events could thus shed some light on how they are related to other auroral processes. Such a study is being conducted on the basis of Astrid-2 microsatellite data [18]. Over 700 auroral crossings have been analyzed from various local time sectors in both the southern and the northern hemisphere. A windowed Fourier transform is carried out on the electric field data, and intervals with enhanced power between 0.5 and 8 Hz are selected. The $\delta E / \delta B$ ratio has been estimated from the ratio of the spectral powers of δE and δB between 0.5 and 8 Hz and found to be consistent with the Alfvén velocity. Regions with fluctuations of different extent are observed, ranging from disturbances shorter than 1 s in the satellite frame to tens of seconds.

Large scale (over 300 km) regions of Alfvénic turbulence occur with the highest probability in the cusp/cleft region (8-14 MLT, above 70° invariant latitude), and less frequently in the premidnight sector (see Fig. 3). The amplitudes of the δE fluctuations in the cusp/cleft region are typically 100-200 mV/m. The cusp region is characterized by magnetosheath ion precipitation and continuous energy input, often being very dynamic and highly structured. The premidnight sector is the place where the strongest precipitation occurs during substorm expansion.

Shorter intervals of electromagnetic activity are rather spred out over the whole auroral zone and have typically smaller amplitudes, probably representing a mixture of different phenomena. Some events are rather narrowband bursts of a few periods of oscillations or localized events similar to those in Fig. 2.

DISCUSSION

Alfvén waves with small transverse scales are observed in the auroral ionosphere. The basic physics governing the low-frequency waves in auroral plasma seems to be well understood. The main questions to be resolved are related to the spatial and temporal structure of the events. Theories often assume some specific space and time dependence for the solutions. It is in general very difficult to derive the spatial and temporal portrait from single satellite observations unambigously, leaving it open to several different interpretations.

The observations indicate the existence of different classes of the phenomena, which probably have somewhat different sources and scales. Both extended and localized regions of wave activity are observed. A common feature which can be

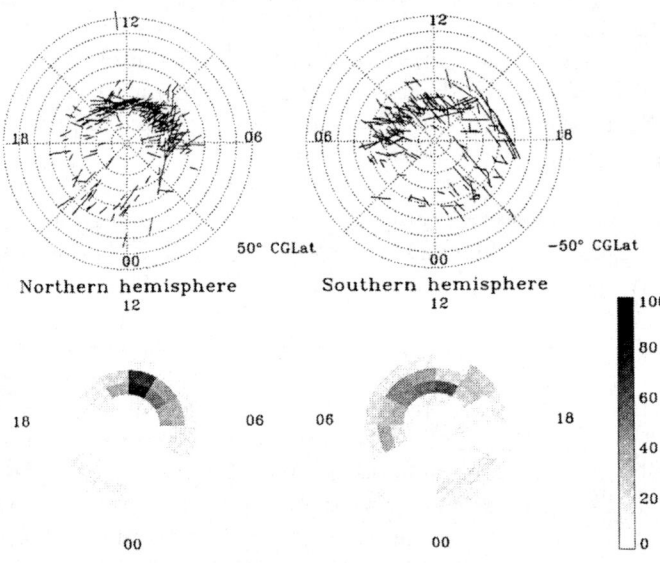

FIGURE 3. Occurrence of the regions of Alfvénic activity observed for more than 40 s on the satellite (spatial extent over 300 km). Top panels show the locations of the observed regions, bottom panels - normalized occurrence rates binned in invariant latitude and magnetic local time.

derived from a careful analysis of the $\delta E/\delta B$ ratio and from multiprobe measurements by Freja Langmuir probes, and multipoint sounding rocket observations is that the waves seem to have a broad wave vector spectrum (a range of the transverse scale sizes or localization with sharp boundaries) with a rather narrow range of frequencies [19]. The regions of the strongest energy input into the ionosphere are often associated with Alfvénic turbulence covering a broad range of wavevectors.

Field aligned electrons accelerated due to $E_\parallel$ of the inertial Alfvén waves, indicate the presence of temporal variations in the wave field. Studies of the field aligned electrons can give a clue to separating the temporal and spatial ambiguity in the measurements. The different velocities of sounding rockets (around 1 km/s) and satellites (7 km/s) provide an opportunity to increase the information on the structure of the events, giving different Doppler shifts of the spatial spectra. Future multipoint missions will provide more insight into the nature of the events.

Theories have to address the question of the boundary conditions along the magnetic field. Some theories are based on infinite and homogeneous models. For low frequency waves it is difficult to rationalize, as the wavelength along the field line becomes very large compared to the scale length of plasma density decrease. This is especially true for non-linear theories, where reflected waves cannot be treated by simple superposition. The effects of the ionospheric Alfvén resonator seen in both ground and satellite data [20,21] should be included in the models.

REFERENCES

1. Aikio, A.T., *et al.*, *J. Geophys. Res.* **26**, 27157 (1996).
2. Stasiewicz, K., *et al.*, *Geophys. Res. Lett.* **27**, 173 (2000)
3. Stasiewicz, K., *et al.*, *Space Science Reviews* in print, (2000).
4. Marklund, G., *ESA* **SP-397**, 387 (1997).
5. Alfvén, H., *Nature* **150**, 405 (1942).
6. Goertz, C.K., *J. Geophys. Res.* **89**, 10847 (1984).
7. Chmyrev, V. M. *et al.*, *Physica Scripta* **38**, 841 (1988).
8. Mishin, V. M. and Förster, M., *Geophys. Res. Lett.* **22**, 1745 (1995).
9. Stasiewicz, K. *et al.*, *J. Geophys. Res.* **102**, 2565 (1997).
10. Knudsen, D. J., *J. Geophys. Res.* **101**, 10761 (1996).
11. Louarn, P. *et al.*, *Geophys. Res. Lett.* **21**, 1847 (1994).
12. Seyler, C.E., J.-E. Wahlund, and Holback B., *J. Geophys. Res.* **100**, 21453 (1995).
13. Polyakov, S.V. and Rapoport, V.O., *Geomagn. Aeron* **21**, 816 (1981).
14. Lysak, R.L., *J. Geophys. Res.* **96**, 1553 (1991).
15. Gelpi, C.G., and Bering, E.A., *Planet. Space Sci.* **32**, 1387 (1984).
16. Ivchenko, N. *et al.*, *Geophys. Res. Lett.* **26**, 3365 (1999).
17. Lynch, K. *et al.*, *Geophys. Res. Lett.* **26**, 3361 (1999).
18. Ivchenko, N. and Marklund G., to be submitted to *Annales Geophysica* (2000).
19. Khotyaintsev, Y., *et al.*, *Physica Scripta* **T84**, 151 (2000).
20. Belayev, P.P., *et al.*, *J. Geophys. Res.* **104**, 4305 (1999).
21. Grzesiak, M., *Geophys. Res. Lett.* **27**, 923 (2000).

Helicon mode driven by O^+ thermal anisotropy

R. L. Mace[a] and Gurbax Lakhina[b]

[a] *Physics Department, University of Durban-Westville, Durban, South Africa*
[b] *Indian Institute of Geomagnetism, Mumbai, India*

Abstract. Preliminary results from an investigation of the helicon instability in a plasma composed of protons, electrons and singly charged oxygen ions, are presented. The velocity distribution function for each plasma component is modelled by a bi-Lorentzian distribution, which allows each particle species to possess a power law tail of arbitrary spectral index. This permits us to model accurately the shape of the power law tails observed on particle species in the plasma sheet region, where the helicon mode is believed to play an important role. The presence of a hard power law tail on the oxygen component is found to dramatically enhance the maximum growth rate of the instability when the oxygen ions possess a small $T_\parallel > T_\perp$ anisotropy. Above a certain value of $T_\parallel/T_\perp$, however, this behaviour is reversed. The growth rate decreases as the spectral index of the protons is decreased. The relevance of these effects to the central plasma sheet region is briefly discussed.

I INTRODUCTION

Lakhina and Tsurutani [1,2] recently showed that oxygen ions (O^+) of ionospheric origin, possessing a $T_\parallel > T_\perp$ thermal anisotropy, can destabilise a low frequency, right hand circularly polarized helicon mode in the near Earth plasma sheet region. The helicon mode occurs at frequencies much lower than the proton gyrofrequency, usually the domain of the MHD modes in a proton-electron plasma. Despite its low frequency the helicon mode is non MHD in character, possessing a non-vanishing Hall current [1].

The most important sources of the magnetospheric O^+ component are the auroral region and the dayside polar cleft [3]. Ionospheric O^+ from these regions can feed the inner plasma sheet on typical substorm timescales. The helicon mode, which can be destabilised by such oxygen ions, is therefore important for our understanding of substorm phenomena. It has been suggested that helicon waves could lead to the fast current and flux penetration across the plasma sheet [4], thus affecting substorm dynamics. The relatively rapid characteristic growth time of the helicon mode instability, on timescales much shorter than the typical duration of enhanced convection events, likely means that the instability attains saturation during these

CP537, *Waves in Dusty, Solar, and Space Plasmas*, edited by F. Verheest, et al.
© 2000 American Institute of Physics 1-56396-962-9/00/$17.00

events [1]. The low frequency electromagnetic turbulence produced by the helicon instability could scatter high energy electrons and help excite ion tearing modes that lead to substorm onset.

In the original papers by Lakhina and Tsurutani [1,2] the helicon mode was investigated in a plasma composed of bi-Maxwellian plasma components. There is, however, substantial evidence that charged particle species in the plasma sheet region possess high energy tails on their velocity distributions [5–8], and that these tails can be accurately fitted with a kappa distribution.

We investigate the helicon mode instability driven by O^+ thermal anisotropy in a model which allows each plasma component to possess a high energy power law tail with variable index κ, in addition to thermal anisotropy ($T_{\parallel} \neq T_{\perp}$). This model provides a more accurate picture of the measured particle distributions in the central plasma sheet (CPS) region, than does one based on bi-Maxwellian distributions. We have omitted particle drifts as we aim to isolate clearly the effects due to the high energy tails in this study. The results are compared and contrasted with those of Lakhina and Tsurutani [1].

II   MODEL AND BASIC EQUATIONS

The kinetic model for particle species in the central plasma sheet is based upon the bi-Lorentzian velocity distribution function

$$f(v_{\perp}^2, v_z) = \pi^{-3/2} \frac{1}{\kappa^{3/2}\theta_{\perp}^2 \theta_{\parallel}} \frac{\Gamma(\kappa+1)}{\Gamma(\kappa-\frac{1}{2})} \left(1 + \frac{v_{\perp}^2}{\kappa\theta_{\perp}^2} + \frac{v_z^2}{\kappa\theta_{\parallel}^2}\right)^{-(\kappa+1)}. \tag{1}$$

The distribution (1) has been normalized so that the integral over velocity space is unity. The constants $\theta_{\perp}$ and $\theta_{\parallel}$ are related to the perpendicular and parallel thermal speeds via $\theta_{\perp} = [(\kappa - \frac{3}{2})/\kappa]^{1/2}(2T_{\perp}/m)^{1/2}$ and $\theta_{\parallel} = [(\kappa - \frac{3}{2})/\kappa]^{1/2}(2T_{\parallel}/m)^{1/2}$. We shall call $\theta_{\perp}$ and $\theta_{\parallel}$ generalised thermal speeds. Distributions similar to (1) have recently been used for the electrons in a study [9] of the destabilization of the parallel whistler mode by electron thermal anisotropy in the electron foreshock.

Employing the distribution (1), the dispersion relation for the parallel propagating R-mode can be written [10,11]

$$\frac{k^2 c^2}{\omega^2} = 1 + \sum_{\alpha=i,e,O^+} \frac{\omega_{p\alpha}^2}{\omega^2} \left\{ A_\alpha + \left[A_\alpha \left(\frac{\omega + \epsilon_\alpha \Omega_\alpha}{k\theta_{\parallel\alpha}} \right) + \frac{\omega}{k\theta_{\parallel\alpha}} \right] \right.$$

$$\left. \times \frac{(\kappa_\alpha - 1)^{3/2}}{\kappa_\alpha^{1/2}(\kappa_\alpha - \frac{3}{2})} Z_{\kappa_\alpha-1} \left[\left(\frac{\kappa_\alpha - 1}{\kappa_\alpha} \right)^{1/2} \frac{\omega + \epsilon_\alpha \Omega_\alpha}{k\theta_{\parallel\alpha}} \right] \right\}, \tag{2}$$

where ω is the complex wave frequency, $k = |k_z|$, and for particle species α the plasma frequency is given by $\omega_{p\alpha} = (n_\alpha q_\alpha^2/\epsilon_0 m_\alpha)^{1/2}$, the gyrofrequency is given by $\Omega_\alpha = |q_\alpha \mathbf{B}_0|/m_\alpha$, charge sign is $\epsilon_\alpha = q_\alpha/|q_\alpha|$, the spectral index of the distribution is given by κ_α, and the temperature anisotropy is given by $A_\alpha = \theta_{\perp\alpha}^2/\theta_{\|\alpha}^2 - 1 = T_{\perp\alpha}/T_{\|\alpha} - 1$. Temperatures are measured in energy units throughout. The uniform magnetic field $\mathbf{B}_0$ is directed along the z axis of our coordinate system, i.e., $\mathbf{B}_0 = B_0\mathbf{e}_z$. The function $Z_\kappa(\zeta)$ appearing in (2) is the modified dispersion function, defined by Summers and Thorne [11] for positive integers $\kappa \geq 2$ and generalized to real positive values of $\kappa > \frac{3}{2}$ by Mace and Hellberg [12]. Equation (2) reduces to that used by Lakhina and Tsurutani [1] in the limit $\kappa \to \infty$ (since $\lim_{\kappa\to\infty} Z_\kappa(\zeta) = Z(\zeta)$, where Z is the plasma dispersion function of Fried and Conte [13]).

Due to the nature of the low frequency helicon mode and the extremely high β of the central plasma sheet it is difficult to approximate the above dispersion relation (2) to obtain analytical formulae for the real part of the frequency and the growth rate. In the following section we solve equation (2) without approximation.

III   NUMERICAL RESULTS

We take as our nominal plasma sheet parameters the following: $n_i = 0.5\,\text{cm}^{-3}$, $T_{\|e} = 0.5\,\text{keV}$, $T_{\|p} = 5\,\text{keV}$, $\beta_{\|O} = 3.5$, $B_0 = 10\,\text{nT}$, $\kappa_e = 5.5$ and $\kappa_p = 5.5$. These parameters are based on those used by Lakhina and Tsurutani [1] in their study of the helicon instability. The values of κ_e and κ_p are representative of those occuring in the plasma sheet [7].

In the calculations we have kept the proton density constant at a value of $0.5\,\text{cm}^{-3}$ and varied the density of O$^+$. This approach mimics the plasma dynamics in the CPS region where the ionospheric oxygen ions (and electrons, to maintain quasineutrality) periodically feed the region, supplementing the already quasineutral plasma there. It should be borne in mind, however, that the addition of oxygen ions increases the parallel electron beta $\beta_{\|e} = 2\mu_0 n_e T_{\|e}/B_0^2$ and the electron plasma frequency ω_{pe} as side effects.

Figure 1 illustrates the low frequency regime $0 < \omega < \Omega_p$ of the weakly unstable R-mode for a range of oxygen densities n_O/n_p. The oxygen temperature anisotropy $A_O = -0.1$, which represents a small $T_{\|O} > T_{\perp O}$ anisotropy, and $\kappa_O = 3.3$. The value of $\beta_{\|O} = 3.5$, which means that $\beta_{\|O} - \beta_{\perp O} = -A_O\beta_{\|O} = 0.35$. The instability in all cases is weakly growing at frequencies below the oxygen gyrofrequency Ω_O. The largest maximum growth rate and highest unstable frequency correspond to the smallest value of n_O/n_p.

Figure 2 illustrates the variation of the real frequency and growth rate as functions of A_O for a fixed oxygen ion density $n_O = 0.1n_p$. It is observed that the real frequency varies only slightly with A_O, whereas γ exhibits a marked dependence on this parameter. In particular, as A_O becomes increasingly negative, and hence $\beta_{\|O} - \beta_{\perp O}$ becomes larger, so the instability growth rate increases and the window of growing wavenumbers increases.

Figure 3 illustrates the variation of the real frequency and growth rate of the helicon mode as a function of the spectral index κ_O, for a small negative value of A_O, i.e., $A_O = -0.1$. The real frequency exhibits no dependence on κ_O to within

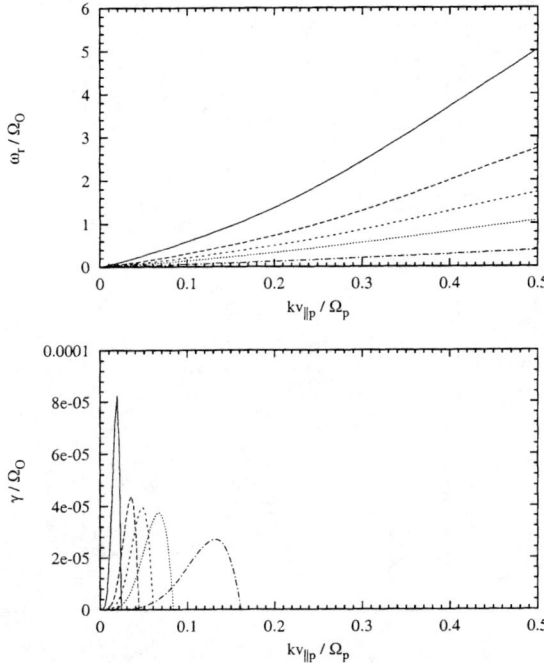

FIGURE 1. The low frequency region of the R-mode instability for a number of values of n_O/n_p: 0.1 (—), 0.5 (– –), 1 (- -), 2($\cdots$), 10(– $\cdot$ –). In all cases $A_O = -0.1$, $A_e = A_p = 0$, $\kappa_e = \kappa_p = 5.5$ and $\kappa_O = 3.3$. Here and in all subsequent figures, unlisted parameters correspond to those given in the text. The curves corresponding to the smallest values of n_O/n_p have the steepest slope.

graphical accuracy. The maximum growth rate, however, increases by well over three orders of magnitude as κ_O is decreased from 10 to 2, i.e., as the power-law tail on the O^+ distribution becomes ever more pronounced. Furthermore, the range of growing wavenumbers extends towards lower values of k as κ_O is decreased. The upper limit of the excited wavenumbers is independent of κ_O.

Figure 4 illustrates the effects of κ_O on the instability for a larger negative value of A_O. Here $A_O = -0.5$ and hence $\beta_{\parallel O} - \beta_{\perp O} = 1.75$, which is a very large β-anisotropy. In this case the effect of κ_O on the growth rate is not as dramatic as was found in figure 3. It is interesting to note that in this regime of larger thermal anisotropy, the effect of κ_O is precisely opposite to that observed in Figure 3, i.e., decreasing κ_O reduces the growth rate of the instability. In other words, a hard tail on the oxygen ions tends to diminish the instability in the regime $\beta_{\parallel O} - \beta_{\perp O} > 1$. However, common to both regimes of A_O is the fact that low values of κ_O excite the largest range of wavenumbers.

We have investigated the effects of varying the the spectral index of the proton

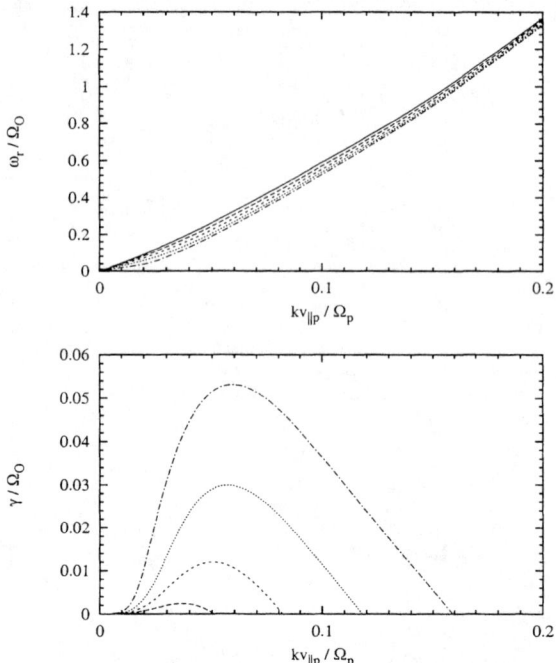

FIGURE 2. The low frequency region of the R-mode instability for a number of values A_O: -0.1 (—), -0.2 (– –), -0.3 (- -), -0.4 ($\cdots$) and -0.5 (– · –). In all cases $n_O = 0.1 n_p$, $A_e = A_p = 0$ and $\kappa_e = \kappa_p = 5.5$, $\kappa_O = 3.3$.

velocity distribution on the dispersion characteristics and growth rate of the helicon instability (not shown). As the proton energy spectrum becomes harder, i.e., as κ_p decreases, so the growth rate of the instability rapidly decreases. However, above $\kappa_p \simeq 10$, when the particle statistics become quasi-Maxwellian, the maximum growth rate varies very slowly with κ_p. The dispersion characteristics of the helicon mode are very weakly dependent on this parameter.

We have investigated the effects of varying κ_e from $\kappa_e = 3\text{--}10$ for $n_O/n_p = 0.5$, $T_{\|p}/T_{\|e} = 10$, $\beta_{\|O} = 3.5$ and $A_O = -0.1$ (not shown). We find little or no dependence on κ_e for these parameters (to within graphical accuracy). Lakhina and Tsurutani [1] found furthermore that the instability growth rate is insensitive to the temperature anisotropy $T_{\|e}/T_{\perp e}$ of the electron distribution. Thus, it appears that the instability growth rate, and indeed the real part of the frequency, is insensitive to the exact details of the electron distribution.

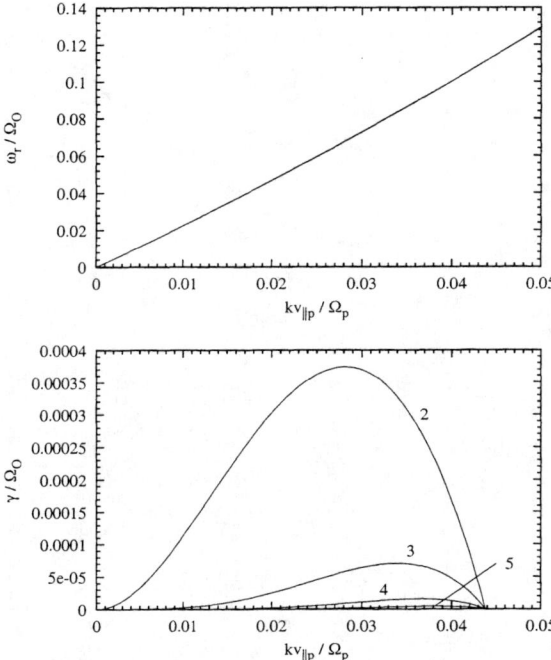

FIGURE 3. The low frequency region of the R-mode instability for a number of values of κ_O: 2, 3, 4, 5 and 10 (shown on curves). In all cases $n_O = 0.5 n_p$, $\kappa_e = \kappa_p = 5.5$, $A_O = -0.1$ and $A_e = A_p = 0$.

IV   DISCUSSION AND CONCLUSIONS

We have investigated numerically the O^+ temperature anisotropy driven instability of the low frequency helicon mode in a plasma that comprises electrons, protons and O^+ ions, each having a power-law tail on their velocity distribution. The presence of power law tails on both the electron and proton species in the central plasma sheet region is now well known [5,7]. To account for the less certain details of the O^+ ions in the central plasma sheet and the variability of their abundance, we have varied their parameters over a correspondingly larger range of values.

This work extends the previous works of Lakhina and Tsurutani [1,2] in two important ways. First, as mentioned, we allow the particle distributions to possess a power law tail, which provides a more accurate statistical model of measured particle velocity distributions in the central plasma sheet. Second, we investigate a larger region of parameter space than was studied by Lakhina and Tsurutani.

The salient findings are as follows. The spectral index of the O^+ velocity distribution dramatically affects the growth rate of the helicon instability in the regime of small $T_{\|O}/T_{\perp O} - 1$, which we tentatively categorise as the regime where $-0.1 < A_O < 0$. In this regime the growth rate rapidly increases as the hardness

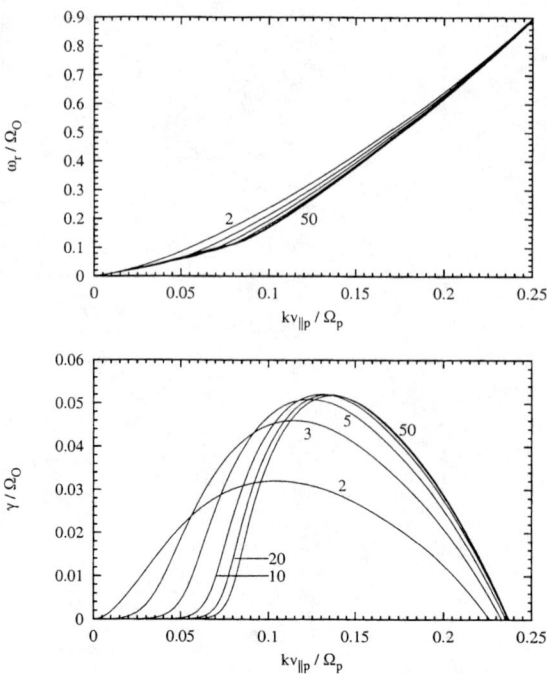

FIGURE 4. The low frequency region of the R-mode instability for a number of values of κ_O: 2, 3, 5 and 10 (shown on curves). In all cases $n_O = 0.5n_p$, $\kappa_e = \kappa_p = 5.5$, $A_O = -0.5$ and $A_e = A_p = 0$.

of the tail spectrum increases, i.e., as κ_O decreases. We observed an increase of over three orders of magnitude in the maximum growth rate of the instability as κ_O decreased from 10 to 2.

When the O^+ ions possess larger thermal anisotropy, i.e., when $T_{\|O}$ is sufficiently larger than $T_{\perp O}$, the growth rate is less sensitive to the value of κ_O. We have found that $A_O \sim -0.5 - -0.3$ characterises this regime. Here the dependence of the maximum growth rate on κ_O is precisely the opposite of that exhibited in the regime of small thermal anisotropy; a harder O^+ tail reduces the maximum growth rate of the helicon mode instability. The range of unstable wavenumbers and frequencies is much larger than it is in the low thermal anisotropy regime, and the maximum growth rate is generally larger.

Since the protons [6] and electrons [7] in the central plasma sheet possess power law tails on their velocity distributions, their effects on the helicon instability are important. It is found that decreasing κ_p, i.e., increasing the hardness of the proton energy spectrum, reduces the growth rate of the helicon instability, whereas the instability is insensitive to variations in the spectral index of the electron distribution κ_e.

The exact nature of the helicon mode in the central plasma sheet will be determined by all of the above parameters, but primarily by the value of A_O, which plays the largest role in determining the maximum growth rate. If $T_{\parallel O}$ is sufficiently larger than $T_{\perp O}$, i.e., A_O is sufficiently negative ~ -0.5, then the maximum growth rate $\gamma_{\max}$ is much larger than it is for small (negative) values of $A_O \sim -0.1$. Accompanying this increased growth rate, the instability develops a more broadband nature, characterised by a larger window of growing wavenumbers. These two facts are generally true independently of the value of κ_O. Consequently, the nonlinearly saturated regime of the instability, that which is almost always observed in space plasmas, would be far more turbulent than the final weakly-turbulent state which would likely occur for small $|A_O|$. The large amplitude helicon turbulence so produced could be responsible for the twisting of the equilibrium magnetic field in the plasma sheet into flux ropes [1]. As discussed by Lakhina and Tsurutani [1], these properties of the helicon instability/turbulence give an indication that it might be playing a role in the processes related to oxygen ion bursts associated with multiple flux ropes in the distant magnetotail (and observed by GEOTAIL [14]).

REFERENCES

1. Lakhina, G. S., and Tsurutani, B. T., *Geophys. Res. Lett.* **24**, 1463–1466 (1997).
2. Lakhina, G. S., and Tsurutani, B. T., "Role of helicon modes in substorm processes", in *Substorms-4*, edited by S. Kokubun and Y. Kamide, Kluwer, 1998, pp. 511–516.
3. Lockwood, M., Waite, J. H., Jr., Moore, T. E., Johnson, J. F. E. and Chappell, C. R., *J. Geophys. Res.* **90**, 4099–4116 (1985).
4. Papadopoulos, K., Zhou, H. B., and Sharma, A. S., *Comments Plasma Phys. Controlled Fusion* **15**, 321 (1994).
5. Sarris, E. T., Krimigis, S. M., Lui, A. T. Y., Ackerson, K. L., Frank, L. A., and Williams, D. J., *Geophys. Res. Lett.* **8**, 349–352 (1981).
6. Lui, A. T. Y., and Krimigis, S. M., *Geophys. Res. Lett.* **8**, 527–530 (1981).
7. Christon, S. P., Mitchell, D. G., Williams, D. J., Frank, L. A., Huang, C. Y., and Eastman, T. E., *J. Geophys. Res.* **93**, 2562–2572 (1988).
8. Williams, D. J., Mitchell, D. G., and Christon, S. P. *Geophys. Res. Lett.* **15**, 303–306 (1988).
9. Mace, R. L., *J. Geophys. Res.* **103**, 14643–14654 (1998).
10. Mace, R. L., *Physica Scripta* **T63**, 207–210 (1996).
11. Summers, D., and Thorne, R. M., *Phys. Fluids B* **3**, 1835–1847 (1991).
12. Mace, R. L., and Hellberg, M. A., *Phys. Plasmas* **2**, 2098–2109 (1995).
13. Fried, B. D., and Conte, S. D., *The plasma dispersion function*, Academic Press, New York, 1961.
14. Wilken, B. Q., Zong, Q. C., Daglis, I. A., Doke, T., Livi, S., Maezawa, K., Pu, Z. Y., Ullaland, S., and Yamamoto, T., *Geophys. Res. Lett.* **22**, 3267–3270 (1995).

AUTHOR INDEX

A

Andries, J., 136
Arber, T. D., 224, 248

B

Ballai, I., 144, 152
Banerjee, D., 160
Berghmans, D., 168, 271
Bharuthram, R., 33, 61, 68
Bhatt, J. R., 84
Boddie, D., 176

C

Čadež, V. M., 91, 295
Califano, F., 126
Campos, L. M. B. C., 184, 192
Chagelishvili, G. D., 200
Cramer, N. F., 23

D

De Groof, A., 208
De Keyser, J., 340
De Moortel, I., 216, 224
De Sterck, H., 232
Doyle, J. G., 160
Dubinin, E., 327

E

Erdélyi, R., 144, 152, 279

F

Farid, T., 41
Fedun, V., 311
Foullon, C., 240

G

Gelinas, L. J., 48
Gerrard, C. L., 248
Gil, P. J. S., 184
Goedbloed, J. P., 109
Goossens, M., 136, 144, 152, 160, 200, 208, 287, 303, 317

H

Hellberg, M. A., 53, 348
Hochedez, J. F., 271
Hood, A. W., 224, 248

I

Ireland, J., 216
Isaeva, N. L., 184
Ivchenko, N., 356

J

Jacobs, G., 53, 91

K

Kelley, M. C., 48
Khotyaintsev, Y., 356
Kim, E. J., 256

L

Lakhina, G., 364
Larsen, M. F., 48

M

Mace, R. L., 348, 364
MacGregor, K. B., 256
Maharaj, S. K., 61

Mamun, A. A., 41
Marklund, G., 356
McKenzie, D., 168
McKenzie, J. F., 327
Mendes, P. M. V. M., 192

N

Nakariakov, V. M., 264

O

Ofman, L., 119
O'Shea, E., 160

P

Pillay, S. R., 33, 53, 68
Poedts, S., 76, 232, 271

R

Rao, N. N., 13, 33, 68, 84
Robbrecht, E., 271
Roberts, B., 176, 240
Rogava, A. D., 76
Ruderman, M. S., 279

S

Sauer, K., 327
Shaikh, A. A., 84
Shukla, P. K., 3, 41, 303

Sirenko, O., 287, 311
Stenflo, L., 303

T

Tevzadze, A. G., 200

V

Van Der Linden, R. A. M., 248
Vanlommel, P., 295
Verheest, F., 53, 91, 348
Verwichte, E., 271
Vladimirov, S. V., 23
Voitenko, Y., 287, 303, 311

W

Walsh, R. W., 216

Y

Yaroshenko, V. V., 53, 99
Yukhimuk, A., 287, 311

Z

Zhugzhda, Y., 317